# Biological and Chemical Systems for Nutrient Removal



## A Special Publication

Prepared by **Task Force on Biological and Chemical Systems for Nutrient Removal**

Movva Reddy, *Chair*

Bob Andryszak
Richard Atoulikian
Britt Bassett
Lee Beetschen
Erik Bundgaard
Jeffrey Chen
Gordon Culp
John Dietz
Peter Dold
Padma Golla
Lawrence Hentz, Jr.
Slawomir Hermanowicz
Samual Jeyanayagam
Ben Koopman
Samuel McClintock
Richard Mines, Jr.
Scott Murphy
J. B. Neethling
Manu Patel
Vikram Pattarkine
Clifford Randall
Roderick Reardon, Jr.
K. Ramesh Reddy
David Refling
Frank Rogalla
Thomas Sadick
Richard Sedlak
Jeannette Semon
Anthony Shemonski
Rao Surampalli
Paul Sutton
Nikolay Voutchkov
Thomas Wilson

Under the Direction of the
**Municipal Subcommittee** of the
**Technical Practice Committee**

**1998**

**Water Environment Federation**
**601 Wythe Street**
**Alexandria, VA 22314–1994 USA**

**IMPORTANT NOTICE**

The contents of this publication are not intended to be a standard of the Water Environment Federation (WEF) and are not intended for use as a reference in purchase specifications, contracts, regulations, statutes, or any other legal document.

No reference made in this publication to any specific method, product, process, or service constitutes or implies an endorsement, recommendation, or warranty thereof by WEF.

WEF makes no representation or warranty of any kind, whether expressed or implied, concerning the accuracy, product, or process discussed in this publication and assumes no liability.

Anyone using this information assumes all liability arising from such use, including but not limited to infringement of any patent or patents.

**Library of Congress Cataloging-in-Publication Data**

Water Environment Federation. Task Force on Biological and chemical systems for nutrient removal : a special publication / prepared by Task Force on Biological and chemical systems for nutrient removal : under the direction of the Municipal Subcommittee of the Technical Practice Committee.

p. cm.

Includes bibliographical references (p. ) and index.

ISBN 1-57278-123-8 (paperback)

1. Sewage—Purification—Nutrient removal. I. Water Environment Federation. Municipal Subcommittee. II. Title.

TD758.5.N87W38 1997

628.3′57—dc21 97-51597

CIP



ISBN 1-57278-123-8

Printed in the USA **1998**

# *Water Environment Federation*

Founded in 1928, the Water Environment Federation is a not-for-profit technical and educational organization. Its goal is to preserve and enhance the global water environment. Federation members number more than 42 000 water quality professionals and specialists from around the world, including engineers, scientists, government officials, utility and industrial managers and operators, academics, educators and students, equipment manufacturers and distributors, and other environmental specialists.

For information on membership, publications, and conferences, contact


Water Environment Federation
601 Wythe Street
Alexandria, VA 22314-1994 USA
1-703-684-2400

# *Special Publication of the Water Environment Federation*

The Water Environment Federation (WEF) Technical Practice Committee (formerly the Committee on Sewage and Industrial Wastes Practice of the Federation of Sewage and Industrial Wastes Associations) was created by the Federation Board of Control (now Board of Directors) on October 11, 1941. The primary function of the committee is to originate and produce, through appropriate subcommittees, special publications dealing with technical aspects of the broad interests of the Federation. These publications are intended to provide background information through a review of technical practices and detailed procedures that research and experience have shown to be functional and practical.

# Contents

# List of Tables

# List of Figures

# Acknowledgments

Principal authors and the chapters for which they were responsible are

Bob Andryszak (1)
Peter Dold (9)
Slawomir Hermanowicz (3)
Mark Hsu* (6)
Ben Koopman (4)
Richard Mines, Jr. (7)
J.B. Neethling (2, 5)
Clifford Randall (8)
Roderick Reardon, Jr. (12)
K. Ramesh Reddy (10)
Movva Reddy (5)
Thomas Sadick (11)
Tom Wilson (6)

*Deceased

Contributing authors are

Patricia Barker
Jeffery Chen
Lawrence Hentz, Jr.
Sam McClintock
Krishna Pagilla

Considerable editing and revision to the the final document were provided by Richard Mines, Jr.

Authors' and reviewers' efforts were supported by the following organizations:

Boyle Engineering Corporation, Orlando, Florida
Brown and Caldwell, Portland, Oregon
Buchart-Horn, Inc., Baltimore, Maryland
Camp Dresser & McKee, Inc., Maitland, Florida
CH2M Hill, Newport News, Virginia
City of Baltimore, Maryland
City of Stamford, Connecticut
HDR Engineering, Inc., Folsom, California
Illinois Institute of Technology
Johnson, Mirmiran, and Thompson, PA, Sparks, Maryland
Krüger Systems AS, Søborg, Denmark
Malcolm Pirnie, Inc., Newport News, Virginia
McMaster University, Hamilton, Ontario, Canada
Metcalf & Eddy, Inc., New York, New York
Montgomery Watson, Cleveland, Ohio

MPR Engineering Corporation, Inc., Longwood, Florida
Pennsylvania State University—Capitol, Middletown, Pennsylvania
P.M. Sutton & Associates, Inc., Bethel, Connecticut
PWT Waste Solutions, Inc., Houston, Texas
Rust E & I, Oak Brook, Illinois
Smith Culp Consulting, Seattle, Washington
Soap & Detergent Association, New York, New York
Stanley Associates, Surrey, British Columbia, Canada
Stearns and Wheler, Cazenovia, New York
Tata Consultancy Services, Pune, India
Tetra Technologies, Inc., Palo Alto, California
University of Central Florida, Orlando
University of California, Berkeley
University of Florida, Gainesville
University of South Florida, Tampa
U.S. Environmental Protection Agency, Kansas City, Kansas
Virginia Polytechnic Institute and State University, Blacksburg, Virginia
Washington Suburban Sanitary Commission, Laurel, Maryland

# Chapter 1
# Nutrients and Their Effects on the Environment



## Overview

This chapter provides a perspective on the need to control nitrogen and phosphorus in treating wastewater. The effects that these nutrients have on the aquatic and land environments are discussed, the typical chemical forms are described, and their sources in raw wastewater are identified. Typical quantities found in wastewater are tabulated. The nitrogen and phosphorus cycles in the environment, as applicable, are presented. Wastewater treatment plant (WWTP) effluent limits are discussed. Analytical techniques that are available to determine forms and quantities are outlined.

**POINT SOURCE CONTROL.** To address the adverse effects of nutrients on water quality, various programs have been undertaken throughout the world to control nutrients in the effluent from municipal and industrial WWTPs. These discharges are classified as "point sources" of nutrients.

Typically, during wastewater treatment, organic nitrogen and ammonia-nitrogen are either converted to nitrates or removed as nitrogen gas to the atmosphere. Phosphorus typically is removed with the sludge following either conversion to a chemical precipitate or incorporation into the biological cell mass.

**NONPOINT SOURCE CONTROL.** Nonpoint sources also contribute significant quantities of nitrogen and phosphorus to the nutrient budget of an aquatic environment. For water quality improvement programs to be successful, it is often necessary to control these nonpoint sources as well. While the means to control nonpoint sources is not addressed in this manual, a brief discussion is included herein to emphasize its importance in meeting water quality objectives.

Attention has focused on the need for establishing a nutrient budget for each body of water. This involves evaluating the amounts contributed by all sources of water entering lakes and reservoirs, namely rivers, creeks, groundwater, atmosphere, rainfall, and wastewater effluent. Streams draining agricultural land where inorganic fertilizers are used only for fertility maintenance receive nutrients through underground and overland drainage carrying silt and nutrients off the land. In the same fashion, both croplands receiving animal manure and grazing lands can also be sources of nitrogen and phosphorus. Agriculturally, it is impossible to attain a complete balance of fertilization with crop consumption of nutrients. Some excess nitrogen must be applied to compensate for losses from denitrification and conversion to nitrates that are purged through the soil and lost in groundwater. In the case of phosphorus, losses are primarily in silt carried away by wind and water erosion.

The establishment of a nutrient budget for a body of water is a time-consuming and expensive proposition. Typically, it requires a year-long study in foul and fair weather. Streams must be gauged, samples must be collected frequently during periods of storm runoff, and nutrient contents of base flows must be established. Analytical procedures should distinguish between readily available and slowly available nutrients. They should also distinguish between insoluble and soluble materials because insoluble materials are apt to become part of bottom deposits.

In some instances, contributions from nonpoint sources may be so great that benefits from removal of nitrogen or phosphorus from point sources become insignificant. The purpose of establishing a nutrient budget is to prevent expenditure of funds for nutrient removal where little or no benefit can be expected. Careful evaluation of data is required because concentration is a factor that must be considered in conjunction with quantity.

**NUTRIENT CONTROL STRATEGIES.** Typically, nutrient control strategies take advantage of a microorganism's necessity for nitrogen and phosphorus. By limiting the availability of either or both of these elements, the growth of undesirable microorganisms can be controlled. The methods used to control nitrogen and phosphorus in wastewater treatment practice are presented in the following chapters.

Nitrogen in the biomass of algae occurs in amounts ranging from 3 to 10%, largely in the form of proteins (smaller amounts are in the form of blue-green algae). However, because several blue-green algae can fix elemental nitrogen dissolved in water, fixed forms of nitrogen available from aqueous sources may not be true growth restraints. Numerous studies have shown that nitrogen can become limiting in the control of algal growth during the summer growing season at a limiting level of 0.05 mg/L of inorganic forms ($NH_3$-N + $NO_3$-N). On this basis, nitrogen removal from wastewater may become necessary when receiving waters are insufficient in quantity to dilute inorganic forms to the limiting level.

Although phosphorus in algal cells occurs in small amounts ranging from 0.5 to 1.0% in the biomass, it has been shown to be a limiting factor in the growth of algae. A value of less than 0.005 mg/L or 5 $\mu$g/L in the ortho form is recognized as a lower growth-limiting concentration (Federal Water Pollution Control Administration, 1968). Its removal from wastewater is highly feasible because most organic and inorganic forms of phosphorus are readily removed by precipitation with the use of alum, ferric salts, or lime. It can also be removed by enhanced biological phosphorus removal processes. Where land is available, it can be removed by application to cultivated, forested, and pasture areas.

In instances where nutrient control for freshwater is deemed desirable, control of phosphorus is considered to be absolutely essential because when nitrogen becomes limiting, any excess of phosphorus can support growth of nitrogen-fixing, blue-green algae. In such cases, the nitrogen budget of a body of water will be increased, thereby materially offsetting any benefits from nitrogen removal. In marine waters, it is considered desirable to control nitrogen inputs because phosphate is in abundant supply.

Considering this information, many authorities believe the eutrophication problem can be controlled by rigid requirements for phosphorus removal alone. Others believe that a high degree of nitrification is also needed to eliminate detrimental side effects of ammonia-nitrogen. The latter proposal meets with some objection on the part of people concerned with public water supplies because current water standards limit nitrate-nitrogen to 10 mg/L. However, blue-green algae do not grow in saline environments. Thus, nitrogen control is a viable strategy in estuarine environments.

# *Effect on the Environment*

**GENERAL.** *Eutrophication* is the term used to describe the natural process by which biological productivity increases with the age of a body of water. This typically is a result of capture by phytoplankton and other aquatic growths of plant nutrients contributed by inflowing waters and of the new growth that results. These resultant organisms and plants eventually die and settle to the bottom, where they decompose to some degree. During decomposition, nutrients are released to the water above. These nutrients reach the upper waters or the euphotic zone in time, and this continual enrichment from external and recycled sources perpetually produces new growth that dies and settles to the bottom. Residue from decomposition and silt carried by inflowing waters gradually fills the lake or reservoir.

*Eutrophic* is the term applied to waters with a high degree of biological productivity. Several parameters are used to determine the trophic condition of a body of water: standing crop of phytoplankton, level of chlorophyll *a*, volume of algae, level of oxygen production, level of oxygen depletion, Secchi disk readings, or a combination of all these factors. In bodies of water that stratify during the summer, a typical method of determination measures dissolved oxygen (DO) levels in the bottom layer, or hypolimnion. If the deep waters remain aerobic during summer stagnation, the body of water is considered oligotrophic. If the dissolved oxygen becomes depleted, the waters are considered eutrophic, and the degree of eutrophy can be estimated by the time it takes for anoxic or anaerobic conditions to develop after the onset of stratification. The usefulness of this parameter requires some judgment based on temperature conditions and the relative volumes of the hypolimnion and the upper water layer, or epilimnion.

**FLOWING WATER BODIES. Ammonia-Nitrogen.** Ammonia-nitrogen entering receiving waters behaves in one of three ways. It may serve as a prime nutrient to support phytoplankton and rooted aquatic growths. Some will be lost to the atmosphere as gaseous ammonia when its partial pressure in the water is greater than that of ammonia in the atmosphere. The degree of loss is a function of pH, which controls the concentration of ammonia (ammonium ion) by its effect on the equilibrium:

$$NH_4^+ \rightleftarrows NH_3 + H^+ \qquad (1.1)$$

The significance of pH in the range of interest in natural waters is shown in Figure 1.1. At pH levels of 9 or above, significant losses can occur, but little loss will occur at pH values below 7. There is a need to control molecular ammonia concentrations to prevent fish toxicity. Molecular or free ammonia

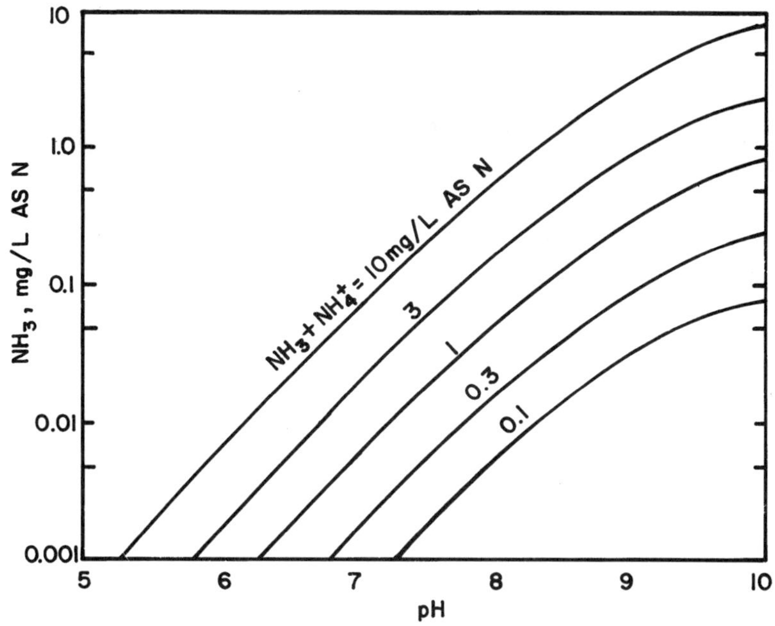


**Figure 1.1** **The effect of pH and ammonia-nitrogen concentration on the concentration of free or molecular ammonia in water.**

in concentrations of more than 0.2 mg/L can cause acute toxicity in several species of fish. The criteria for ambient water quality have been set for un-ionized ammonia as a function of PH and temperature (U.S. EPA, 1985a).

The most significant effect of ammonia-nitrogen is related to its effect on the oxygen resources of a stream or river, typically referred to as *nitrogenous oxygen demand* (NOD). Ammonia-nitrogen can be readily converted to nitrite and then nitrate, provided the necessary nitrifying bacteria are present:

$$2NH_4^+ + 3O_2 \xrightarrow{Nitrosomonas} 2NO_2^- + 2H_2O + 4H^+ \qquad (1.2)$$

$$2NO_2^- + O_2 \xrightarrow{Nitrobacter} 2NO_3^- \qquad (1.3)$$

The manner in which these reactions occur is displayed in Equations 1.2 and 1.3, and the rate of the reaction is in direct proportion to the population of nitrifying bacteria. Thus, it is possible that the reaction may be so slow that the rate of stream reaeration offsets the demand for oxygen and no noticeable oxygen sag will occur. On the other hand, where significant populations do occur, the demand for oxygen may be in excess of natural reaeration capacity and a distinct oxygen sag will develop, as demonstrated in Figure 1.2.

The reason for the difference in the behavior of ammonia in rivers is related to differences in nitrifying populations. Typically, extensive and rapid nitrification has been noted in shallow streams, and very low rates have been noted in deep ones.

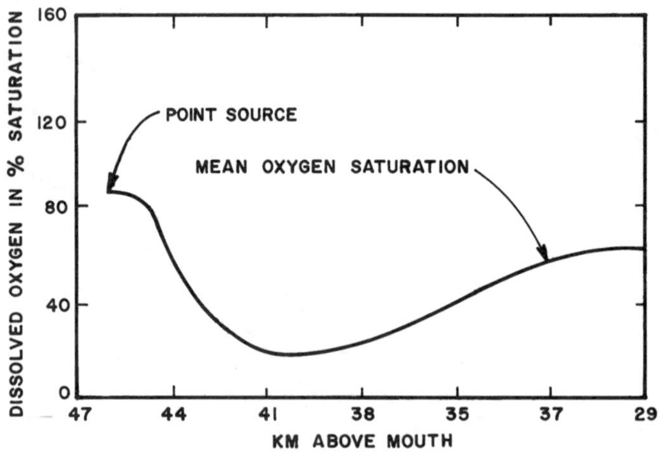


**Figure 1.2 Dissolved oxygen sag caused by nitrification (Clinton River, Michigan, August 23–25, 1960).**

One explanation for this phenomenon may be that in natural waters, flowing or otherwise, the population of suspended nitrifying bacteria is small. Because nitrifying bacteria are slow-growing organisms, it is impossible for them to reproduce fast enough to create conditions as shown in Figure 1.2 within any short distance downstream of a point source. Because the normal habitats for nitrifying bacteria are attached growths, they occur in great populations in slimes on rocks and other surfaces. For this reason, the nitrification rate in a stream is a function of the ratio of water volume to surface contact area. This explains why nitrification is rapid in shallow streams and very slow in deep ones. In effect, nitrification rate, like reaeration, is inversely related to depth. This hypothesis finds strong confirmation in a study conducted on the Willamette River, Oregon (USGS, 1985). Besides nitrification bacteria habitat, two other factors related to nitrification rate in streams are water temperature and oxygen. In shallow streams, both average water temperature and oxygen are significantly higher than in deep streams. As a result, larger water volume is involved in the nitrification in shallow streams. In deep streams, only the surface layer sustains nitrification bacteria because it is oxygen rich and warm enough. Problems involving NOD can be solved by the conversion of all ammonia to nitrates in the WWTP.

**Nitrogen and Phosphorus.** The fertilizing effects of nitrogen and phosphorus in receiving waters vary in different depths. In deep rivers, there is stimulation of phytoplankton. Rooted aquatics are stimulated in shallow waters, along shorelines, and in backwaters.

Typically, the effects are not serious in deep waters because of widely fluctuating water levels and flushing actions that occur at high river stages. However, the situation is different in shallow streams where rooted aquatics gain enough sunlight to thrive. Stimulation of growths can be so severe that natural flow may be impeded and backflooding may occur. If this happens in a stream draining from a lake, water levels in the lake can become uncontrollable. Dredging to remove the aquatic growths may become necessary. In some instances, severe diurnal fluctuations in dissolved oxygen can occur in stream waters passing through weed beds, as shown in Figure 1.3.

The data in Figure 1.3 were obtained at three stations downstream of a WWTP. Station A was first in a line of flow and received the full impact of residual biochemical oxygen demand (BOD) plus ammonia and phosphorus. Stations B and C were further downstream and were spared the full load of BOD that Station A received. During the day, submerged, in-stream growths produced oxygen in excess of the demands imposed by the BOD and NOD of the treatment plant effluent, particularly at B and C. At night, the demands

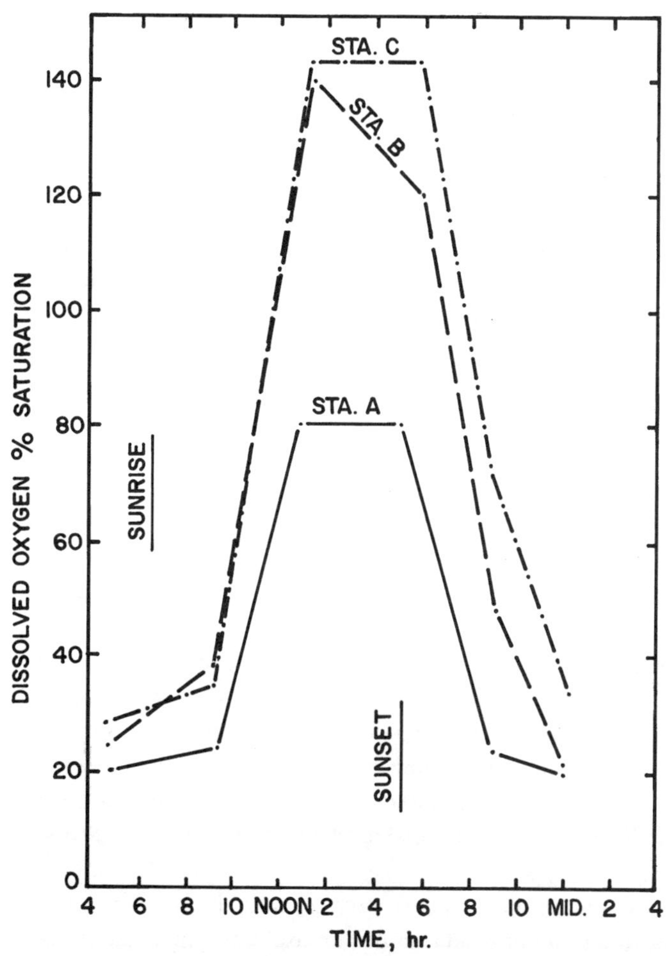


**Figure 1.3** **Dissolved oxygen profiles (Spring Creek, Pennsylvania).**

imposed by the effluent were increased by respiration of the plants, and dissolved oxygen levels fell so low that they would not support fish life. This caused a catastrophe at a downstream fish hatchery that used diverted stream waters.

**STATIONARY WATER BODIES.** The most damaging and obvious effects of nutrients in lakes, ponds, and reservoirs are stimulation of algal growths that typically occur in pulses referred to as *blooms.* Of particular concern are algae such as *Spirogyra, Anabaena, Aphanizomenon, Gleotrichia,* and others that tend to float to the surface at certain stages in their life cycles and are wafted about by gentle breezes to form accumulations in coves and along shorelines. Such aggregations can interfere with normal recreational activities such as swimming and cause serious nuisances because of algal decay, subsequent odorous bacterial decomposition, and unsightliness. Odor problems can be particularly offensive when high-protein blue-green algae, such as *Anabaena, Aphanizomenon,* and *Gleotrichia,* are involved.

Algal bloom nuisances have sparked legal controversies in the past. Perhaps the classic case was one in Madison, Wisconsin, involving lake property owners, the state legislature, and the courts. Deliberations went on for 18 years before the Metropolitan Sewerage District finally was forced to divert its WWTP effluent away from downstream lakes.

The algal bloom phenomenon is a simple biological process. Death is the ultimate fate of photosynthetic plants such as algae and rooted aquatics that thrive on plant nutrients. At the end of their lives, they settle in with previous crops in bottom sediments, where they then undergo bacterial decomposition. In decomposing under aerobic conditions they consume oxygen, which eventually becomes depleted. This condition is quite normal in lakes that are deep enough to stratify. When oxygen is reduced below certain levels, the effects on fish life and other oxygen-dependent organisms are disastrous. Nitrates in the water may temporarily prevent the formation of anaerobic conditions, but their subsequent depletion will allow anaerobic conditions to develop. Release of hydrogen sulfide can also result from the reduction of sulfates.

Bacterial decomposition under both aerobic and anaerobic conditions then leads to a phase that perpetuates the blooming process. The decaying bacteria release nitrogen as ammonia and phosphorus as phosphates. Small amounts also are released in soluble organic compounds, and a fraction of the nutrients always remains as undigestible residues in bottom deposits. However, the predominant release is of prime nutrients, which are then recycled to waters above.

In bodies of water that do not stratify, all water is considered in complete circulation as a result of wind action. Because of this mixing, nutrients released from bottom deposits are quickly carried back to waters in the eutrophic zone, where they become free to support new growths of photosyn-

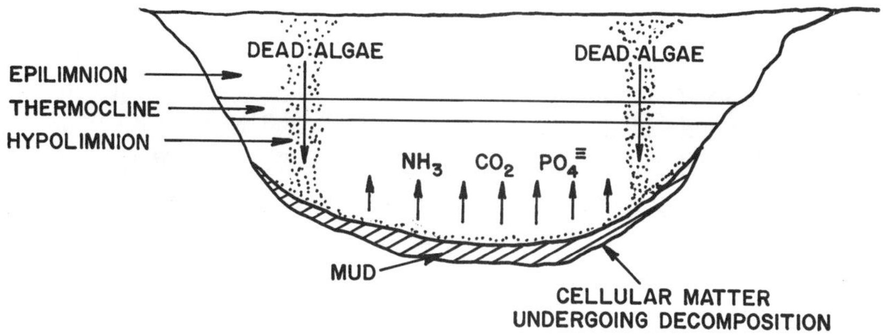


**Figure 1.4 Backfeed of nutrients from bottom.**

thetic organisms. In deep lakes that do not stratify during the summer, the nutrients released are circulated and stored within the hypolimnion (see Figure 1.4). Small amounts may diffuse through the thermocline to the epilimnion and reach the euphotic zone. However, the major part of the nutrients released remains effectively sealed off until the fall overturn when the hypolimnion is disrupted by falling water temperatures and circulation patterns caused by strong winds. The nutrients then become distributed throughout the entire body of water. This seasonal enrichment of waters in the euphotic zone often results in fall blooms of algae.

The most damaging potential effect of a nutrient release like this will occur in lakes, ponds, and reservoirs with long retention periods. Only ponds without outlets can be considered strictly stationary; however, all lakes and reservoirs have finite flows over time that range from a few days to many years. Typically, at least an average 20-day detention time is required for released nutrients to develop serious algal growths. At shorter retention times, bodies of water assume characteristics similar to rivers.

**LAND ENVIRONMENT.** Wastewaters bearing nutrients are introduced to the land environment for any one of four reasons:

- Irrigation water,
- Nutrient removal,
- Removal of BOD and suspended solids, or
- Groundwater recharge.

**Irrigation.** Use of treated wastewater for irrigation purposes has been widely practiced around the world, particularly in semiarid or arid regions where rainfall is insufficient to interfere with land application as required by effluent quantities. In other areas, wastewater augments natural waters from pumped or stored supplies. In the U.S., it has been used on a wide variety of cultivated crops, pastures, and forest lands where disease transmission is not a factor.

Wastewater from domestic sources is highly prized by agriculturists because of nutrient content, primarily nitrogen, phosphorus, and potassium. Wastewater has been used for irrigation purposes for several years in California and Florida without apparent damage to the land. Numerous studies have indicated that crop, pasture, and forest lands effectively remove phosphorus and a large part of nitrogen from the water. The benefits derived from irrigation with wastewater of domestic origin are considerable with respect to nutrient removal. This form of wastewater treatment constitutes the highest order of practice of returning to the land the things that came from the land.

**Nutrient Removal.** In some instances, nutrient removal from wastewater is necessary to protect receiving waters. This can be accomplished by advanced or land treatment methods. Two notable cases where land treatment has been used for nutrient removal are State College, Pennsylvania, and Muskegon County, Michigan.

At State College (Penneypacker *et al.*, 1967), serious degradation of Spring Creek had occurred (see Figure 1.3). To relieve this situation, controlled experiments were conducted using the WWTP effluent to irrigate crop and forest lands in the Spring Creek watershed. Studies conducted on the groundwater draining to Spring Creek from the experimental area showed it to be effectively free of phosphorus and containing a significant fraction of nitrogen. The nitrogen passing through the soil was converted to nitrates.

At Muskegon, which is located on the eastern shore of Lake Michigan, a large irrigation system was developed to remove nutrients from wastewater as a means of protecting the lake. Because of location and climatic conditions, wastewater storage is needed for approximately 6 months of flow. Percolate from the Muskegon County system is collected in underdrains and discharged to a nearby stream. Phosphorus removal has averaged 95 to 98% with percolate total phosphorus concentrations of 0.05 to 0.1 mg/L. Nitrogen concentrates in the applied wastewater have averaged 8.2 mg/L, and the percolate total nitrogen has been 2.5 mg/L.

**Groundwater Recharge.** In many areas of the world, groundwater sources have been pumped out in excess of natural recharge. As a result, water levels have fallen and pumping costs have increased because of the greater lifts that must be overcome. In some cases, wells have gone dry and, in locations near the ocean, saltwater intrusions have damaged aquifers. In some situations, such as in Peoria, Illinois, recharge has been accomplished by diverting river water to recharge pits during the winter time when water temperatures are low. In this way, an adequate domestic water supply and suitable water for cooling purposes in industrial operations are kept available.

To correct the groundwater situation in many areas, particularly in semi-arid areas and in some densely populated places, such as Nassau County on Long Island, New York, projects involving recharge with WWTP effluent have been initiated (Asano, 1985). Groundwater recharge is accomplished by either of two methods: deep well injection or surface spreading with resulting percolation.

**Deep Well Injection.** In deep well injection, nitrogen, phosphorus, and the remaining suspended solids present in the WWTP effluents are pumped into the aquifer along with the water. The Nassau County system and the Orange County, California, Water Factory No. 21 are two WWTPs designed to produce renovated wastewater suitable for well injection (Asano, 1985). In each case the primary purpose of injection is to arrest seawater intrusion.

**Surface Spreading and Percolation.** Where appropriately sloped land is available at reasonable cost, surface spreading and percolation often is the preferred method of groundwater recharge. Natural recharge, however, involves rainwater or runoff after contact with the soil. Typically, it is low in total dissolved solids (TDS) and carries little, if anything, that can be altered during passage through the soil. Because of the natural solvent powers of the water and carbon dioxide it can carry, natural water can dissolve considerable amounts of mineral matter while passing through soils. On the other hand, wastewater is typically high in TDS and does contain both phosphates and nitrogenous compounds that can undergo significant changes in the soil.

One of the first large-scale facilities for groundwater recharge by surface spreading of wastewater effluent built in the U.S. was constructed in 1962 at the Whittier Narrows plant in Los Angeles County, California (Crook *et al.*, 1990). Smaller installations of an experimental or permanent nature exist in other locations. Most of these have operated with a reasonable degree of success. However, stricter requirements for recharge are becoming increasingly evident.

# *NUTRIENT FORMS, SOURCES, AND CYCLES*

**NITROGEN COMPOUNDS.** The nitrogen cycle shown in Figure 1.5 identifies the various chemical forms of nitrogen and illustrates generalized pathways of transformation. As shown, nitrogen can exist in many states of oxidation, ranging from reduced ammonia/ammonium to highly oxidized nitrates. In all, nitrogen can exist in seven states of oxidation, as shown in Table 1.1.

However, only ammonia (and organic derivates), nitrogen gas, nitrites, and nitrates have significance in aquatic environments. These compounds are considered the major nitrogen sources for most microorganisms in soil and water. Microorganisms use these compounds for amino acid production, protein synthesis, cell growth, and energy transfer. In the process, the nitrogen compounds are transformed and changed to other oxidative states. Figure 1.6

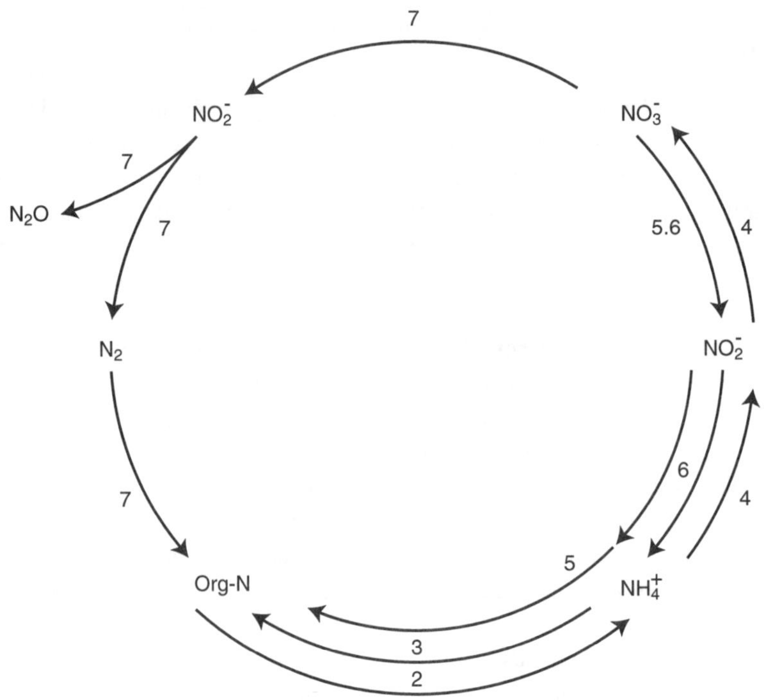


**Figure 1.5** **Schematic view of the biotic part of the nitrogen, biogeochemical cycle (1 = nitrogen fixation, 2 = mineralization, 3 = immobilization [ammonia assimilation], 4 = nitrification, 5 = immobilization [nitrate and nitrite assimilation], 6 = nonassimilatory nitrate reduction, and 7 = denitrification).**

illustrates the many microbiological processes that use and transform nitrogen compounds.

**NITROGEN SOURCES.** Because nitrogen compounds are found throughout the environment, the potential sources are infinite, though urea and fecal matter are the primary sources in wastewater. Biological fixation of atmospheric nitrogen serves as a significant source of natural nitrogen input to wastewater. Chemical fertilizers, runoff, and industrial discharges top the list of human-made sources. Industrial discharges typically high in nitrogen include feedlots, rendering plants, milk and tobacco processors, petroleum refineries, dye manufacturers, meat and poultry processors, certain synthetic fiber plants, and industries that clean with ammonia-containing compounds. Table 1.2 illustrates the variety of nitrogen content originating from industrial discharges.

By far, the single largest source of nitrogen in wastewater is human fecal matter. Raw wastewater typically contains only nitrogen and ammonium. Organic nitrogen comes from proteins, amines, nucleic acids, peptides, amino

**Table 1.1 The seven states of oxidation in which nitrogen can exist.**

| Compound | Formula | Valence |
|---|---|---|
| Ammonia/ammonium | $NH_3$-N/$NH_4$-N | −3 |
| Nitrogen gas | $N_2$ | 0 |
| Nitrous oxide | $N_2O$ | +1 |
| Nitric oxide | NO | +2 |
| Nitrite | $NO_2$-N | +3 |
| Nitrogen dioxide | $NO_2$ | +4 |
| Nitrate | $NO_3$-N | +5 |

acids, and other cellular constituents that are excreted in fecal matter. Ammonium originates from bacteria mineralization of proteins and urea, which is the principal component of urine. Nitrites and nitrates typically are not found in raw wastewater because most raw wastewater does not contain free oxygen for nitrification. Nitrates are only found in raw wastewater that has significant contributions from industries or from drinking water with high nitrate concentrations.

Ammonium constitutes approximately 60% of the total nitrogen in raw wastewater. Deamination of proteins and hydrolysis of urea are the primary

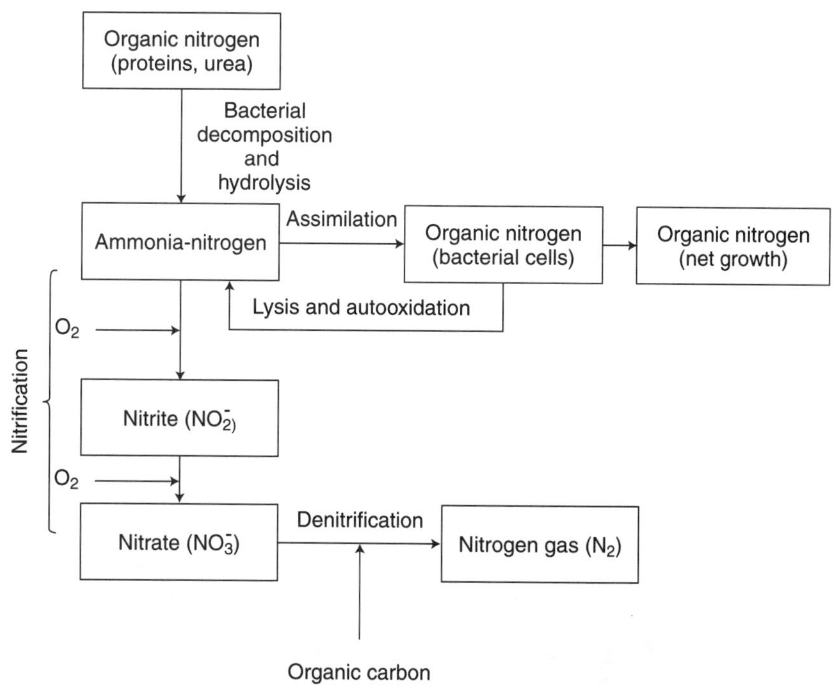


**Figure 1.6 Nitrogen transformations in biological treatment processes.**

**Table 1.2 Nitrogen concentrations in industrial wastewater.**

| Industry | Organic nitrogen, mg/L | | Ammonia-nitrogen, mg/L | | Total nitrogen, mg/L | |
|---|---|---|---|---|---|---|
| | Mean | Range | Mean | Range | Mean | Range |
| Meat packing | | | | | 52 | |
| Slaughter house | | | | | 115 | |
| Packing house | | | | | 76 | |
| Processing plant | | | | | | |
| Fish and seafood | | | | | | |
| Catfish | 32.7 | 20.7–49.8 | 0.8 | 0.5–1.4 | | |
| Crab | 94 | 58–138 | 6 | 4–9 | | |
| Shrimp | 215 | 164–266 | 6.9 | 4.4–9.5 | | |
| Tuna | 30 | | 7.6 | 2.2–13.0 | | |
| Combined poultry | | | | | | 15–300 |
| Fruit and vegetable processing | | | | | | |
| Apples | | | | | 3.88 | |
| Citrus fruits | | | | | 20.8 | |
| Potatoes | | | | | 5.5 | |
| Winery wastewater | | | | | 20 | 5–40 |
| | | | | | 40 | 10–50 |
| Chemical industries | | | | | | |
| $NH_3$-N, mg/L | | | | | 1 270 | |
| $NO_3$-N, mg/L | | | | | 550 | |

sources of this ammonium. However, ammonium may ionize to ammonia, depending on the pH and temperature of the wastewater. At pH 8 and below, the proportion of ammonium typically exceeds 99%. The equilibrium shifts to 50% at approximately pH 9.5. At pH 11, ammonia predominates. The ammonia and ammonium equilibrium is described by the following expression:

$$K_a = [NH_3][H^+]/[NH_4^+] \qquad (1.4)$$

Where

$[NH_3]$ = ammonia concentration, moles/L;
$[H^+]$ = hydrogen ion concentration, moles/L;
$[NH_4^+]$ = ammonium ion concentration, moles/L; and
$K_a$ = ionization constant with the following values.

The following parameters apply:

- At 0°C, ionization constant = $8.26 \times 10^{-11}$;
- At 10°C, ionization constant = $1.86 \times 10^{-10}$;
- At 15°C, ionization constant = $2.73 \times 10^{-10}$;
- At 20°C, ionization constant = $3.98 \times 10^{-10}$; and
- At 25°C, ionization constant = $5.68 \times 10^{-10}$.

During wastewater treatment most of the organic nitrogen is metabolized to ammonium as shown in Figure 1.6. The ammonium can be assimilated into bacteria (immobilization) or further oxidized to nitrates (nitrification). Biological nitrogen removal processes ensure conversion of organic nitrogen and ammonium to nitrate-nitrogen. In addition, these processes provide oxygen-deficient environments that allow the nitrates to be converted to molecular nitrogen (denitrification). These principles are discussed in Chapter 4. However, none of these processes removes all of the influent nitrogen. A small portion of the organic ammonia is not metabolized to ammonium and passes through these treatment processes as nonbiodegradable total Kjeldahl nitrogen (TKN). It is believed that this portion of the TKN consists of nitrogen heterocycles such as purine, pyridine, and pyrimidine. The nitrogen in these compounds is contained within a ring structure and is not easily degraded. However, the total quantity of all the nonbiodegradable TKN typically is 1 mg/L, with a range of 0.5 to 2.5 mg/L.

Other sources of nitrogen that are often overlooked include wastewater sludge treatment processes and septage disposal. Anaerobic digesters and sludge dewatering processes recirculate a significant amount of nitrogen back to the plant influent. Odor scrubbers, dewatering sidestreams, and thickener overflows also contribute to the nitrogen recycle. It is not unusual for the recycled nitrogen to increase the raw wastewater nitrogen concentrations by 20 or 30%. In addition, each litre of septage contains 25 times more nitrogen than most raw wastewater. Therefore, it is important to account for recycled nitrogen and septage inputs when developing nutrient control strategies.

**TYPICAL NITROGEN CONCENTRATIONS.** Typical nitrogen concentrations for raw domestic wastewater are listed in Table 1.3. The nitrogen quantities are expressed in terms of mass per unit volume (milligrams per litre and pounds per million gallons). Influent TKN typically averages 30 mg/L with ammonium at approximately 18 mg/L.

Table 1.4 provides typical removal efficiencies for progressive treatment processes. Nitrogen removal by primary clarification is mostly because of settlement of gross solids containing organic nitrogen. Secondary treatment efficiencies are primarily because of bacteria assimilation and subsequent sludge wasting. Advanced nutrient removal processes are required to increase nitrogen removal efficiencies beyond 30%. These nutrient removal processes are discussed in later chapters.

**PHOSPHORUS.** Unlike nitrogen, phosphorus only exists in aquatic environments in the +5 valence state. In essence, this limits the types of naturally occurring phosphorus compounds to salts and esters of phosphoric acid. Phosphoric acid is easily ionized to inorganic phosphates such as orthophosphates, which are readily available for biological metabolism. Bacteria and many other organisms use orthophosphates to create organic phosphates and build solid-phase structures. Calcium phosphate is the basic component of bone. Organic phosphates are found in nucleic acids, phospholipids, adenosine triphosphate (ATP), hormones, and many other compounds that are essential to energy and reproduction.

Adenosine triphosphate is probably the most important organic phosphate. It is an energy-rich compound that is essential to energy exchange in both plant and animal cells. Adenosine triphosphate is synthesized by processes that supply energy to the microorganism (for example, sunlight and organic metabolism). Processes that require energy degrade ATP by cleaving a phosphate bond. Cleaving this high-energy bond releases a relatively large

**Table 1.3 Nitrogen in domestic wastewater.**

| | Concentration (as N)[a] In mg/L | In lb/ mil. gal[b] | Generation, kg/cap/a (lb/cap/yr)[c] |
|---|---|---|---|
| Total | 20–85 | 167–709 | 3.4–5.0 |
| Organic | 8–35 | 67–292 | (7.5–11) |
| Free ammonia/ammonium | 12–50 | 100–417 | (4.8–8.8) |

[a] Typical concentrations in fresh domestic wastewater.
[b] lb/mil. gal × 0.119 8 = $g/m^3$.
[c] lb/cap/yr × 0.453 6 = kg/cap/a.

**Table 1.4 Typical removal efficiencies, total nitrogen.**

| Treatment level | Removal of influent concentration, % |
|---|---|
| Primary | 5–10 |
| Secondary | 10–20 |
| Secondary with nitrification | 20–30 |

amount of energy that is used for synthesis of proteins, reproduction, and movement.

Besides orthophosphates and organic phosphorus, polyphosphates are the third type of phosphorus compound with pollution-related significance. Polyphosphates are synthetic compounds that are condensed and dehydrated versions of complex orthophosphates. Polyphosphates typically are used in water systems to protect distribution pipes from corrosion. All polyphosphates eventually hydrolyze in aqueous solution and revert to orthophosphate. The speed of reversion typically is slow but can be increased at lower pH and higher temperatures.

Because phosphates are nonvolatile, their occurrence is limited to soil and aquatic environments. Figure 1.7 illustrates the various paths that phosphorus compounds travel in the aquatic environment. As shown, sediment in water bodies is a sink for organic and inorganic phosphorus. Fertilizer-laden runoff and erosion contribute to this phosphorus deposition. Anaerobic activity at the bottom releases orthophosphates back to the standing water, where it is consumed by bacteria and algae. The amount of phosphorus in the water depends on which phase of this cycle is dominant and the amount of phosphorus added by runoff.

The most significant sources of phosphorus in wastewater are human excrement, food wastes, synthetic laundry detergents, household cleaners, and industrial and commercial discharges. Polyphosphate added to drinking water is an additional source, but its contribution varies with the utility. Those utilities that use polyphosphates typically apply it at small doses that rarely exceed 0.25 mg/L. Phosphorus excreted by humans has been estimated at 0.2 to 1.0 kg/cap·a, with an annual mean of 0.6 kg phosphorus/cap·a. The next most significant phosphorus source is synthetic laundry detergents. The amount of phosphorus in wastewater attributed to detergents has been estimated at 0.3 kg/cap·a. This contribution has been determined by monitoring wastewater phosphorus concentrations before and after phosphate detergent bans. Because household cleaners and commercial detergents typically are excluded in phosphate detergents ban, their contributions remain undetermined. However, the soap and detergent industry has proposed 0.1 kg/cap·a as an estimate of its contribution to domestic wastewater (Jenkins and Hermanowicz, 1991). As usual, industrial and commercial contributions are highly variable, with fertilizer manufacturers, feed lots, and meat, milk, and food processors among the significant sources.

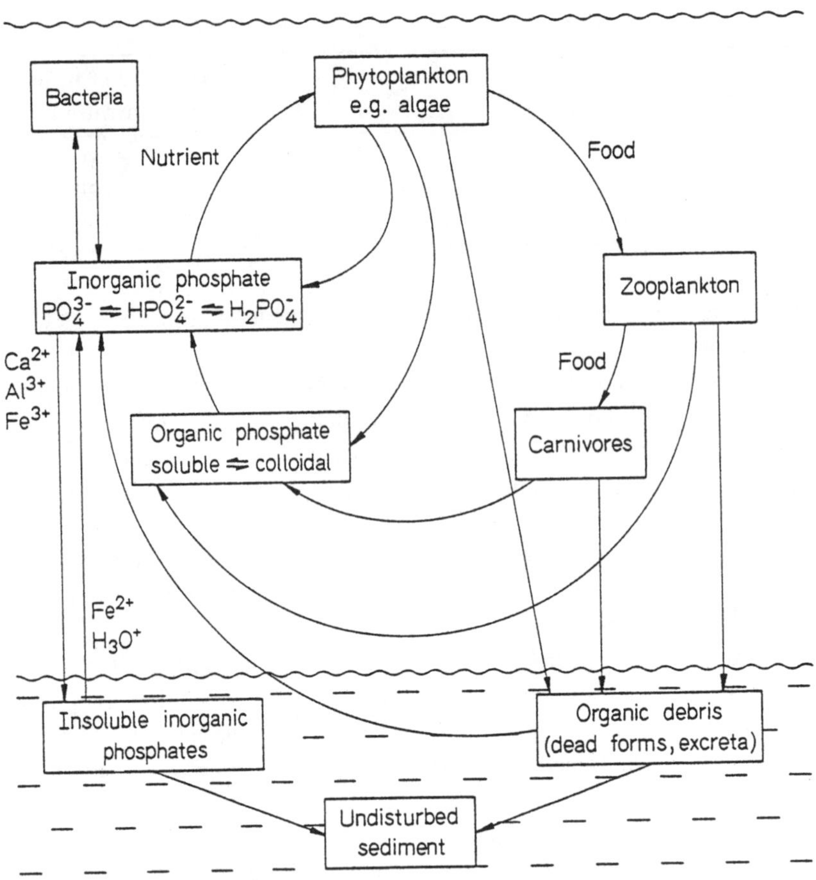


**Figure 1.7 The water-based phosphate biocycle.**

**TYPICAL PHOSPHORUS CONCENTRATIONS.** The concentration of phosphorus in domestic wastewater is affected by each of the previously cited sources. In general, phosphorus contributions from human excrement, food wastes, household cleaners, and commercial detergents remain relatively consistent throughout the U.S. This is also true for industrial discharges, which are required to meet pretreatment standards. Therefore, the formulation of detergents within the wastewater collection area represents a significant variable in determining the wastewater phosphorus concentration. Table 1.5 illustrates typical wastewater phosphorus concentrations for several facilities in Virginia and Maryland before and after phosphate detergent bans. The phosphate detergent bans went into effect on December 1, 1985, in Maryland and on January 1, 1988, in Virginia. The average total phosphorus concentration was 8 mg/L before the phosphate bans and 5.4 mg/L after the bans. The average orthophosphate concentration was 6 mg/L before and 3.7 mg/L after the phosphate bans. Total phosphorus reductions at individual plants ranged from 15 to 50%, with an average of 32%. Almost all of this reduction was a result of orthophosphate reduction.

**Table 1.5 Typical phosphorus concentrations before and after phosphate bans.**[a]

| Plant influent data | Preban, mg/L | | Postban, mg/L | | Reduction, % | |
|---|---|---|---|---|---|---|
| | Total phosphorus | Orthophosphate | Total phosphorus | Orthophosphate | Total phosphorus | Orthophosphate |
| Alexandria | 5.22 | | 4.46 | | 14.6 | |
| Arlington | 6.17 | | 4.80 | | 22.2 | |
| Army Base | 6.20 | 4.70 | 4.18 | 2.68 | 32.6 | 43.0 |
| Boat Harbor | 7.30 | 5.52 | 5.24 | 3.80 | 28.2 | 31.2 |
| Chesapeake/Elizabeth | 9.60 | 7.78 | 5.93 | 4.62 | 38.2 | 40.6 |
| Falling Creek | 6.84 | 4.56 | 4.68 | 3.07 | 31.6 | 32.7 |
| James River | 9.20 | 7.36 | 6.10 | 3.78 | 33.7 | 48.6 |
| Lower Potomac | 7.51 | | 7.01 | | 6.7 | |
| H L Mooney | 9.90 | | 5.50 | | 44.4 | |
| Nansemond | 9.30 | 8.20 | 4.94 | 4.06 | 46.9 | 50.5 |
| Quantico | 8.05 | | 6.36 | | 21.0 | |
| Williamsburg | 7.26 | 4.30 | 6.22 | 3.77 | 14.3 | 12.3 |
| York River | 8.80 | | 4.88 | | 44.5 | |
| Horsepen Branch | 9.40 | | 5.94 | | 36.8 | |
| Parkway | 7.22 | | 5.48 | | 24.1 | |
| Damascus | 12.04 | | 7.20 | | 40.2 | |
| Western Branch | 8.15 | | 5.12 | | 37.2 | |
| Seneca | 7.32 | | 5.35 | | 26.9 | |
| Piscataway | 6.49 | | 3.96 | | 39.0 | |
| Average | 8.00 | 6.06 | 5.44 | 3.68 | 32.0 | 39.2 |

[a] Maryland phosphate ban initiated December 1, 1985; Virginia phosphate ban initiated January 1, 1988.

# EFFLUENT LIMITS

**CONCENTRATION AND MASS LOADINGS CRITERIA.** In the U.S., the effluent limits for nutrient removal are not uniform. The limits vary based on the receiving body and its water use, water quality, and relative pollutant loads to each water body. Regional water quality criteria for nutrients may also be established by state and local agencies. Recently, the European community also initiated programs to control the discharge of nitrogen and phosphorus. The trend for limiting the discharge of nutrients to more stringent levels is continuing.

**AMMONIA TOXICITY.** According to U.S. Environmental Protection Agency (U.S. EPA) water quality criteria, concentrations of total ammonia, which contain an un-ionized ammonia concentration of 0.020 mg/L of $NH_3$, are considered toxic for freshwater aquatic life. When ammonia dissolves in water, some of the ammonia reacts with the water to form ammonium ions. The toxicity of ammonia depends on pH, the concentration of ammonia, and temperature and ionic strength. Percent of ammonia for aqueous ammonia solutions at different pH and temperature values can be used to determine the amount of ammonia that is toxic. Many states have adopted similar water quality standards and use them in conjunction with receiving water quality data and establishing discharge limits. U.S. EPA has proposed specific toxicity test procedures to determine the toxicity of a given solution. The test procedures are useful in identifying the toxicity of a given wastewater for specific wastewater samples (U.S. EPA, 1985b).

# ANALYTICAL TECHNIQUES

**NITROGEN ANALYSES.** It is typical to report the concentration of nitrogen compounds in terms of mg/L as nitrogen. For example, 1.00 mg/L nitrate should be reported as 0.23 mg/L nitrate-nitrogen. Part 4500-N of the 19th edition of *Standard Methods for the Examination of Water and Wastewater* (APHA *et al.*, 1995) provides detailed instructions for determining the concentrations of nitrogen compounds in wastewater samples.

Ammonia-nitrogen and organic nitrogen compounds are both determined by the TKN procedure. This is why the sum of ammonia-nitrogen and organic nitrogen is often reported as TKN. Both macro- and micro-Kjeldahl methods can be used, depending on the amount of nitrogen in the sample. Micro-Kjeldahl methods should be limited to samples containing concentrations of organic nitrogen in the range of 0.2 to 2 mg total nitrogen. In either method, ammonia is first determined by distilling the sample and capturing the free ammonia in a boric acid solution. The amount of ammonia is then determined by colorimetric analysis, titration, or an ammonia-selective electrode. The titration method is used for ammonia concentrations greater than 5 mg/L ammonia-nitrogen. The electrode method is applicable in a range of 0.03 to 1 400 mg/L. Colorimetric analyses include nesslerization and phenate methods, which are limited to lower concentrations.

Following ammonia distillation, organic nitrogen compounds are digested and converted to ammonia. The sample is digested with heat, sulfuric acid, potassium sulfate, and a mercuric sulfate catalyst. Ammonia is immobilized as a nonvolatile sale (ammonium sulfate) in the acid medium. The digested solution is then made alkaline and the ammonia is distilled and captured in boric acid. Again, the amount of ammonia is determined by the same methods previously described.

The determination of nitrate- and nitrite-nitrogen concentrations involve relatively complex procedures. These procedures have limited ranges of color and turbidity. The ion chromatograph method determines nitrate and nitrite concentrations directly. The cadmium reduction method reduces nitrate to nitrite and determines nitrite concentrations colorimetrically. Determining nitrite concentrations before cadmium reduction differentiates the quantities of each compound. This method is limited to nitrite concentrations between 0.01 and 1.0 mg/L. Nitrate electrodes can be used to measure nitrate concentrations between 0.14 and 1 400 mg/L.

**PHOSPHORUS ANALYSES.** It is typical to segregate orthophosphates, polyphosphates, and organic phosphorus into dissolved (soluble) and suspended (particulate) fractions. Orthophosphates and polyphosphates are mostly soluble. Organic phosphorus may be dissolved or suspended. This is because organic phosphorus can be found in a variety of cellular materials. Dissolved phosphorus typically is determined by filtering the sample through a 0.45-$\mu$m pore diameter membrane filter. Although this procedure does not constitute a true separation of suspended and dissolved forms of phosphorus, it is accepted as a convenient demarcation. Most data reported in the literature are based on using this technique. Particulate (suspended) phosphorus can be determined by analyzing an unfiltered sample (total phosphorus) and a filtered sample, and subtracting the difference between the two results. Methods for analysis of phosphorus are presented in part 4500-P of *Standard Methods* (APHA *et al.*, 1995).

# REFERENCES

American Public Health Association, American Water Works Association, and Water Environment Federation (1995) *Standard Methods for the Examination of Water and Wastewater.* 19th Ed., Washington, D.C.

Asano, T. (1985) *Artificial Recharge of Groundwater.* Butterworth Publishers, Boston, Mass.

Crook, J., *et al.* (1990) Groundwater Recharge with Reclaimed Water in California. *Water Environ. Technol.*, **2**, 8, 42.

Federal Water Pollution Control Administration (1968) Water Quality Criteria. Superintendent of Documents, U.S. Gov. Printing Office, Washington, D.C.

Jenkins, D., and Hermanowicz, S.W. (1991) Principles of Chemical Phosphate Removal. In *Phosphorus and Nitrogen Removal from Municipal Wastewater: Principles and Practices.* 2nd Ed., R.I. Sedlak (Ed.), Lewis Publishers, Boca Raton, Fla.

Penneypacker, S.P., *et al.* (1967) Renovation of Wastewater Effluent by Irrigation of Forest Land. *J. Water Pollut. Control Fed.*, **39**, 285.

U.S. Environmental Protection Agency (1985a) Ambient Water Quality Criteria for Ammonia. EPA-440/5-85-001, Washington, D.C.

U.S. Environmental Protection Agency (1985b) Methods for Measuring the Acute Toxicity of Effluents to Freshwater and Marine Organisms. EPA-600/4-85-013, Cincinnati, Ohio.

U.S. Geological Survey (1985) Methodology for River-Quality Assessment with Application to the Willamette River Basin, Oregon. Geological Survey Circular 715M, Arlington, Va.

# Chapter 2 Overview of Nutrient Removal Processes



This chapter presents an overview of various nutrient removal processes. The reader is referred to other chapters for a more detailed description of the theory and specific application of the various processes. The processes described in this chapter are grouped as follows:

- Target nutrient removed: nitrogen and phosphorus.
- Mechanism of removal: physical, chemical, and biological.

Many of the processes shown here are protected by U.S. patents. While some patents have expired, a number still remain in effect.

# NITROGEN REMOVAL

Table 2.1 summarizes the various options for removing or converting nitrogen from one species to another. To completely remove nitrogen from the system, five processes are available:

- Conversion of nitrogen to nitrogen gas, $N_2(g)$, which escapes into the atmosphere. This is achieved in biological treatment systems through nitrification, followed by denitrification (breakpoint chlorination).
- Biological uptake of nitrogen for growth of biomass.
- Stripping of ammonia from the water because $NH_3(g)$ can be achieved at high pH.
- Ion exchange to chemically exchange nitrogen ions as $NH_4^+$ or as $NO_3^-$ using a cation or an ion-exchange resin, respectively. For example, the naturally occurring zeolite clinoptilolite will selectively remove ammonium.
- Processes that will remove essentially all pollutants from water, such as reverse-osmosis membranes, can be used to remove nitrogen. Efficiency varies with the type of membrane and nitrogen species.

Figure 2.1 summarizes the various processes that can be used to remove nitrogen from wastewater.

**Table 2.1 Nitrogen removal and conversion processes.**

| Initial species | Ultimate species | Process |
|---|---|---|
| Organic-N | Ammonia, $NH_4^+$-N | Ammonification<br>Biological conversion of organic nitrogen to ammonia |
| Ammonia, $NH_4^+$-N | Nitrate, $NO_3^-$-N | Biological nitrification |
| | Ammonia gas, $NH_3(g)$ | Stripping |
| | Organic nitrogen | Biological uptake during bacterial growth |
| | Nitrogen gas, $N_2(g)$ | Breakpoint chlorination |
| | Resin | Ion exchange will exchange ammonia for another cation |
| Nitrate, $NO_3^-$-N | Nitrogen gas, $N_2(g)$ | Biological denitrification |
| | Resin | Ion exchange will exchange nitrate for another anion |

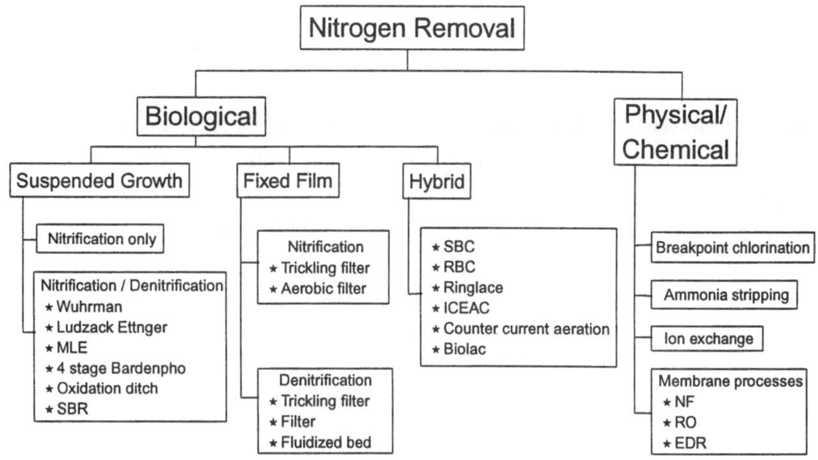


**Figure 2.1 Processes used for nitrogen removal.**

**BIOLOGICAL NITROGEN REMOVAL.** Biological nitrogen removal is achieved through a series of biochemical reactions that transform nitrogen from one form to another. The key transformations are

- Nitrification—the conversion of ammonium, $NH_4^+$, to nitrite, $NO_2^-$, and finally to nitrate, $NO_3^-$. The conversion to nitrite is catalyzed by *Nitrosomonas* sp. and the conversion to nitrate by *Nitrobacter* sp. In practice, the kinetics of the overall process are limited by *Nitrosomonas* sp. and nitrite is rapidly converted to nitrate by *Nitrobacter* sp. One therefore typically finds little (<0.2 mg/L) nitrite in solution if the reactions occur under optimum conditions.
- Denitrification—in the absence of dissolved oxygen, bacteria will use nitrate as a terminal electron acceptor and convert nitrate to nitrogen gas. Denitrification occurs under two distinct conditions:

  — Rapid denitrification is achieved when an external substrate (such as wastewater or methanol) is available for bacterial growth. Substrate level denitrification is relatively rapid and proceeds typically at 0.03 to 0.11 kg $NO_3^-$-N/kg VSS·d (SDA, 1989). The concentration of available substrate (food-to-microorganism ratio) and type of substrate will affect the denitrification rate.

  — Slow denitrification occurs when bacteria use nitrate under conditions without an external substrate source, also known as *endogenous level denitrification*. Endogenous level denitrification is slow and proceeds typically at rates between 0.01 and 0.03 kg $NO_3^-$-N/kg VSS·d (Stensel, 1981). Endogenous-level denitrification rates are related to sludge age, or the active mass fraction.

See Chapter 4 for a detailed discussion of the stoichiometrics, reaction rates, and other factors affecting biological nitrogen removal.

**Suspended-Growth Processes.** The activated-sludge process is a typical suspended-growth system, where bacteria are kept in suspension under appropriate conditions to allow the bacteria to grow and consume pollutants from the water. Earlier, nitrification was separated from carbon oxidation (biochemical oxygen demand [BOD] removal) in activated-sludge plants in a two-stage process by using a high-rate, first-stage activated-sludge process for BOD removal, followed by a low-rate nitrification stage (see Chapter 4). However, carbon oxidation and nitrification can proceed readily in the same aeration basin under favorable conditions.

For a suspended-growth system, it is important to maintain a sufficiently high sludge age to allow the slow-growth nitrifiers to reproduce in the aeration basin and prevent them from being wasted out of the system. The minimum required sludge age for nitrification is determined by temperature, pH, and other operating conditions as discussed in Chapter 3. Longer sludge ages are often used to provide a stable sludge and for other process considerations. When estimating the sludge age of nitrifier sludge mass, one must reduce the traditional sludge age by considering only the sludge mass that is kept under aerobic conditions, because nitrifiers are obligate aerobes and do not grow under anoxic or anaerobic conditions. Those zones used for denitrification or anaerobic zones should therefore not be included when calculating the nitrifier sludge age. See Chapter 3 for a detailed discussion.

After nitrification is established, effluent ammonia concentrations below 1 mg $NH_4^+$-N/L can be readily achieved using activated sludge at a high sludge age. The design and operation of a nitrification activated-sludge system is similar to a "normal" activated-sludge system operated for BOD removal only. However, the oxygen demand is significantly higher, and the operator must consider the effects of decreases in alkalinity and a possible drop in pH. Because the nitrifier growth is affected by pH, alkalinity supplementation may be required. Denitrification in suspended-growth plants can be readily achieved by creating zones depleted of oxygen by replacing aeration with mixing in some areas. If these zones, without dissolved oxygen, are created at the feed end of the aeration basin, the presence of BOD in the feed water provides substrate for denitrification, and substrate level denitrification will occur. If the zone is created toward the end of the basin where the substrate level is low, endogenous level denitrification will occur. An external substrate, such as methanol, can also be added to the denitrification basin to increase the denitrification rate.

Denitrification in activated-sludge secondary clarifiers will lead to the flotation of solids, because the gaseous $N_2(g)$ produced in the process becomes entrapped in the activated-sludge flocs and floats to the tank surface. This can be controlled by reducing $NO_3$ before the clarifier by denitrifying in the reactor, by a long solids retention time to produce a stable sludge, or by increasing the mixed liquor dissolved oxygen before the water enters the clarifier.

**Fixed Film Processes.** Fixed film processes, such as trickling filters and rotating biological contactors (RBCs), have been used effectively for nitrification. In this case, the process analysis is less straightforward than in activated sludge because bacteria are in direct competition for oxygen with heterotrophic bacteria (Okey and Albertson, 1989a and 1989b). Because the heterotrophic bacteria typically outgrow the nitrifiers, heterotrophic activity tends to dominate areas where the BOD is high, such as the top of a trickling filter and front end of an RBC. However, after the BOD loading decreases, oxygen becomes more available for nitrifiers and nitrification starts.

Trickling filters have been used successfully to achieve nitrification. Heterotrophic growth will typically dominate the top of the trickling filter, and nitrification will occur at the lower end of the trickling filter. Nitrifying trickling filters typically are tall, and forced ventilation is provided to improve oxygen supply. To reduce the heterotrophic zone, the U.S. Environmental Protection Agency (U.S. EPA, 1975) suggested maintaining the organic loading rate on trickling filters for combined BOD removal and nitrification at 0.16 to 0.19 g $NH_4^+$-N/m$^2$·d. Gullicks and Cleasby (1986) suggested that oxygen limitations may occur when the ammonia concentration exceeds 4 mg $NH_4^+$-N/L and that zero-order kinetics may be used to design trickling filters for loading rates to 1.17 g $NH_4^+$-N/m$^2$·d. This loading rate is affected by the liquid loading rate, which is between 0.41 and 118 m$^3$/m$^2$·d. Similarly, Okey and Albertson (1989a and 1989b) determined the effects of oxygen limitations on nitrification in trickling filters used to nitrify secondary effluent and concluded that nitrification rates showed oxygen limitations when the loading rate exceeded 1.2 g $NH_4^+$-N/m$^2$·d for tertiary-treated wastewater.

Trickling filter effluent ammonia concentrations of 1 mg/L are achievable under favorable conditions. However, below this concentration the confidence level of the process decreases and additional facilities may be required to ensure that effluent limits are consistently met. Pilot testing for specific applications should be considered to determine the process performance under site-specific conditions. At the Reno-Sparks, Nevada, wastewater treatment facility, effluent ammonia levels below 1 mg $NH_4^+$-N/L were consistently achieved, even when the temperature of the water fell to 2°C (Parlin and Peel, 1990). After they are established, nitrification remains efficient in trickling filter plants.

Because denitrification requires the absence of oxygen, denitrification can be achieved in fixed film reactors in a deep film or by operating in a submerged mode. Both upflow and downflow modes of operation can be used. An external carbon source, such as methanol, is often added to increase the denitrification rate and reduce the reactor size.

Denitrification can be achieved in filters and fluidized-bed reactors following any nitrification process. An external substrate such as methanol typically is added to increase denitrification rates. Koopman *et al.* (1990) found that complete denitrification (effluent nitrate plus nitrite <1 mg/L) can be achieved with methanol addition of 3.3 to 3.5 g methanol/g N. Madireddi *et al.* (1994) achieved nitrate concentrations below 1 mg/L in an upflow fluidized-bed reactor receiving nitrified trickling filter effluent and adding methanol as a carbon source. Temperature ranged from 10 to 20°C with a bed hydraulic retention time of 10 minutes.

**Biological Nitrogen Removal Schemes.** Figure 2.2 shows several of the biological nitrogen removal schemes used to take advantage of the biological nitrogen transformations. This is by no means a complete listing of all possi-

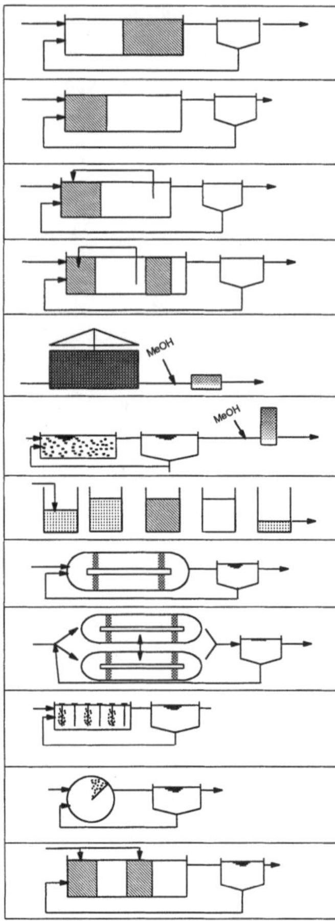


| | Substrate Level Denitrification | Endogenous Level Denitrification | Separate Stage Nitrification | Mixed Nitrification/Denitrification | Fixed Film | Chemical (methanol) Addition | Internal Recycle |
|---|---|---|---|---|---|---|---|
| 1. Wuhrman | | X | | | | | |
| 2. Ludzack-Ettinger | X | | | | | | |
| 3. MLE | X | | | | | X | X |
| 4. Bardenpho | X | X | | | | X | X |
| 5. Trickling Filter Filter | | | X | | X | X | |
| 6. Activated Sludge Fluidized Bed | | | X | | X | X | |
| 7. SBR | | | | X | | | |
| 8. Oxidation Ditch | | | | X | | | |
| 9. Biodenitro | | | | X | | | |
| 10. Biolac | | | | X | | | |
| 11. Counter Current Aeration | | | | X | | | |
| 12. Step Feed Denitrification | X | X | | | | | |

**Figure 2.2 Typical biological nitrogen removal schemes.**

ble process schemes, but it demonstrates the variety of possible ways to achieve nitrogen removal. In general, the various schemes achieve nitrogen removal by

- Sequential nitrification followed by denitrification—schemes 1, 4, 5, 6, and 12 show these types of arrangements. Because denitrification rates are slow, large reactors are required or a chemical substrate is added, as shown for the postdenitrification fluidized bed or filter in schemes 5 and 6.
- Denitrification using influent organics to achieve substrate level denitrification, as shown in schemes 2, 3, 4, and 12—these schemes rely on the return activated sludge and often an internal recycle to return nitrate from the nitrification zone to the denitrification zone.
- Mixed systems such as shown in schemes 7, 8, 9, 10, and 11, where nitrification and denitrification occur within the same reactor—two types of schemes are used:

    — Change the environment and feed pattern to achieve nitrification and denitrification at various times. Schemes 7 and 9 are examples of the feed and aeration cycles being manipulated to achieve nitrification and denitrification.

    — Operate the system to achieve simultaneous nitrification and denitrification at different locations within the same reactor by controlling environmental (such as dissolved oxygen) and hydrodynamic (mixing) conditions. Schemes 8, 10, and 11 are examples of aerobic and anoxic regions being created within one system.

**PHYSICAL/CHEMICAL NITROGEN REMOVAL PROCESSES.** Nitrogen removal can be removed chemically by converting the nitrogen to nitrogen gas, $N_2(g)$; by stripping $NH_3(g)$; by an ion-exchange process; or by using membrane separation. Physical/chemical nitrogen removal processes typically are not used and many of the earlier full-scale systems have been decommissioned. Biological processes have essentially replaced physical/chemical nitrogen removal systems. A brief overview of these processes is included in the following sections.

**Breakpoint Chlorination.** In breakpoint chlorination, ammonium is oxidized to nitrogen gas. The process is efficient and follows a series of complex reactions where ammonia reacts with chlorine to form monochloramine ($NH_2Cl$), dichloramine ($NHCl_2$), and ammonium trichloride ($NCl_3$). In addition to nitrogen gas, other nitrogen oxidation species also form, primarily nitrate, $NO_3^-$. While the actual mechanisms for these reactions are quite complex, the stoichiometry of the breakpoint reaction is simply

$$2NH_4^+ + 3Cl_2 \rightarrow N_2(g) + 6Cl^- + 8H^+ \tag{2.1}$$

According to Equation 2.1, it requires 7.6 mg $Cl_2$/mg $NH_4^+$-N to oxidize ammonium to nitrogen gas. Because some ammonia is converted to nitrate and other oxidized nitrogen species, the actual practical chlorine dose is often approximately 10 mg $Cl_2$/mg $NH_4^+$-N. It is also important to note that the breakpoint process produces a significant amount of acid (hydrochloric acid [HCl]), which will consume alkalinity at a ratio of 10.7 mg as $CaCO_3$/mg $NH_4^+$-N. For a typical wastewater treatment plant, oxidizing a 20-mg $NH_4^+$-N/L stream by breakpoint chlorination will therefore require 152 mg $Cl_2$/L and consume 214 mg/L as calcium carbonate of alkalinity.

Chlorine is a very reactive oxidant and produces other byproducts of concern. Ammonium trichloride typically is produced during breakpoint, resulting in the release of noxious ammonium trichloride gas. Reactions with organic compounds in the water lead to the formation of many halogenated organics, principally trihalomethanes such as chloroform, $CHCl_3$. Chlorinated organic compounds have long been recognized as potential human carcinogens.

Safety aspects of storage and dispensing of chlorine gas are major health and safety concerns, making the use of large quantities of chlorine gas unacceptable for many public utilities. Fire and safety code requirements have placed severe restrictions on its use.

Full-scale applications of breakpoint chlorination as the primary means of nitrogen removal have largely been discontinued. Some utilities maintain the ability to breakpoint chlorinate as as emergency backup process.

**Air Stripping.** Ammonia nitrogen can occur in water as the ammonium ion, which is in equilibrium with its conjugate base, ammonia. The equilibrium has a $pK_a$ value of approximately 9.5, as follows:

$$NH_4^+ = NH_{3(eq)} + H^+ \qquad pK_a = 9.5 \tag{2.2}$$

Aqueous ammonia exists in equilibrium with its gaseous counterpart in accordance to Henry's law:

$$NH_{3(eq)} \rightleftharpoons NH_3(g) \tag{2.3}$$

The equilibrium in Equation 2.2 is shifted toward the right-hand side as the pH increases, with significant amounts of un-ionized ammonia [$NH_{3(eq)}$] forming when the pH exceeds the $pK_a$ value (pH > 9.5). This means that to remove ammonium ions, the ammonium nitrogen is first converted to un-ionized ammonia by raising the pH and then stripped from the liquid according to Henry's law (Equation 2.3).

Ammonia stripping can be achieved using these principles. It typically requires a large air-to-water (A:W) ratio to achieve a high degree of nitrogen removal, with a minimum A:W ratio for ammonia removal on the order of 1 600 $m^3/m^3$. A high safety factor typically is used with practical attainable levels of 1 to 3 mg $NH_3$-N/L.

Full-scale ammonia-stripping towers at South Lake Tahoe, California; at Water Factory 21 of Orange County Water District, California; and in South Africa have been decommissioned because of operational problems and cost. The most significant problems with air-stripping plants follow:

- The process is highly temperature dependent and has failed during cold weather; freezing has been a problem.
- Scaling occurrs on the tower packing, specifically with the use of lime to raise the pH.
- The process has failed to consistently achieve a low ammonia concentration.

Although some stripping plants remain in operation (Central New York State, for example), most plants have been replaced by more economical biological nitrogen removal systems.

# PHOSPHORUS REMOVAL PROCESSES

Unlike nitrogen, there is no gaseous form of phosphorus through which it can be removed from wastewater. Consequently, phosphorus must be converted to a particulate (solid) form and removed as a particulate by sedimentation, filtration, or some other solids removal process or be concentrated into a sidestream using membrane treatment. Table 2.2 summarizes the vari-

**Table 2.2 Phosphorus species and reactions.**

| Species | Comment |
|---|---|
| Organic-P | Organic phosphorus can be converted to orthophosphate and polyphosphate |
| Orthophosphate | Most abundant phosphorus species<br>Reactive species in chemical reactions and consumed in biological growth |
| Polyphosphates | Condensed orthophosphates<br>Possibly reacts with metal salts<br>Can be used for biological growth |
| Chemical phosphorus | Precipitated phosphates formed by reacting orthophosphate with metal salts, or precipitates as phosphate hydroxides |
| Biological phosphorus | Phosphorus incorporated into the biomass for growth<br>Excess phosphorus may accumulate under certain conditions |

ous options for removing or converting phosphorus species. To remove phosphorus from the system, three options are available.

- Convert the phosphorus to a chemical species by adding a metal salt or lime. The efficiency of phosphorus removal is dependent on two factors: the chemical equilibrium between the phosphorus liquid and solid phases, and the efficiency of the solids removal process. Typically, the latter process controls the removal efficiency.
- Incorporate the phosphorus into the biomass. Typically, biomass contains 1.5 to 2.5% (w/w) phosphorus per volatile solids. Under certain conditions, the biomass will accumulate phosphorus levels far in excess of the nutritional requirements to 6 to 8% phosphorus—a process referred to as *enhanced biological phosphorus removal* (EBPR). The phosphorus removal efficiency for biological systems depends on the phosphorus content of the sludge removed and the efficiency of the solids separation process.
- Processes that will remove essentially all pollutants from water, such as reverse osmosis, EDR, or nanofilters, can be used to remove phosphorus. Membrane treatment is expensive and not used for mainstream phosphorus removal. However, membranes used for another objective (such as total dissolved solids removal) will also remove phosphorus.

The different chemical species must be considered when considering National Pollutant Discharge Elimination System permit requirements and evaluating process options. Typically, phosphorus limits are stated in terms of total phosphorus, which includes all phosphorus species, dissolved and soluble. In such cases, removal of the chemical or biological bound solid-phase phosphorus is critical in meeting effluent standards. If the discharge limit is based on orthophosphate only, suspended solids removal is less critical.

Phosphorus removal processes can be grouped into two basic groups: chemical processes and biological processes. The following sections describe typical phosphorus removal processes. Figure 2.3 gives a basic overview of phosphorus removal processes. The selection of the specific process must be based on a case-by-case evaluation of the system economics, including both capital and operating costs.

**BIOLOGICAL PHOSPHORUS REMOVAL.** In the 1950s, Greenburg *et al.* (1955) proposed that activated sludge could take up phosphate at a level beyond its normal microbial growth requirements. Srinath *et al.* (1959) reported batch experiments in which soluble phosphorus concentrations could be reduced to below 1 mg/L following vigorous aeration. However, Levin and Shapiro (1965) were the first to report EBPR, in the District of Columbia activated-sludge plant. Follow-up work in the U.S. and South Africa (Barnard, 1974) clearly demonstrated that EBPR can occur. Typically, biomass in the activated-sludge process will contain 1.5 to 2.5% phosphorus

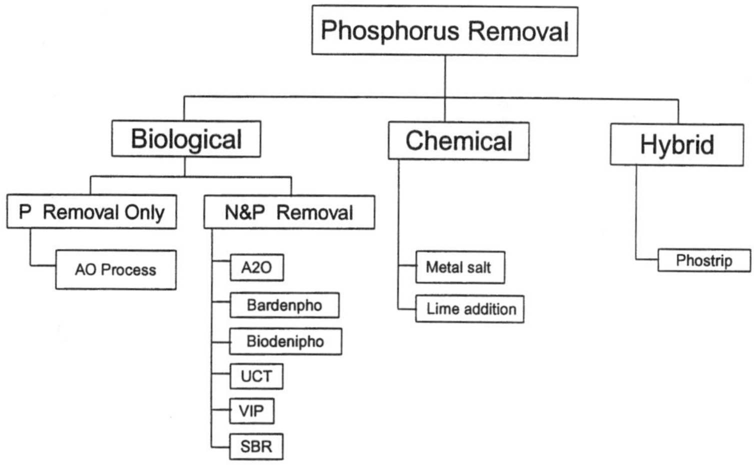


**Figure 2.3 Overview of phosphorus removal processes.**

based on dry weight. However, under proper conditions, certain bacteria will accumulate phosphorus in the cell in excess of typical nutritional requirements.

A proposed biological phosphorus removal mechanism is shown in Figure 2.4 (U.S. EPA, 1987). Acetate and other fermentation products (mostly volatile fatty acids [VFAs]) are produced from fermentation reactions catalyzed by facultative organisms in the anaerobic zone. The model assumes that these products are derived from soluble portions of the influent BOD and that there is insufficient time in the anaerobic zone for hydrolysis and fermentation of particulate BOD. The VFAs can also be added directly to the anaerobic basin. The fermentation products are readily assimilated by these microorganisms capable of enhancing phosphorus removal and stored as intracellular hydrocarbons. The assimilation and storage is aided by the energy made available from the hydrolysis (and release) of polyphosphates previously stored in the cells, thus causing an increase in the soluble phosphorus concentration. This accumulation of storage products gives the *Acinetobacter,* and other phosphorus removal organisms, a competitive edge for growth and survival, which it needs because of its extremely low growth rate under normal activated-sludge conditions. Thus, the anaerobic stage serves two important purposes: it provides a fermentation zone to produce simple hydrocarbons used by phosphorus removal bacteria, and it provides an environment that gives these organisms a competitive edge to ensure their survival in the system.

During the aerobic phase, polyhydroxybutyrates stored inside the cells are depleted and soluble phosphorus is taken up, with excess amounts stored as polyphosphates inside the cells. The phosphate-using bacterial population increases during this time because of substrate use and growth. According to this model, the level of phosphate accumulation in the cells is therefore re-

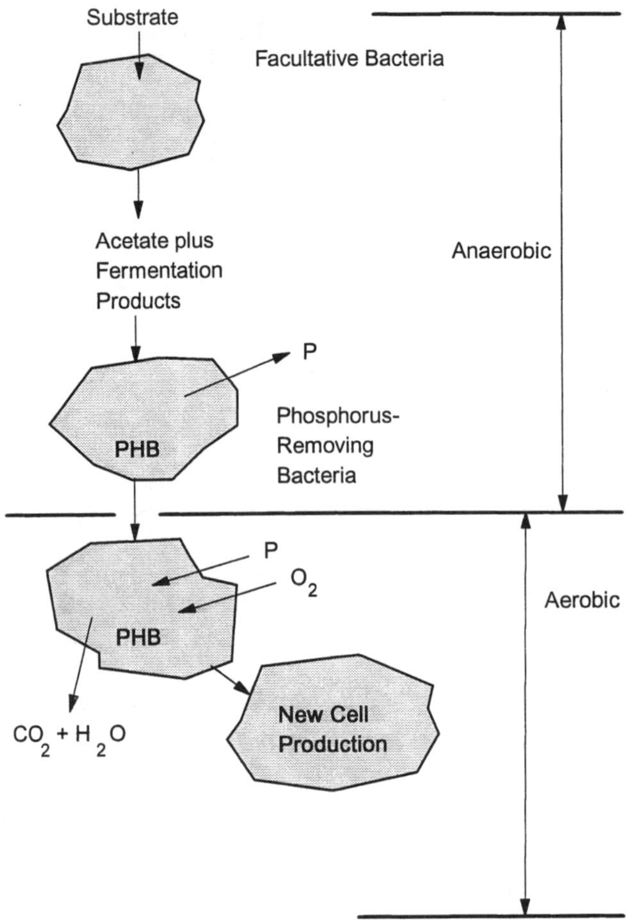


**Figure 2.4 Proposed mechanism for phosphorelease and uptake (U.S. EPA, 1987).**

lated to the amount of substrate assimilated and stored in the anaerobic phase.

The conditions required for EBPR therefore include the following:

- Provide an *anaerobic* zone, that is, a zone free of dissolved oxygen and devoid of nitrate. In this zone, phosphate is released from the cell, with accompanying substrate uptake. Koch and Oldham (1984) suggested that redox potential be used to determine the degree of anaerobiosis. Redox potential is between that controlling $NO_3^-$ reduction to $N_2$ (denitrification) and $SO_4^=$ reduction to $HS^-$ (sulfate reduction).
- Soluble substrate such as VFA must be present in significant quantities. The initial phosphorus release is accompanied by soluble chemical oxygen demand uptake. It is now known that simple organics such as VFAs are most effective in contributing to phosphorus release and uptake (Gerber *et al.,* 1986).

All biological phosphorus removal schemes include an anaerobic zone in the process. The proper functioning of the anaerobic zone requires the absence of dissolved oxygen and nitrate. Consequently, the various processes use different approaches to eliminate oxygen and nitrate from the anaerobic zone.

Some process schemes include a fermentation step to generate VFAs and feed the VFA-rich stream to the anaerobic zone. These VFAs can also directly be supplemented by chemical addition.

Figure 2.5 shows a summary of the various biological phosphorus removal processes.

**CHEMICAL PHOSPHORUS REMOVAL.** Phosphorus removal by chemical addition is attractive for its simplicity of operation and ease of implementation but results in increased sludge production and additional opera-

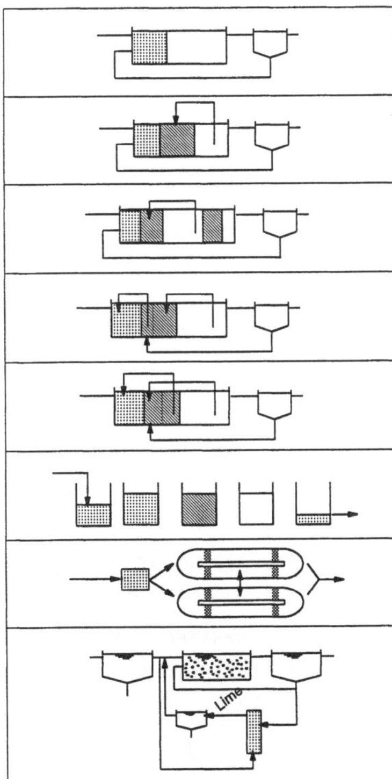


| Process | Nitrification | Nitrogen Removal | Internal Recycle | Chemical Addition |
|---|---|---|---|---|
| AO Process | | | | |
| A20 Process | X | X | X | |
| Bardenpho | X | X | X | |
| UCT | X | X | X | |
| VIP | X | X | X | |
| SBR | X | X | | |
| Biodenipho | X | X | | |
| PhoStrip | | | | X |

**Figure 2.5 Biological phosphorus removal processes.**

tion and maintenance costs. Chemicals are added to the wastewater at a well-mixed location, followed by flocculation and solids removal by sedimentation, filtration, or similar processes. The equilibrium of the chemical reactions favors phosphorus precipitation and will produce low residual phosphorus levels. Phosphorus levels less than 0.1 mg P/L can consistently be achieved with chemical addition and well-designed filtration facilities. See Chapter 8 for a detailed description of the chemistry and design of chemical phosphorus removal systems.

Metal salts, specifically iron and aluminum salts, can be readily added to wastewater for phosphate removal. The stoichiometry of the reactions for chemical phosphorus removal with aluminum and ferric salts is as follows:

$$Al^{3+} + H_2PO_4^- \rightarrow AlPO_{4(s)} + 2H^+ \tag{2.4}$$

$$Fe^{3+} + H_2PO_4^- \rightarrow FePO_{4(s)} + 2H^+ \tag{2.5}$$

In addition to metal phosphate precipitants, metal complexes also form with hydroxide ions as shown below:

$$Al^{3+} + 3H_2O \rightarrow Al(OH)_{3(s)} + 3H^+ \tag{2.6}$$

$$Fe^{3+} + 3H_2O \rightarrow Fe(OH)_{3(s)} + 3H^+ \tag{2.7}$$

The formation of metal hydroxides adds to the complexity of predicting chemical reactions and their results. First, the formation of hydroxy–metal complexes adds a competing reaction for the added metal. The dose must therefore be adjusted to account for the metal hydroxide reactions. Second, the reactions produce a significant amount of sludge (2.9 mg solids/mg Al for alum and 1.9 mg solids/mg Fe for ferric) that must be processed through dewatering and disposal. Third, the reactions consume a significant amount of alkalinity (5.8 mg as $CaCO_3$/mg Al and 2.7 mg as $CaCO_3$/mg Fe).

Dosage requirements for phosphate removal can be estimated based on the stoichiometric requirements (see Equations 2.4 and 2.5) until the phosphate level decreases to approximately 1 mg P/L when the reactions become equilibrium controlled and competing reactions start to dominate. To achieve phosphate levels below 1 mg/L, predictions based on stoichiometry are no longer accurate and excess metal salt dosing is required. For example, reducing the phosphate levels in secondary effluent from approximately 1 mg P/L to 0.27 mg P/L requires an alum dose of 20 mg/L. This represents an alum:phosphorus ratio of 20, or approximately twice the stoichiometric requirements (Neethling *et al.*, 1991).

Lime can also be used to precipitate phosphorus; however, lime is messy to handle and difficult to use compared with metal salts. A variety of calcium phosphates will form when calcium reacts with phosphate and hydroxide.

Snoeyink and Jenkins (1980) showed that the precipitation of hydroxyapatite [$Ca_5(PO_4)_3OH$], the thermodynamically stable calcium–phosphate precipitant, predicts equilibrium phosphate levels far lower than those found in natural waters. Equilibrium with ß-tricalcium phosphate shows orders-of-magnitude higher phosphate concentrations, but they still are very low. These calculations indicate that the solid precipitant that forms in practical situations is often not the thermodynamically most stable form, but some other chemical compound. In addition, the kinetics of precipitation are very slow and some reactions may never quite reach equilibrium. Even so, for practical application, equilibrium calculations can be used to predict the ultimate equilibrium phosphorus concentration and provide a boundary for dose requirements to meet the treatment objectives.

Practical experiences with lime addition demonstrate that the treatment becomes effective when the pH is raised to approximately 10.0 or higher, with pH levels of 10.5 used to achieve low phosphate levels. Lime dose requirements typically are determined by the bicarbonate alkalinity of the water because, in effect, the wastewater must be titrated to a pH of 10.5 as follows:

$$Ca(OH)_2 + HCO_3^- \rightarrow CaCO_{3(s)} + H_2O + OH^- \qquad (2.8)$$

Successful chemical phosphorus removal requires efficient solids removal processes and often includes filtration to achieve low phosphorus concentrations. Because the chemical sludge quantities are high, sedimentation typically is added to reduce the solids loading onto the filters. Chemical clarification processes such as contact clarifiers, sludge blanket clarifiers, and claricones have been used successfully in chemical phosphorus removal schemes.

# REFERENCES


Barnard, J.L. (1974) Cut P and N Without Chemicals—Parts 1 and 2. *Water Wastes Eng.,* **11,** 33/41.

Gerber, A., *et al.* (1986) The Effect of Acetate and Carbon Compounds in the Kinetics of Biological Nutrient Removal. *Water SA,* **12,** 7.

Greenburg, A.E., *et al.* (1955) The Effect of Phosphorus Removal on the Activated Sludge Process. *Sew. Ind. Wastes,* **27,** 227.

Gullicks, H.A., and Cleasby, J.L. (1986) Design of Trickling Filter Nitrification Towers. *J. Water Pollut. Control Fed.,* **58,** 60.

Koch, F.A., and Oldham, W.K. (1984) ORP—A Tool for Monitoring, Control, and Optimization of Biological Nutrient Removal Systems. Paper presented at IAWPRC Post-Conf. Seminar Enhanced Biological Phosphorus Removal Wastewater, Paris, Fr.

Koopman, B., *et al.* (1990) Denitrification in a Moving Bed Upflow Sand Filter. *Water Environ Res.,* **62,** 239.

Levin, G.V., and Shapiro, J. (1965) Metabolic Uptake of Phosphorus Removal Process. *J. Water Pollut. Control Fed.,* **37,** 800.
Madireddi, K., *et al.* (1994) Denitrification in a Fluidized Sand Column: Lake Arrowhead Water Reclamation Pilot Study. Paper presented at 67th Annu. Water Environ. Fed. Tech. Exposition Conf., Chicago, Ill.
Neethling, J.B., *et al.* (1991) Chemical Phosphorus Removal in a Trickling Filter Plant. Paper presented at 64th Annu. Conf. Water Pollut. Control Fed., Toronto, Ont., Can.
Okey, R.W., and Albertson, O.A. (1989a) Evidence of Oxygen Limiting Conditions During Tertiary Fixed-Film Nitrification. *J. Water Pollut. Control Fed.,* **61,** 5100.
Okey, R.W., and Albertson, O.A. (1989b) Diffusions' Role in Regulating Rate and Masking Temperature Effects in Fixed-Film Nitrification. *J. Water Pollut. Control Fed.,* **61,** 500.
Parlin, L., and Peel, J. (1990) Biological Ammonia Removal at Reno-Sparks, Nevada: Nitrification Utilizing High Density Plastic Media Towers. Paper presented at Annu. Calif. Water Pollut. Control Assoc. Meet., South Lake Tahoe, Calif .
Soap and Detergent Association (1989) *Principles and Practice of Phosphorus and Nitrogen Removal from Municipal Wastewater.* R.I. Sedlak (Ed.), Soap and Detergent Assoc., New York, N.Y.
Snoeyink, V.L, and Jenkins, D. (1980) *Water Chemistry.* John Wiley and Sons, New York, N.Y.
Srinath, E.G., *et al.* (1959) Rapid Removal of Phosphorus from Sewage by Activated Sludge. *Experientia* (Switz.), **15,** 339.
Stensel, H.D. (1981) *Biological Nitrogen Removal System Design.* Am. Inst. Chem. Eng., New York, N.Y., 327.
U.S. Environmental Protection Agency (1975) *Process Design Manual for Nitrogen Control.* EPA-625/1-77-007, Washington, D.C.
U.S Environmental Protection Agency (1987) *Design Manual: Phosphorus Removal.* EPA-625/1-87-001.

# Chapter 3 Chemical Phosphorus Removal



## PRINCIPLES OF CHEMICAL PHOSPHORUS REMOVAL

The basic principle of chemical phosphorus (P) removal relies on the transformation of soluble phosphorus to a particulate form and the removal of this form (together with any phosphorus already present in a particulate form), typically by sedimentation. Some phosphorus is removed in a conventional secondary wastewater treatment plant (WWTP), both in the primary and secondary treatment processes (see Chapter 1). Removal during primary settling is limited to a fraction of the particulate form and depends on the efficiency of the primary clarifiers. In a secondary treatment process, phosphorus is incorporated to biomass and removed from wastewater through secondary sedimentation with waste biomass. As such, the quantity of phosphorus removed by a conventional secondary treatment process is a function of biomass yield and production.

Total effluent phosphorus concentration, $C_{TP,eff}$, can be estimated as

$$C_{TP,eff} = C_{SP,eff} + X_{eff} m_{PX} \tag{3.1}$$

Where

$C_{SP,eff}$ = concentration of soluble phosphorus (mainly soluble orthophosphate), mg/L;
$X_{eff}$ = effluent suspended solids concentration, mg/L; and
$m_{PX}$ = phosphorus content in suspended solids dry mass, mg P/mg suspended solids.

For conventional activated sludge, $m_{PX}$ is 20 to 25 mg P/g volatile suspended solids (VSS) (2 to 2.5%). For chemical phosphorus removal, $m_{PX}$ varies between 40 and 100 mg P/mg suspended solids (4 to 10%). Equation 3.1 emphasizes the contribution of suspended solids to total effluent phosphorus concentration and the need for effective solids removal. For example, if effluent suspended solids concentration is 20 mg/L with a phosphorus content of 5% (50 mg P/g suspended solids), a total phosphorus concentration below 1 mg P/L cannot be achieved.

The following cations typically are used for the precipitation of phosphorus from wastewater:

- Aluminum,
- Iron, and
- Calcium.

Under the right conditions, all three cations form insoluble precipitates with orthophosphate. For this reason soluble orthophosphate is the primary phosphorus species affected by chemical removal. Any accurate analysis of its removal can be carried out only if its concentrations are measured and reported. Unfortunately, soluble orthophosphate is rarely determined in practice. All other phosphorus species (such as condensed polyphosphates, colloids, and particulates containing phosphorus) are removed by secondary mechanisms including adsorption, coagulation, sedimentation, filtration, or biologically mediated removal. These mechanisms will not be discussed in this chapter.

The nature of the precipitates formed during chemical phosphate removal is not well known. Table 3.1 lists examples of chemical solids that may be formed during phosphate removal. Exact stoichiometric composition of

**Table 3.1 Possible precipitates formed during phosphate removal.**

| Cation | Precipitate |
|---|---|
| Al(III) | Aluminum phosphate [$Al_r(H_2PO_4)(OH)_{3r-1}$] |
| | Aluminum hydroxide [$Al(OH)_3$] |
| Fe(II) | Ferrous phosphate [$Fe_3(PO_4)_2$] |
| | Ferrous hydroxide [$Fe(OH)_2$] |
| Fe(III) | Ferric phosphate [$Fe_r(H_2PO_4)(OH)_{3r-1}$] |
| | Ferric hydroxide [$Fe(OH)_3$] |
| Ca(II) | Tricalcium phosphate [$Ca_3(PO_4)_2$] |
| | Hydroapatite [$Ca_5(OH)(PO_4)_2$] |
| | Dicalcium phosphate [$CaHPO_4$] |
| | Calcium carbonate [$CaCO_3$] |

phosphate precipitates is not fully known, but it is likely that it will vary from the simple forms such as $AlPO_4$ or $FePO_4$ . Some authors (Arvin and Petersen, 1980, and Hsu, 1973) suggested that all three cations (aluminum, iron, and calcium) can be present in a precipitate together with other ions such as magnesium, sulfate, and bicarbonate. Variability in precipitate composition led to suggestions (Leckie and Stumm, 1970) that phosphate ion is also removed by adsorption on other chemical precipitates.

From an engineering point of view, three parameters are of particular importance for design, operation, and analysis of chemical phosphorus removal:

- Dose requirements,
- Minimum achievable phosphate concentration, and
- Effects of pH.

These factors will be discussed in the subsequent sections for the three cations involved.

**FERROUS IRON.** The main source of ferrous iron [Fe(II)] is spent pickle liquor containing mostly $FeSO_4$ and originating from metal-processing operations. It is a potentially convenient and economical source of precipitating agent, but it may contain hazardous materials (such as heavy metals) that can either pass through with the effluent or accumulate in the sludge. It should be noted, however, that commercially available technical-grade iron salts [both Fe(II) and ferric iron, or Fe(III)] may also contain a certain amount of heavy metals.

Recht and Ghassemi (1970) conducted an extensive study of phosphate precipitation with Fe(II) in the absence of dissolved oxygen. They concluded that optimum removal occurred at pH 8, with residual orthophosphate at 0.4 mg P/L. On either side of the optimum pH, orthophosphate residual concentrations were much greater: 8 mg P/L at pH 7 and 3.5 mg P/L at pH 9. At pH

8, a reaction time of 2 hours was required to achieve maximum removal. Although the initial iron:phosphorus molar ratio was 1:1, not all Fe(II) was removed from the solution. The precipitate formed in their experiments was identified as vivianite, $Fe_2(PO_4)_3 \cdot 8H_2O$. Despite theoretically more advantageous stoichiometry for orthophosphate precipitation with Fe(II), the results of two full-scale applications of Fe(II) to raw wastewater (Mentor, Ohio, and Texas City, Texas) cited by Recht and Ghassemi indicated that the majority of removal occurred in the aeration tanks. Low phosphate removal in primary treatment was attributed to inadequate reaction time of Fe(II), formation of poorly settling precipitate, or complexation of Fe(II) by organic matter.

If exposed to aerobic conditions (for example, when dosed to an aeration tank), Fe(II) oxidizes rapidly to Fe(III). Leckie and Stumm (1970) indicated that ferric iron Fe(III) formed by oxidation of Fe(II) is more effective for phosphate precipitation than Fe(III) added directly from a stock solution. Recht and Ghassemi (1970) found that the speed of phosphate precipitation increased substantially in the presence of oxygen as Fe(II) was oxidized to Fe(III). Similarly to Leckie and Stumm (1970), they claimed that phosphate removal efficiency with oxidized Fe(II) was better than for an equivalent dose of Fe(III) from stock solution. However, the practical efficiency of phosphorus removal is also affected by the settling characteristics of the precipitate. Leckie and Stumm (1970) reported that the precipitate formed by oxidized Fe(II) in clean water was inferior to those from Fe(III).

**FERRIC IRON AND ALUMINUM.** Because there are many similarities between the chemistry of orthophosphate precipitation with Fe(III) and aluminum [Al(III)], these two agents will be discussed together. Ferric iron is used as ferric chloride ($FeCl_3$) or ferric sulfate [$Fe_2(SO_4)_3$ ]. Aluminum for phosphate removal is used as either alum [$Al_2(SO_4)_3 \cdot 18H_2O$], sodium aluminate ($NaAlO_2$), or polyaluminum chloride (PAC). The latter chemical is often used when enhanced solids removal is also a treatment objective. Regardless of the form used, $Fe^{3+}$ or $Al^{3+}$ cations are the precipitating agents. Besides formation of phosphate precipitate, ferric iron and aluminum can react in an aquatic environment to produce a range of other compounds, soluble or not, including hydroxides [for example, $Fe(OH)_3(s)$ or $Al(OH)_3(s)$] and a variety of polymeric species. Major chemical reactions are listed in Table 3.2.

When an excess dose of metal salt [Fe(III) or Al(III)] is added, metal hydroxide will precipitate in addition to metal phosphate precipitating (Ferguson and King, 1977, and Kavanaugh *et al.*, 1978). In this case, residual phosphate concentration can be calculated from equilibrium equations if the appropriate equilibrium constants are known. Table 3.2 lists the equilibrium constants compiled from a number of sources. Of those, only solubility products for $Al(OH)_3$ and $Fe(OH)_3$ are known with any accuracy. The values of solubility products for metal phosphates and stability constants for soluble

**Table 3.2** **Chemical reactions of Al(III) and Fe(III) (Feitknecht and Schindler, 1963; Gates *et al.,* 1990; Hogfeldt, 1983; Smith and Martell, 1976; and Stumm and Morgan, 1970).**

| Reaction | | Equilibrium constants Al(III) | Equilibrium constants Fe(III) |
|---|---|---|---|
| 1. $Me^{3+} + 3\ H_2O \rightleftarrows Me(OH)_3(s)$ | $pK_{MeOH}$ = | 9.1 | −0.5 |
| 2. $r\ Me^{3+} + H_2PO_4^- + (3r\text{-}1)\ OH^-$ | $pK_s$ = | 25.8 | 67.1 |
| $\rightleftarrows Me_r(H_2PO_4)(OH)_{3r-1}(s)$ | r = | 0.8 | 1.6 |
| 3. $Me^{3+} + H_2PO_4^- \rightleftarrows MeH_2PO_4^{2+}$ | $pK_{MP1}$ | | −21.3 |
| 4. $Me^{3+} + HPO_4^{2-} \rightleftarrows MeHPO_4^+$ | $pK_{MP2}$ | −12.1 | |
| 5. $H_3PO_4 \rightleftarrows H^+ + H_2PO_4^-$ | $pK_1$ = | 2.1 | |
| 6. $H_2PO_4^- \rightleftarrows H^+ + HPO_4^{2-}$ | $pK_2$ = | 7.2 | |
| 7. $HPO_4^{2-} \rightleftarrows H^+ + PO_4^{3-}$ | $pK_3$ = | 12.3 | |

metal–phosphate complexes were estimated from field and laboratory data on phosphate removal (Gates *et al.*, 1990).

When both metal hydroxide and metal phosphate precipitates are present, the residual orthophosphate concentration can be calculated as follows:

1. Calculate metal cation concentration ($Me^{3+}$) from the solubility product of its metal hydroxide (reaction 1 in Table 3.2):

$$\log[Me^{3+}] = pK_{MeOH} - 3\ pH \tag{3.2}$$

2. Calculate $H_2PO_4$ concentration from the solubility product of metal phosphate (reaction 2 in Table 3.2):

$$\log[H_2PO_4^-] = -pK_s - r\log[Me^{3+}] + (3r - 1)(pK_w - pH) \tag{3.3}$$

3. Calculate remaining orthophosphate species (reactions 5 through 7 in Table 3.2):

$$\log[H_3PO_4] = \log[H_2PO_4^-] - pH + pK_1 \tag{3.4}$$

$$\log[HPO_4^{2-}] = \log[H_2PO_4^-] + pH - pK_2 \tag{3.5}$$

$$\log[PO_4^{3-}] = \log[HPO_4^{2-}] + pH - pK_3$$

At pH values typically found in wastewater treatment, Equation 3.5 is the most important.

4. Calculate soluble metal–phosphate complexes (reactions 3 and 4 in Table 3.2).

$$\log[MeH_2PO_4^{2+}] = \log[Me^{3+}] + \log[H_2PO_4^-] + pK_{MP1} \quad (3.7)$$

$$\log[MeHPO_4^+] = \log[Me^{3+}] + \log[HPO_4^{2-}] + pK_{MP2} \quad (3.8)$$

It appears that for Al(III) addition, soluble complex $AlHPO_4^+$ (Equation 3.8) is the dominant of the two complexes, while for Fe(III) addition, soluble complex $FeH_2PO_4^{2+}$ (Equation 3.7) predominates.

5. Calculate total residual soluble phosphate concentration (in mol/L):

$$C_{P,res} = [H_3PO_4] + [H_2PO_4^-] + [HPO_4^{2-}] + [PO_4^{3-}] + [MeH_2PO_4^{2+}] + [MeHPO_4^+] \quad (3.9)$$

The residual phosphate concentration of Equation 3.9 is the lowest theoretically achievable concentration because of phosphate precipitation at high doses of Fe(III) or Al(III). It is a function of pH as shown in Figure 3.1 for Al(III) and Figure 3.2 for Fe(III). These theoretical curves delineate minimum residual soluble orthophosphate concentrations observed in laboratory-

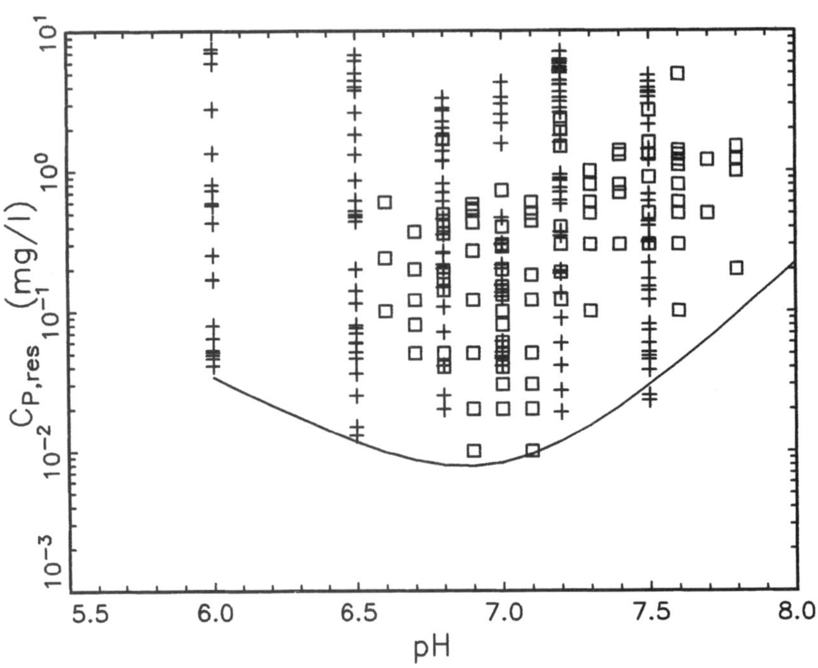


**Figure 3.1 Observed residual soluble orthophosphate concentrations and theoretical solubility limit for Al(III) (+ = laboratory [Gates *et al.,* 1990], and □ = full-scale data [Trenton, N.J., WWTP]).**

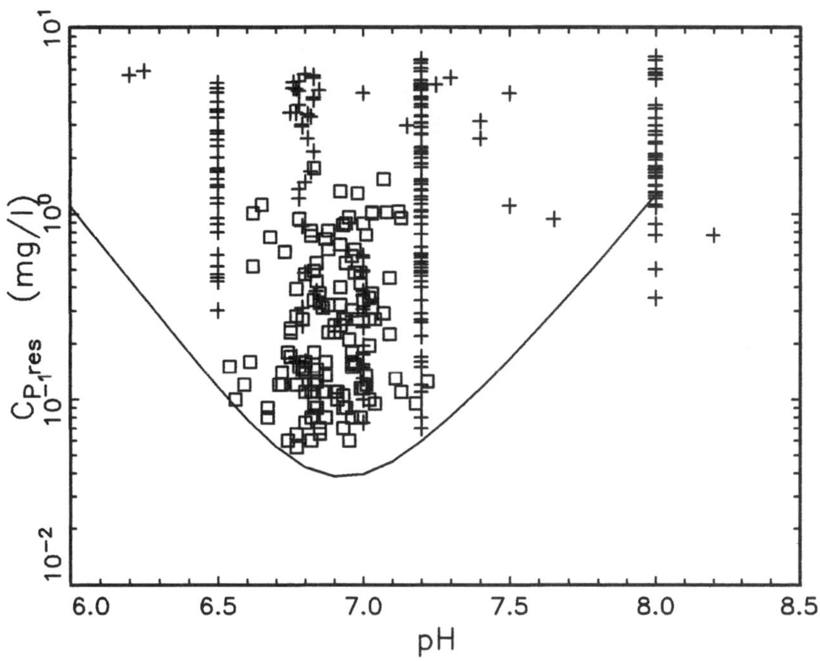


**Figure 3.2 Observed residual orthophosphate concentrations and theoretical solubility limit for Fe(III) (+ = laboratory data [Luedecke *et al.*, 1988], and □ = full-scale data [Blue Plains WWTP, Washington, D.C.]).**

and full-scale activated-sludge systems. It is worth noting that for both Al(III) and Fe(III), optimum pH (corresponding to minimum metal phosphate solubility) is approximately 6.8 and is relatively broad. In contrast, Recht and Ghassemi (1970) reported an optimum pH of 6.0 for phosphate precipitation from distilled water with Al(III) and a pH of 3.5 to 4.0 for precipitation with Fe(III). The difference in optimum pH can be attributed to the influence of other chemical species present in wastewater. Hsu (1973) reported that the addition of $Ca^{2+}$ brought about a change from a well-defined optimum pH of 4 without $Ca^{2+}$ to a broad pH range of 4 to 8 at 2 mmol/L $Ca^{2+}$. Similar calcium effects were also reported by Grohman *et al.* (1984), who studied ferric phosphate precipitation from "clean" solutions. Arvin and Petersen (1980) incorporated the effects of calcium and bicarbonate to a complex semiempirical model of phosphate precipitation. Despite these studies, the mechanism of calcium effect is not fully understood.

Composition of metal phosphate precipitate has a significant effect on the required metal dose. The ratio of metal dose to soluble orthophosphate removed closely approximates the overall precipitate (or precipitate mixture) composition because the residual soluble metal concentrations are very small. Figure 3.3 shows the ratio as a function of residual phosphate concentration for batch and continuous experiments with aluminum for a pH in the range 6 to 7.5. Figure 3.4 presents a similar relationship for Fe(III) for laboratory- and full-scale activated-sludge systems for pH 6.5 to 7.5. For both metals, the relationships corroborate the chemical mechanism of phosphate precipitation. At high residual phosphate concentrations (that is, at low metal doses) the ratio in the precipitate remains essentially constant, indicating stoichiometric precipitation. At high metal doses, the phosphate solubility limit is approached, and the excess of Al(III) or Fe(III) precipitates as metal hydroxide, resulting in a dramatic increase of the Me/P ratio. At residual phosphate concentrations of less than approximately 1 mg P/L, the Me/P ratio slightly increases as a result of either additional phosphate adsorption (Luedecke *et al.,* 1988, and Goldshmid and Rubin, 1978) or substitution of orthophosphate for OH in the precipitate (Hsu, 1973). Although the exact mechanism of precipitate formation is not fully understood, Figures 3.3 and 3.4 can be used to calculate the metal dose necessary to achieve required residual soluble orthophosphate:

$$Me_{dose} = (Me/P)(C_{P,in} - C_{P,res}) \tag{3.10}$$

where (Me/P) can be read off the graphs for required residual phosphate concentration. For biological treatment systems, influent concentration includes

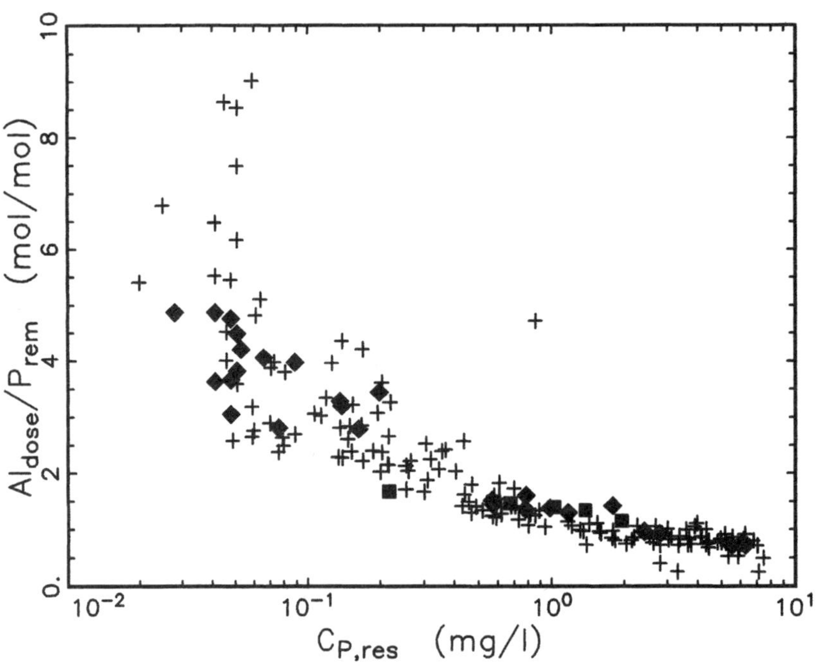


**Figure 3.3** **Ratio of Al(III) dose to phosphorus removed as a function of residual soluble orthophosphate concentration (+ = laboratory batch experiments [Gates, 1991], ◆ = laboratory continuous-flow experiments [Gates, 1991], and ■ = laboratory University of Cape Town system [Rabinowitz *et al.,* 1987]).**

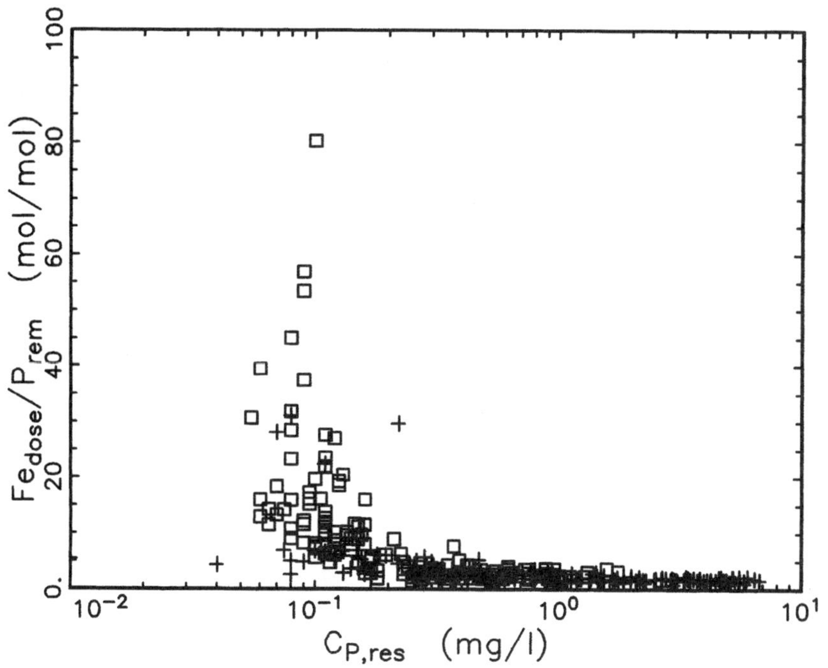


**Figure 3.4 Ratio of Fe(III) dose to phosphorus removed as a function of residual soluble orthophosphate concentration (+ = laboratory experiments [Luedecke *et al.*, 1988], and □ = full-scale data [Blue Plains WWTP, Washington, D.C.]).**

soluble orthophosphate concentration and a portion of soluble nonorthophosphate and particulate phosphorus concentration. The latter two fractions of total phosphorus are partially hydrolyzed during the treatment and converted to orthophosphate requiring an additional metal dose.

**CALCIUM.** Phosphate precipitation with lime was the earliest method of phosphorus removal. Calcium forms several insoluble compounds with phosphate (see Table 3.2), among which hydroxyapatite $Ca_5(PO_4)_3(OH)$ seems to be the most important (Menar and Jenkins, 1972). Additionally, calcium carbonate can form depending on pH, wastewater alkalinity, and calcium dose. Figure 3.5 shows that significant phosphate removal can only be achieved at higher pH values. Most reports (Buzzell and Sawyer, 1967; Menar and Jenkins, 1972; and Spiegel and Forrest, 1969) indicated that removal of phosphate to values below 1 mg P/L requires values of pH of 10.5 to 11. For this reason lime is used either in primary treatment or following biological treatment. The dose of lime and the amount of solids produced primarily are functions of total alkalinity of wastewater according to the following reaction:

$$Ca(OH)_2 + HCO_3^- \leftrightharpoons CaCO_3 + H_2O + OH^- \qquad (3.11)$$

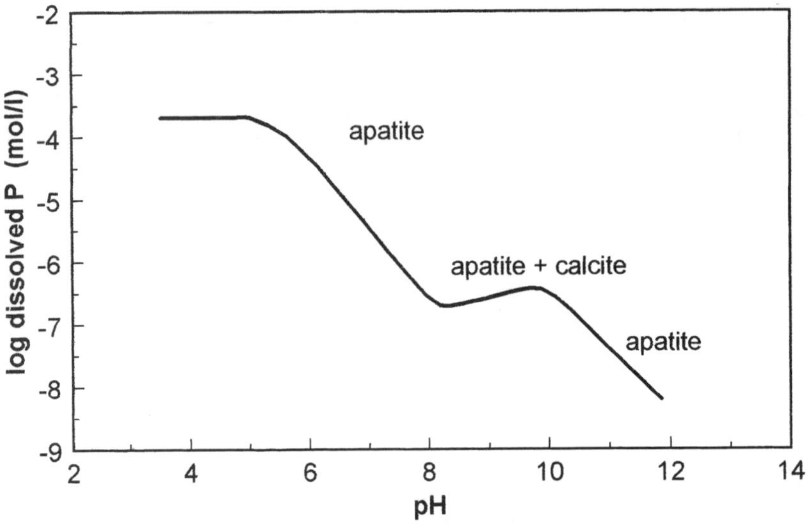


**Figure 3.5** **Equilibrium solubility for calcium phosphate system. (Adapted with permission from Jenkins, D., and Hermanowicz, S.W. [1991] Principles of Chemical Phosphorus Removal. In *Phosphorus and Nitrogen Removal from Municipal Wastewater: Principles and Practice*. R.I. Sedlak [Ed.]. Copyright CRC Press, Boca Raton, Florida © 1994.)**

These factors do not significantly depend on phosphate concentration. According to Equation 3.11, the required lime dose [in mg $Ca(OH)_2$/L] is approximately 1.5 times the total alkalinity (as mg $CaCO_3$/L).

# DESIGN EXAMPLES

**CHEMICAL PHOSPHORUS REMOVAL PROCESS DESIGN.** The process design of chemical phosphorus removal is dependent on several factors. The most important factors are

- Wastewater characteristics such as phosphorus concentrations, total suspended solids (TSS), pH, and alkalinity;
- Chemicals used for precipitation, such as aluminum, calcium, or iron; and
- The point of chemical addition, for example primary treatment, secondary treatment, or tertiary treatment.

Wastewater characterization involves determining constituent parameters under varying diurnal and seasonal loading conditions such that chemical type, chemical requirements, and feed rates can be designed to satisfy both peak and minimum requirements for best phosphorus removal. Because the chemicals added to remove phosphorus may also be involved in other reactions such as alkalinity consumption, reaction with sulfides, and coagulation of suspended solids, chemical doses in excess of those estimated using wastewater characteristics and precipitation chemistry of phosphorus must often be provided.

The selection of the chemical used for phosphorus removal depends on the cost of the chemical, alkalinity consumption, quantities of sludge generated, and safety in handling and use of the chemical (Jenkins and Hermanowicz, 1991). Typical forms of the three cations used for phosphorus removal are aluminum sulfate (alum) and sodium aluminate for aluminum; ferric chloride, ferrous chloride, and ferrous sulfate for iron; and lime for calcium. However, lime addition for phosphorus removal from wastewater is not typical in current designs because of drawbacks such as large sludge production rates, high pH requirement, and large investments in equipment, operation, and maintenance (Jenkins and Hermanowicz, 1991). Hence, use of calcium for phosphorus removal from wastewater is not discussed further in this section.

The ability to achieve the required effluent phosphorus limits is strongly dependent on the point of chemical addition in the treatment process. The potential points of chemical addition in the treatment process are before primary treatment (pre-precipitation), before secondary treatment prior to either the aeration basins or the secondary clarifiers (simultaneous precipitation), and postsecondary treatment or post-precipitation. The addition of chemicals before primary treatment requires provision of mixing tank for dissolution of chemicals, and dosages of chemicals required for phosphorus removal are much higher than stoichiometric requirements because of chemical consumption for suspended solids coagulation. Similarly, postsecondary addition of chemicals to remove phosphorus requires a mixing tank for chemical dissolution and tertiary clarification of filtration for solids removal. Jenkins and Hermanowicz (1991) stated that either a combination of primary and secondary chemical addition or tertiary treatment is necessary to achieve effluent phosphorus concentrations lower than 1.0 mg P/L.

The following examples provide the procedures for process design of phosphorus removal using metal salts addition before primary clarification or to aeration basin. The chemical doses are determined using Figures 3.3 and 3.4 and Equation 3.10. These design examples can be used as a starting point to either design a new chemical phosphorus removal system or to retrofit an existing WWTP. The final design should be based on pilot-scale trials, and possibly full-scale trials of these example process designs to account for the variations in the characteristics of the wastewater being treated.

**Phosphorus Removal by Alum Addition to Raw Wastewater and Aeration Basin.** This example presents the basic steps in the design of phosphorus removal by alum addition during primary and secondary treatment stages. Because secondary treatment by activated-sludge processes or other biological processes requires phosphorus for biomass growth, it is undesirable to remove very high levels of total phosphorus in the primary clarifiers.

The wastewater characteristics and effluent requirements for this design example are as follows:

- Wastewater average flow = 5 mgd ($1.89 \times 10^7$ $m^3/d$),
- Wastewater peak flow = 12.5 mgd ($4.73 \times 10^7$ $m^3/d$),
- Biochemical oxygen demand ($BOD_5$) = 250 mg/L,
- Total suspended solids = 270 mg/L,
- Total phosphorus = 8 mg P/L,
- Orthophosphate = 5 mg P/L,
- pH = 7.0,
- Alkalinity = 250 mg/L,
- Effluent total phosphorus limit ≤ 0.5 mg P/L,
- Effluent TSS limit ≤ 15 mg/L, and
- Effluent $BOD_5$ limit ≤ 15 mg/L.

*ALUM DOSE DETERMINATION.* The aluminum dose to be added can be determined by using Equation 3.10 and Figure 3.3. Because the soluble orthophosphate form is removed by chemical precipitation, the dose of aluminum required to remove orthophosphate in primary treatment is as follows:

$$Al_{dose} = (Al/P)\,(C_{P,in} - C_{P,res}) \tag{3.12}$$

From Figure 3.3, we can see that for $C_{P,res} < 0.1$ mg/L, the Al/P ratio increases steeply. It is economical to keep the Al/P ratio low; therefore, $C_{P,res} = 0.1$ mg/L can be selected. For $C_{P,res} = 0.1$ mg/L, Al/P ratio = 3 (from Figure 3.3) and $C_{P,in}$ (ortho P) = 5 mg/L (given). Therefore,

$$Al_{dose} = 3\,(5 - 0.1) = 14.7 \text{ mg/L}$$

Use $Al_{dose} = 15$ mg/L.

For secondary treatment, the influent $C_{P,in}$ includes all the remaining phosphorus because the nonorthophosphate and particulate phosphorus are partially hydrolyzed and converted to orthophosphate during biological treatment. Some of the phosphorus is used for biomass growth; however, determining $C_{P,in}$ as all the remaining phosphorus in the primary effluent is a conservative approach:

$$C_{P,in} = 8 - (5 - 0.1) = 3.1 \text{ mg/L} \tag{3.13}$$

Similar to primary treatment, using Equation 3.10 and Figure 3.3, we can determine aluminum dose to be added to secondary treatment to obtain effluent soluble P = 0.1 mg/L. It is required to achieve effluent soluble P = 0.1 mg/L because the TSS in the effluent contains particulate phosphorus, approximately 2 to 2.5%. Therefore, at effluent TSS = 15 mg/L and phosphorus content of 2.5% in the TSS, the particulate phosphorus concentration in the effluent = 0.375 mg/L.

To achieve an effluent total of P $\leq$ 0.5 mg/L, the effluent soluble phosphorus must be less than (0.5 – 0.375) = 0.125 mg/L.

- $C_{P,in}$ = 3.1 mg P/L;
- $C_{P,res}$ = 0.1 mg P/L;
- Al/P = 2 (from Figure 3.3);
- $Al_{dose}$ = 2 (3.1 − 0.1) = 6.0 mg/L; and
- Total $Al_{dose}$ = 15 + 6 = 21 mg/L.

*CHEMICAL REQUIREMENTS AND STORAGE.* The amount of aluminum required, in kilograms per day, is equal to

$$\text{Flow (L/d)} \times \text{Dose (mg/L)} \times 10^{-6}\ \text{kg/g} \qquad (3.14)$$

$$= (1.89 \times 1.7\ \text{L/d})\ (21\text{mg/L})\ (10^{-6}\ \text{kg/g}) = 397\ \text{kg/d}$$

Aluminum typically is added as alum, $Al_2(SO_4)_3$. Because 14 $H_2O$ = 594.3 g/mole,

$$\text{Percent aluminum in dry alum} = (2 \times 27 \times 100)/594.3 = 9.08\%$$

Therefore,

$$\text{Amount of alum required} = (397 \times 100)/9.08 = 4\ 377\ \text{kg/d}$$

Using 50% alum by weight, unit weight = 1.33 kg/L (11.1 lb/gal). The volume of 50% alum solution required at average flow is equal to

$$4\ 377/(0.5 \times 1.33) = 6\ 575\ \text{L/d}$$

The volume of 50% alum solution required at peak flow is equal to

$$6.575 \left(\frac{4.73 \times 10^7}{1.89 \times 10^7}\right) = 16\ 455\ \text{L/d}$$

Alum storage tanks are sized either as 1.5 times the largest shipment or 10-day storage at peak flow rates.

$$\text{Provide 10-day storage at peak levels} = 10(16\ 455\ \text{L/d}) = 165\ 000\ \text{L}\ (43\ 830\ \text{gal})$$

Provide a 19 000-L (50 000-gal) storage tank that can accommodate alum solution storage from 12 tank trucks (4 000 gal/tank truck) with a 7 570-L (2 000-gal) free space. The storage tank should be provided with temperature control higher than 30°F (below which alum crystallizes), recirculation pumps, and secondary containment for possible spills during unloading from tank trucks and overflow spillage.

*SLUDGE GENERATION FROM ALUM ADDITION.* Addition of alum to both primary treatment and secondary treatment for phosphorus removal results in more primary sludge production and less secondary sludge production. More sludge production in primary treatment is because of additional chemical sludge, sludge generated by enhanced removal of influent suspended solids, and soluble total organic carbon after chemical addition. Lesser sludge production from secondary treatment may be because of lower BOD loads in primary effluent that has been chemically treated.

In primary treatment, TSS and BOD removal without chemical addition are approximately 50% and 30%, respectively. Addition of alum improves the TSS removal to greater than 75%, and BOD removal to 50% (U.S. EPA, 1987). Additional organic carbon removal is also possible if polymer is added along with metal salts in the primary treatment, which will not be considered in this design example. The amount of primary sludge generated can be estimated as follows. Additional primary sludge removed because of alum addition (improved TSS removal from 50% to 75%) is equal to

$$\begin{aligned}&(0.75 - 0.5) \times \text{Influent TSS} \times \text{Flow} \times 10^{-6}\\&= 0.25 \times 270 \times 1.89 \times 10^{7} \times 10^{-6} \qquad (3.15)\\&= 1\,277 \text{ kg/d}\end{aligned}$$

Chemical sludge because of alum addition is from two components: aluminum phosphate precipitate, and aluminum hydroxide precipitate formed from aluminum in excess of stoichiometric amounts.

According to reaction 2 in Table 3.2, $Al_{0.8}(H_2PO_4)(OH)_{1.4}$ (MW = 142.4) is the form of precipitate formed after aluminum addition, and according to reaction 1, $Al(OH)_3$ (MW = 78) is the excess aluminum hydroxide formed.

Al dose = 15 mg Al/L = 15/27 = 0.555 mmole Al/L
P removed = 4.9 mg P/L = 4.9/31 = 0.158 mmole P/L
Stoichiometric Al required, $r$ = (0.8 mmole Al/mmole P removed) × 0.158 = 0.126 mmole Al/L
Excess Al added = 0.555 − 0.126 = 0.428 mmole Al/L
$Al_{0.8}(H_2PO_4)(OH)_{1.4}$ sludge = 0.158 × 142.4 = 22.5 mg/L
$Al(OH)_3$ sludge = 0.428 × 78 = 33.4 mg/L
Total chemical sludge produced = 22.5 + 33.4 = 55.9 mg/L

Chemical sludge because of alum addition (55.9 mg/L) = Flow × 55.9 × $10^{-6}$ = 1.89 × $10^7$ × 55.9 × $10^{-6}$ = 1 058 kg/d

Total increase in primary sludge production = 1 277 + 1 058 = 2 335 kg/d

Waste activated sludge (WAS) from biological treatment is reduced by alum addition in the primary treatment because BOD removal in the primary treatment is increased from approximately 30% to 50% (U.S. EPA, 1987).

Secondary sludge production = WAS + Mass of effluent TSS (3.16)

WAS $P_x$, kg VSS/d (without alum) = $Y_{obs}$ × Flow × BOD removal × $10^{-6}$

Assume observed yield, $Y_{obs}$ = 0.5 g VSS/g BOD

Primary effluent BOD (without alum) = 250 × (1 − 03) = 175 mg/L

Primary effluent BOD (with alum) = 250 × (1 − 0.5) = 125 mg/L

Secondary effluent BOD = 15 mg/L

WAS $P_x$, kg VSS/d (without alum) = 0.5 × 1.89 × $10^{-7}$ ×$10^{-6}$ = 1 514

WAS $P_x$, kg VSS/d (with alum) = 0.5 × 1.89 × $10^7$ × (125 − 15) × $10^{-6}$ = 1 041

Decrease in WAS because of alum addition to primary treatment = 1 514 – 1 041 = 473 kg VSS/d

Decrease in WAS because of alum addition to primary treatment = 473/0.85 (assume TSS/VSS ratio = 0.85) = 556 kg TSS/d

Similar to primary sludge, the chemical sludge composition is because of two components: aluminum phosphate sludge and aluminum hydroxide sludge.

Al dose = 6 mg Al/L = 6/27 = 0.222 mmole Al/L

P removed = 3.0 mg P/L = 3.0/31 = 0.097 mmole P/L

Stoichiometric Al required, *r* = (0.8 mmole Al/mmole P removed) × 0.097 = 0.077 mmole Al/L

Excess Al added = 0.222 − 0.077 = 0.145 mmole Al/L

$Al_{0.8}(H_2PO_4)(OH)_{1.4}$ sludge = 0.097 × 142.4 = 3.8 mg/L

$Al(OH)_3$ sludge = 0.0145 × 78 = 11.3 mg/L

Total chemical sludge produced = 13.8 + 11.3 = 25.1 mg/L

Chemical sludge because of alum addition (25.1 mg/L) = Flow × 25.1 × $10^{-6}$ = 1.89 × $10^7$ × 25.1 × $10^{-6}$ = 475 kg/d

Reduction in secondary sludge produced = 556 − 475 = 81 kg TSS/d

Therefore, the net effect of chemical addition to both primary and secondary treatment processes is an increase in primary sludge production by 2 335 kg/d (5 146 lb/d), and a decrease in secondary sludge production of 81 kg/d (179 lb/d).

**Phosphorus Removal by Ferric Chloride Addition to Raw Wastewater before Primary Treatment.** The following example illustrates the steps involved in process design for phosphorus removal in a typical WWTP (described in the previous design example), with the addition of ferric chloride to the wastewater before the primary treatment process. This process design involves the use of Figure 3.4 and Equation 3.10 to determine the dose of ferric iron required.

*FERRIC IRON DOSE DETERMINATION.* Because soluble orthophosphate form is removed by chemical precipitation, the dose of iron required to remove orthophosphate in the primary treatment is as follows:

$$\text{Fe(III)}_{\text{dose}} = (\text{Fe/P})\,(C_{\text{P,in}} - C_{\text{P,res}}) \qquad (3.17)$$

From Figure 3..4, we can see that for $C_{\text{P,res}} < 0.2$ mg/L, the Fe/P ratio increases steeply. It is economical to keep the Fe/P ratio low; hence, $C_{\text{P,res}} =$ 0.2 mg/L can be selected. For $C_{\text{P,res}} = 0.2$ mg/L, Fe/P ratio = 5 (from Figure 3.4). If $C_{\text{P,in}}$ (orthophosphate) = 5 mg/L (given), then $\text{Fe}_{\text{dose}} = 5 \times (5 - 0.2)$ = 24 mg/L.

$$\text{Primary effluent phosphorus concentration} = 8 - 4.8 = 3.2 \text{ mg P/L}$$

It is possible to achieve lower effluent phosphorus concentrations after the secondary treatment involving biological processes because a certain amount of phosphorus present in the primary effluent would be incorporated to the biomass for growth. Phosphorus removal by secondary treatment will not be discussed in this design example.

*CHEMICAL REQUIREMENTS AND STORAGE.* The amount of ferric iron required, in kilograms per day, is

$$\text{Flow (L/d)} \times \text{Dose (mg/L)} \times 10^{-6} = 1.89 \times 10^{7} \times 24 \times 10^{-6} = 454 \text{ kg/d}$$

Ferric iron typically is added as $FeCl_3$; formula weight = 162.3 g/mole. Therefore,

$$\text{Percent ferric iron in dry FeCl}_3 = (55.85 \times 100)/162.3 = 34.4\%$$

Therefore,

$$\text{Amount of FeCl}_3 \text{ required} = (454 \times 100)/34.4 = 1\,319 \text{ kg/d}$$

Using 30% $FeCl_3$ solution by weight, unit weight = 1.34 kg/L (11.2 lb/gal).

Volume of 30% $FeCl_3$ solution required at average flow =
$1\ 319/(0.3 \times 1.34)$
$= 3\ 280$ L/d
Volume of 30% $FeCl_3$ solution required at peak flow
$= (3\ 280 \times 4.73 \times 10^7)/1.89 \times 10^7 = 8\ 209$ L/d

$FeCl_3$ solution storage tanks are sized either as 1.5 times the largest shipment or as 10-day storage at peak flow rates.

Provide 10-day storage at peak levels = $10 \times 8\ 209 = 82\ 100$ L (21 640 gal)

Provide a 95 000-L (25 000-gal) storage tank that can accommodate $FeCl_3$ solution storage from six tank trucks (15 000 L [4 000 gal]/tank truck) with 3 800 L (1 000 gal) of free space. The storage tank should be provided with temperature control above −58°F (below which 30% $FeCl_3$ solution freezes), recirculation pumps, and secondary containment for possible spills during unloading from tank trucks and overflow spillage.

*SLUDGE GENERATION FROM FERRIC CHLORIDE ADDITION.* The addition of a $FeCl_3$ solution for phosphorus removal results in higher sludge production from primary treatment because of additional chemical sludge, sludge generated by enhanced removal of influent suspended solids, and soluble total organic carbon after chemical addition. In primary treatment, TSS and BOD removals without chemical addition are approximately 50% and 30%, respectively. Addition of coagulant such as $FeCl_3$ improves the TSS removal to more than 75%, and BOD removal improves it to 50% (U.S. EPA, 1987). The amount of primary sludge generated can be estimated as follows:

Additional primary sludge removed because of $FeCl_3$ addition
(improved TSS removal from 50% to 75%) =
$(0.75 - 0.5) \times \text{Influent TSS} \times \text{Flow} \times 10^{-6}$ (3.18)
$= 0.25 \times 270 \times 1.89 \times 10^7 \times 10^{-6}$
$= 1\ 277$ kg/d

Chemical sludge resulting from $FeCl_3$ addition is caused by two components: ferric phosphate precipitate and ferric hydroxide precipitate formed from ferric iron in excess of stoichiometric amounts.

According to reaction 2 in Table 3.2, $Fe_{1.6}(H_2PO_4)(OH)_{3.8}$ (MW = 2.51 g/mole) is the form of precipitate formed after ferric iron addition, and according to reaction 1, $Fe(OH)_3$ (MW = 106.8) is the excess ferric hydroxide formed.

Fe dose = 24 mg Fe/L = 24/55.85 = 0.430 mmole Fe/L
P removed = 4.8 mg P/L = 4.8/31 = 0.155 mmole P/L
Stoichiometric Fe required, $r$ = (1.6 mmole Fe/mmole P removed) ×
0.155 = 0.248 mmole Fe/L

Excess Fe added = 0.430 − 0.248 = 0.182 mmole Fe/L
$Fe_{1.6}(H_2PO_4)(OH)_{3.8}$ sludge = 0.155 × 251 = 38.9 mg/L
$Fe(OH)_3$ sludge = 0.182 × 106.8 = 19.4 mg/L
Total chemical sludge produced = 38.9 + 19.4 = 58.3 mg/L
Chemical sludge resulting from $FeCl_3$ addition (58.3 mg/L)
= Flow × 58.3 × $10^{-6}$ = 1.89 × $10^7$ 5 × 58.3 × $10^{-6}$
= 1 103 kg/d
Total increase in primary sludge production = 1 277 + 1 103
= 2 380 kg/d

Ferric chloride addition to raw wastewater before primary treatment for phosphorus removal increases the primary sludge production by 2 380 kg/d (5 146 lb/d).

# References


Arvin, E., and Petersen, G. (1980) A General Equilibrium Model of Phosphate precipitation. *Prog. Water Technol.*, **12**, 283.

Buzzell, J., and Sawyer, A. (1967) Removal of Algal Nutrients from Raw Wastewater with lime. *J. Water Pollut. Control Fed.*, **39**, R16.

Feitknecht, W., and Schindler, P. (1963). *Solubility Constants of Metal Hydroxides and Metal Hydroxide Salts in Aqueous Solutions.* Butterworths, London, G.B.

Ferguson, J.F., and King, T. (1977) A Model for Aluminum Phosphate Precipitation. *J. Water Pollut. Control Fed.*, **49**, 646.

Gates, D.D. (1991) Simultaneous Precipitation of Orthophosphate in Activated Sludge Systems with Al(III). Ph.D. thes., Dep. Civ. Eng., Univ. Calif., Berkeley.

Gates, D.D., *et al.* (1990) Mechanism of Chemical Phosphorus Removal in Activated Sludge with Al(III) and Fe(III). *Proc. 1990 Am. Soc. Civ. Eng. Natl. Conf. Environ. Eng.*, Arlington, Va.

Goldshmid, T., and Rubin, A.J. (1978) Aqueous Chemistry and Precipitation of Aluminum Phosphate. In *Chemistry of Wastewater Technology*. A.J. Rubin (Ed.), Ann Arbor Science Publisher, Mich., 59.

Grohman, A., *et al.* (1984) Kinetics of Phosphate Precipitation by Means of Iron(III) Salts and the Influence of Calcium Ions. *Vom Wasser* (Ger.), **62**, 171.

Hogfeldt, E. (1983) Stability Constants of Metal-Ion Complexes. IUPAC Chemical Data Series No. 21, Pergamon Press, Oxford, G.B.

Hsu, P.H. (1973) Complementary Role of Iron(II), Sulfate and Calcium in Precipitation of Phosphate from Solution. *Environ. Lett.*, **5**, 115.

Jenkins, D., and Hermanowicz, S.W. (1991) Principles of Chemical Phosphorus Removal. In *Phosphorus and Nitrogen Removal from Municipal Wastewater: Principles and Practice.* R.I. Sedlak (Ed.), Lewis Publishers, Chelsea, Mich.

Kavanaugh, M.C., *et al.* (1978) Phosphorus Removal by Post-Precipitation with Fe(iii). *J. Water Pollut. Control Fed.*, **50**, 216.

Leckie, J., and Stumm, W. (1970) Phosphate Precipitation. In *Water Quality Improvements by Physical and Chemical Processes*. Univ. Texas, Austin, 237.

Luedecke, C., *et al.* (1988) Precipitation of Ferric Phosphate in Activated Sludge. *Water Sci. Technol.*, **21**, 325.

Menar, A., and Jenkins, D. (1972) Calcium Phosphate Precipitation in Wastewater Treatment. U.S. EPA Report No. R2-72-064, Washington, D.C.

Rabinowitz, B., *et al.* (1987) Phosphorus Removal in Activated Sludge Systems. Paper presented at IAWPRC Conf., Port Elizabeth, S. Afr.

Recht, H.L., and Ghassemi, M. (1970) Kinetics and Mechanisms of Precipitation and Nature of the Precipitate Obtained in Phosphate Removal from Wastewater Using Aluminum(iii) and Iron(iii) Salts. Federal Water Quality Administration Report No. 17010 EKI 04/70, U.S. Dep. Interior, Washington, D.C.

Smith, R.M., and Martell, A.E. (1976) *Critical Stability Constants*. Vol. 4, Plenum Press, N.Y.

Spiegel, M., and Forrest, T. (1969) Phosphate Removal: Summary of Papers. *J. Am. Soc. Civ. Eng.*, **95**, 803.

Stumm W., and Morgan, J. (1970) *Aquatic Chemistry*. Wiley-Interscience, New York, N.Y.

U.S. Environmental Protection Agency (1987) *Design Manual: Phosphorus Removal*. EPA-625/1-87-001, Cincinnati, Ohio.

# Chapter 4
# Principles of Biological Nitrogen Removal



The biological reduction of nitrate and nitrite to gaseous products (primarily nitrogen gas, $N_2$), a process known as *denitrification,* is one of the most important biological means for removing nitrogen from wastewater. Because most of the nitrogen in municipal wastewater is present in reduced forms (for example, ammonium or organic nitrogen), a stage in which the reduced nitrogen compounds are oxidized to nitrite and nitrate (nitrification) is a necessary precursor to denitrification. This chapter presents the microbiology and biochemistry of nitrification and denitrification, especially as they relate to nitrogen removal from wastewater.

## Nitrification

The term *nitrification* typically is applied to the biological oxidation of ammonia ($NH_4^+$-N, which as used here refers to the total concentration of ammonia-nitrogen, including the dissociated and undissociated forms) to nitrite ($NO_2$-N) and the further oxidation of nitrite to nitrate ($NO_3^-$-N). Species of ammonia-oxidizing bacteria isolated from wastewater treatment environments most typically belong to the genus *Nitrosomonas*. Among the nitrite oxidizers isolated from wastewater, members of the genus *Nitrobacter* are the most common.

Both *Nitrosomonas* and *Nitrobacter* are autotrophic, meaning they obtain the carbon for cell synthesis from inorganic compounds (such as carbon dioxide or $HCO_3^-$). Oxidation of the ammonia or nitrite provides the energy needed for cell synthesis. These bacteria are obligate aerobes, meaning they can grow only in environments in which dissolved oxygen (DO) is present. The absence of DO for prolonged periods, however, is not lethal (Painter, 1970).

**STOICHIOMETRY.** The reaction for ammonia oxidation by *Nitrosomonas* can be written as

$$NH_4^+ + 1.5\ O_2 \rightarrow NO_2^- + 2\ H^+ + H_2O \qquad (4.1)$$

This is termed an energy reaction because it does not reflect the production of cell mass. The nitrite produced is, in turn, oxidized by *Nitrobacter*, as shown in Equation 4.2:

$$NO_2^- + 0.5\ O_2 \rightarrow NO_3^- \qquad (4.2)$$

The overall energy reaction is

$$NH_4^+ + 2\ O_2 \rightarrow NO_3^- + 2\ H^+ + H_2O \qquad (4.3)$$

Based on the stoichiometry of the overall energy reaction, 2 moles of oxygen are required to oxidize 1 mole of ammonium-nitrogen to nitrate. This is equivalent to the consumption of 4.57 g of oxygen per gram of $NH_4^+$-N oxidized.

Two equivalents of $H^+$ (used as shorthand notation for the hydronium ion, $H_3O^+$) are produced from the oxidation of 1 mole of ammonium. The $H^+$ will, in turn, react with two equivalents of bicarbonate in the wastewater. The result is that 7.14 g of alkalinity (as calcium carbonate) will be destroyed per gram of ammonium-nitrogen oxidized.

Equation 4.3 will be altered somewhat when biosynthesis is considered because some ammonia and bicarbonate will be incorporated into cell mass. The oxygen is obtained from the fixation of carbon dioxide and nitrogen into biomass. The overall reaction with biosynthesis will vary depending on the yield of bacterial mass. An overall reaction that is based on the representative yield of 0.17 g of nitrifying bacteria oxidizing ammonium all the way to nitrate is given in Equation 4.4 (Gujer and Jenkins, 1974):

$$1.02\ NH_4^+ + 1.89\ O_2 + 2.02\ HCO_3^-$$
$$\rightarrow 0.021\ C_5H_7O_2N + 1.06\ H_2O + 1.92\ H_2CO_3 + 1.00\ NO_3^- \quad (4.4)$$

The oxygen requirement and alkalinity consumption in nitrification change little when biosynthesis is considered because of the low yield of bacterial mass in the reaction. The oxygen requirement decreases to 4.3 g $O_2$/g $NH_4^+$-N oxidized, whereas alkalinity consumption increases to 7.2 g as calcium carbonate per gram $NH_4^+$-N oxidized. In design, the values derived from the energy reactions (4.57 g $O_2$ consumed and 7.14 g alkalinity destroyed per gram $NH_4^+$-N oxidized) typically are used.

**KINETICS.** Kinetic studies typically have demonstrated the growth rate of *Nitrosomonas* to be lower than that of *Nitrobacter.* This is supported by the observation that nitrite-nitrogen concentrations in biological treatment processes for nitrification can be low (0.1 g/$m^3$ or less) (circumstances leading to higher nitrite concentrations will be discussed later in this chapter). The oxidation of ammonia to nitrite can be considered to be the rate-limiting step in the overall nitrification process. Accordingly, nitrification can be modeled satisfactorily as a one-step reaction, carried out by a group of autotrophic bacteria that includes both ammonia oxidizers and nitrite oxidizers.

The Monod equation typically is used to describe the effect of limiting substrates on microbial growth (Monod, 1949). A limiting substrate is one of the compounds necessary for growth that is present at a concentration low enough to affect the rate of substrate removal. Both ammonium and DO are substrates for nitrification. Either or both of these substrates can be low enough in concentration to limit the specific growth rate of nitrifying bacteria in wastewater treatment systems.

Because there are two limiting substrates in nitrification, a dual form of the Monod equation is appropriate (Henze *et al.*, 1986). Using this approach, growth rate of nitrifying bacteria can be expressed as

$$\mu_A = \hat{\mu}_A \left[\frac{S_{NH}}{S_{NH} + K_{NH}}\right]\left[\frac{S_O}{S_O + K_{OA}}\right] \quad (4.5)$$

Where

$\mu A$ = specific growth rate of nitrifying (autotrophic) biomass (grams of cells formed per day per gram of nitrifying biomass present), $d^{-1}$;
$\hat{\mu} A$ = maximum specific growth rate of nitrifying biomass, $d^{-1}$;
$S_{NH}$ = ammonia-nitrogen concentration, mg N/L;
$K_{NH}$ = ammonia-nitrogen half-saturation coefficient for nitrification, mg N/L;
$S_o$ = DO concentration of bulk mixed liquor or wastewater, mg $O_2$/L; and
$K_{OA}$ = oxygen half-saturation coefficient for nitrification, mg $O_2$/L.

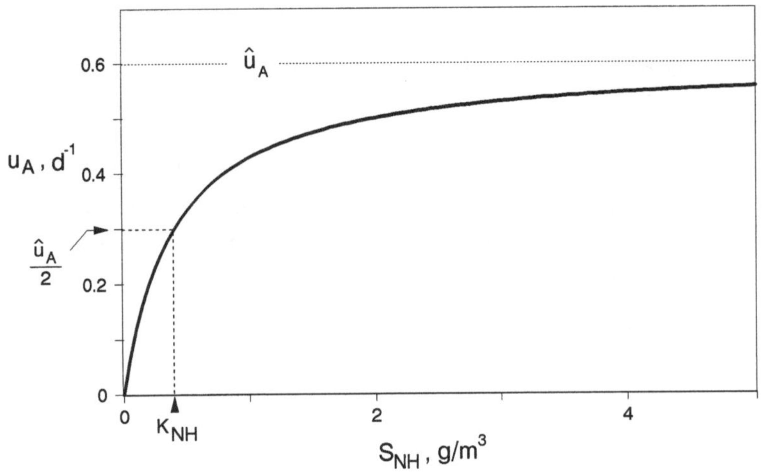


**Figure 4.1 Relationship between specific growth rate of ammonia-oxidizing bacteria and ammonia-nitrogen concentration as predicted by the Monod equation (dissolved oxygen is assumed to be nonlimiting).**

Figure 4. 1 shows a plot of specific growth in relation to ammonia-nitrogen concentration. The plot is drawn for the case in which the DO concentration is great enough that it does not limit growth. Specific growth rate increases in approximately linear fashion with respect to ammonia-nitrogen concentration near the origin of the plot, where the ammonia-nitrogen concentration is low. At higher ammonia-nitrogen concentrations, the specific growth rate approaches an asymptotic value (the maximum specific growth rate), thus exhibiting zero-order behavior with respect to ammonia-nitrogen concentration. The ammonia-nitrogen half-saturation coefficient for nitrification is the ammonia-nitrogen concentration at which the specific growth rate is one-half its maximum value.

A similar plot could be drawn to depict the relationship between specific growth rate and DO concentration, with ammonia-nitrogen at nonlimiting concentrations. The oxygen half-saturation coefficient is the DO concentration where the specific growth rate is equal to one-half its maximum value.

The specific growth rate of nitrifying bacteria is related to the specific rate of ammonia oxidation in a wastewater treatment process by the following expression:

$$q_A = \frac{1}{Y_A} \mu_A \tag{4.6}$$

Where

$q_A$ = specific ammonia oxidation rate, grams ammonia oxidized per day per gram nitrifier biomass present; and

$Y_A$ = autotrophic yield, grams nitrifying biomass produced per gram ammonia-nitrogen oxidized.

The yield coefficient in Equation 4.6 is the true growth yield, representing the quantity of nitrifying biomass that would be formed if all of the energy captured by the bacterial cells were used in cell synthesis. A decay coefficient ($b_A$) is used to account for the consumption of cell energy reserves for maintenance requirements and the effects of predation and cell lysis. An overall mass balance on the nitrifying biomass results in the following equation:

$$\theta_x = \frac{1}{\mu_A - b_A} \tag{4.7}$$

Where

$\theta_x$ = mean cell residence time, $d^{-1}$; and
$b_A$ = decay coefficient for autotrophs, $d^{-1}$.

Nearly all available values for the maximum specific growth rate reported in the literature represent a net value that incorporates decay, and $b_A$ in Equation 4.7 typically is ignored.

**VALUES OF BIOKINETIC COEFFICIENTS.** The reported values for nitrification growth rates and half-saturation coefficients include a wide range. Several reasons for this variability beyond the effects of pH, temperature, and toxins have been postulated (Atkinson and Rahman, 1979; Baillod and Boyle, 1970; Bakti and Dick, 1992; LaMotta and Shieh, 1978; Mueller *et al.*, 1968; Stenstrom and Poduska, 1980; and Stenstrom and Song, 1991). These include the effect of organic loading in single-sludge systems, the concentration of DO within the mixed liquor floc, and simultaneous double substrate limiting kinetics. Work by a number of these researchers has shown that the resistance of the floc particle to the mass transfer of oxygen into the floc particle can make the rate of oxygen diffusion the rate-limiting step to the overall nitrification process. A number of parameters can affect the concentration of DO within the floc. These include the effect of floc shape and size, mixing intensity, and growth rate of the bacteria within the floc. According to Hanaki *et al.* (1990), the heterotrophic bacteria in a single-sludge system may assimilate ammonia faster than nitrifiers, thus reducing the ammonia available for the nitrifiers. The heterotrophic biomass may also hinder the transport of ammonia and DO within the floc. This would help explain why nitrification is slower in combined systems for carbon and ammonia oxidation. This concept is illustrated in Figure 4.2. Regardless of the specific mechanism, the apparent effect of mass-transfer limitations is to increase the half-saturation coefficient and increase the minimum DO required for nitrification.

Table 4.1 and Figure 4.3 summarize values of the nitrifier maximum specific growth rate reported in the literature. These values of $\hat{\mu}_A$ range from

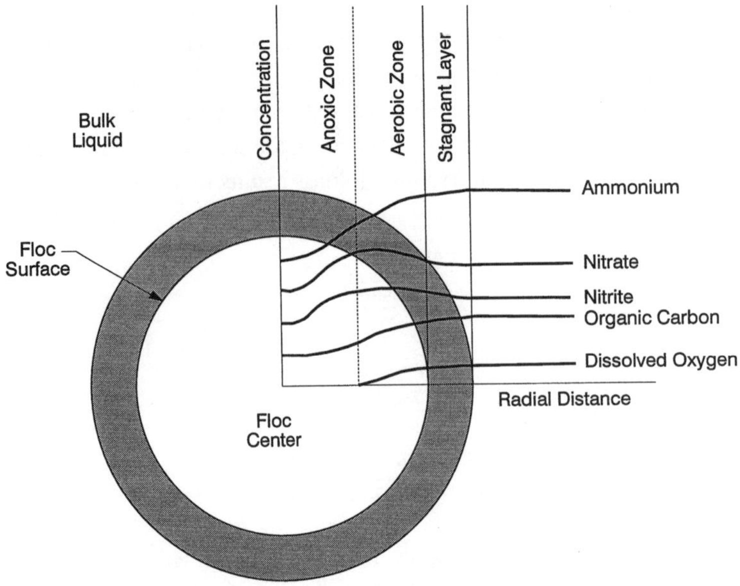


**Figure 4.2** **Illustration of substrate concentration profiles within a microbial floc showing simultaneous nitrification and denitrification. (Reprinted from *Water Research,* Vol. 26, Bakti, N.A.K., and Dick, R.I., A Model for a Nitrifying Suspended-Growth Reactor Incorporating Intraparticle Diffusional Limitation, p. 1681, Copyright 1992, with kind permission from Elsevier Science Ltd, The Boulevard, Langford Lane, Kidlington OX5 1GB, UK.)**

**Table 4.1** **Nitrification maximum specific growth rates. (Reprinted from *Water Science and Technology,* Vol. 25, Randall, C.W., *et al.,* Nitrification Kinetics in Single-Sludge Biological Nutrient Removal Activated Sludge Systems, p. 195, Copyright 1992, with kind permission from Elsevier Science Ltd, The Boulevard, Langford Lane, Kidlington, OX5 1GB, UK.)**

| Source | $\mu n$max vs. temperature, °C | $\mu n$, max ($d^{-1}$) | | |
|---|---|---|---|---|
| | | 10° | 15° | 20° |
| Downing *et al.* (1964) | $(0.47)e^{0.098(T-15)}$ | 0.29 | 0.47 | 0.77 |
| Downing and Hopwood (1964) | $(0.18)e^{0.116(T-15)}$ | 0.10 | 0.28 | 0.32 |
| Hultman (1971) | $(0.50)10^{0.033(T-20)}$ | 0.23 | 0.34 | 0.50 |
| Barnard (1975) | $0.33(1.127)^{T-20}$ | 0.10 | 0.28 | 0.37 |
| Painter and Loveless (1983) | $(0.18)e^{0.0729(T-15)}$ | 0.12 | 0.18 | 0.26 |
| Beccari *et al.* (1979) | | | | 0.27 |
| Hall and Murphy (1980) | | | | 0.46 |
| Lawrence and Brown (1976) | | | | 0.50 |

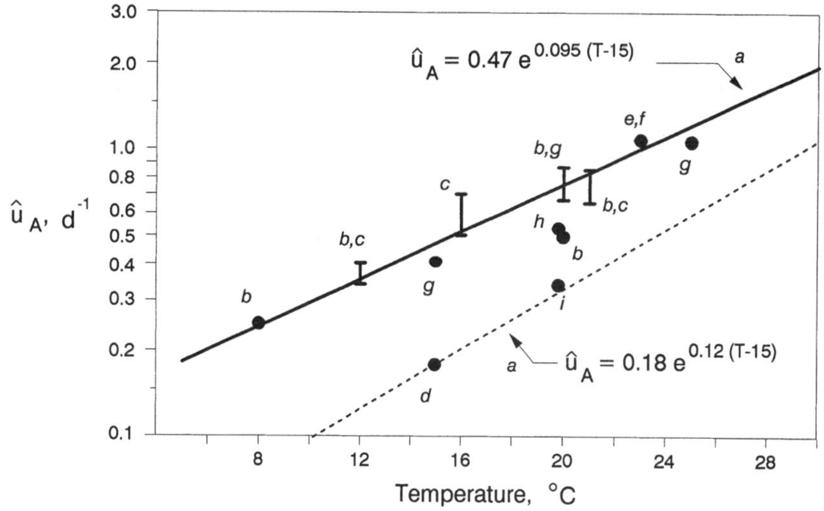


**Figure 4.3** **Effect of temperature on maximum specific growth rate of ammonia-oxidizing bacteria, $\hat{\mu}_A$ (a, Knowles *et al.* [1965]; b, U.S. EPA [1975]; c, Gujer and Jenkins [1974]; d, Benefield and Randall [1980]; e, Poduska and Andrews [1974]; f, Poduska and Andrews [1975]; g, Antoniou *et al.* [1990]; h, Stratton and McCarty [1967]; I, Downing and Hopwood [1964]).**

0.10 to 1.1 $d^{-1}$ and are strongly influenced by temperature. Nitrifier maximum specific growth rates are one to two orders of magnitude less than maximum specific growth rates for heterotrophic bacteria in wastewater treatment systems. Hence, systems intended to nitrify must be designed to operate at growth rates that are much lower than those allowable for removal of organic matter.

Caution should be taken in using any published value of the maximum specific growth for ammonia-oxidizing bacteria because toxic chemicals present in wastewater can substantially inhibit the growth of these bacteria. Wherever possible, experimental studies should be carried out to determine wastewater-specific values of this coefficient.

The optimal temperature for nitrification is between 30 and 36°C, with growth possible between 4 and 50°C (Focht and Chang, 1975, and Painter, 1970). The top expression of Figure 4.3 may be considered reliable between 8 and 30°C. Kinetic data for temperatures beyond this range should be obtained from experimental studies, using the wastewater of concern.

The effect of pH on nitrification has been reported in terms of the ratio of the specific growth rate or activity of a culture at a particular pH to the specific growth rate or activity of the same culture or a parallel culture at the optimum pH. This ratio is symbolized as $f_{pH}$ in the present chapter. Values of $f_{pH}$ from a number of investigations show a broad optimum near pH 8.0, with the nitrification rate approaching zero at pH below 6 or above 10 (Figure 4.4). Most of these data are from studies in which pure or mixed cultures of nitrifying bacteria have been grown at one pH and then subjected in short-term experiments to environments having other pH values. Because mixed cultures have been observed to acclimate to suboptimal pH (U.S. EPA, 1975), these data may be somewhat conservative.

Proposed relationships between $f_{pH}$ and pH are plotted as lines in Figure 4.4. Expressions (1) and (2) are limited to pH ranges on the low side of the optimum. Much of the reported data lies above these lines, suggesting that the expressions may be overly conservative, especially considering that most of the data are for unacclimated cultures. Expression (3) is the Michaelis pH function for the fraction of enzyme in the active state (Bailey and Ollis, 1986).

Values of the DO half-saturation coefficient for nitrification ($K_{OA}$) have been reported to range from 0.15 to 2.0 mg/L. These are summarized in Table 4.2. Reported values of the ammonia-nitrogen half-saturation coefficient for nitrification are summarized in Figure 4.5 and range from 0.54 to 10 mg/L. The half-saturation coefficients have been found to increase with tem-

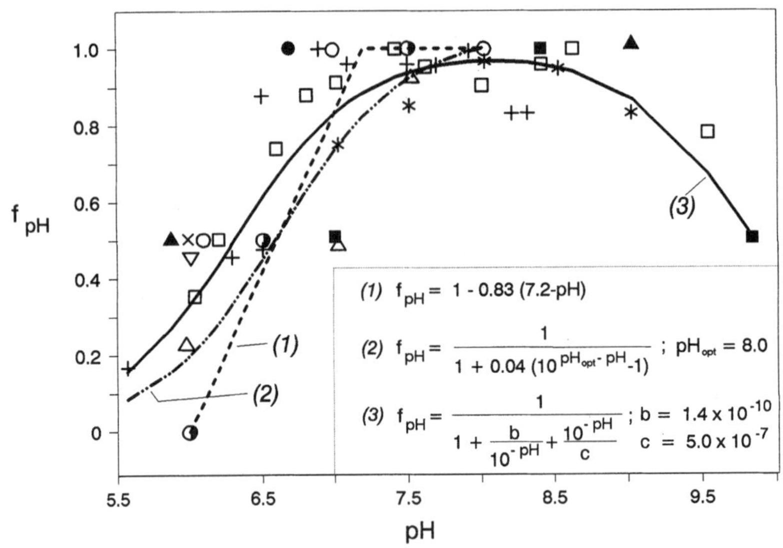


References:

○ Haug and McCarty (1971), △ Laudelot et al. (1976), □ Engel and Alexander (1958),
+ Clayfield (1974), ✱ Suzuki et al. (1974), ◑ Painter and Loveless (1983),
● Barritt (1933), ▲ Forster (1974), ■ Wild et al. (1971), × Kawai et al. (1965),
▽ Stratta et al. (1982)

*(1)* Downing and Knowles (1967), *(2)* Benefield and Randall (1980),
*(3)* Nonlinear regression on data shown

**Figure 4.4 Effect of pH on the relative growth rate or activity of ammonia-oxidizing bacteria ($f_{pH}$).**

Table 4.2 **Reported half-saturation values describing effect of DO on nitrification rates. (Reprinted from *Water Research,* Vol. 14, Stenstrom, M.K., and Poduska, R.A., The Effect of Dissolved Oxygen Concentration on Nitrification, p. 643, Copyright 1980, with kind permission from Elsevier Science Ltd, The Boulevard, Langford Lane, Kidlington, OX5 1GB, UK.)**

| Organism | $K_{OA}$, mg/L | Reference |
|---|---|---|
| *Nitrosomonas* | 0.3 | Loveless and Painter (1968) |
| | 0.25 | Peeters *et al.* (1969) |
| | 0.50 | Laudelot *et al.* (1974) |
| *Nitrobacter* | 1.84 | Peeters *et al.* (1969) |
| | 0.72 | Laudelot *et al.* (1976) |
| Activated sludge | 2.0 | Nagel and Haworth (1969) |
| | 0.45,0.56 | Stenstrom and Song (1991) |

perature (Knowles *et al.*, 1965; Painter, 1970; Painter, 1977; Parker *et al.*, 1975; and Sharma and Ahlert, 1977). Knowles *et al.* (1965) proposed a relationship between $K_{NH}$ and temperature for *Nitrosomonas* in river water, which can be written as follows:

$$K_{NH} = 0.4\ \theta^{0.118(T-15)} \qquad (4.8)$$

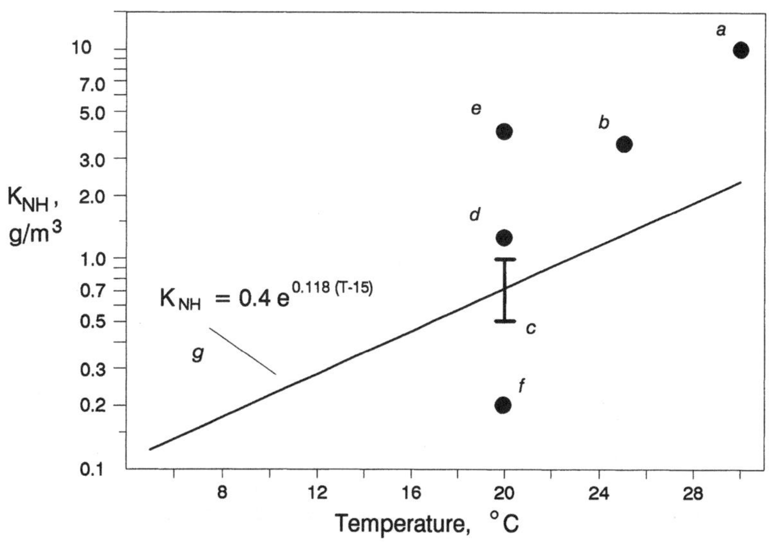


Figure 4.5 **Effect of temperature on the ammonia half-saturation coefficient for ammonia-oxidizing bacteria ($K_{NH}$) (a, Painter [1970]; b, Painter [1977]; c, Sharma and Ahlert [1977]; d, Downing *et al.* [1964]; e, Stratton and McCarty [1967]; f, Antoniou [1989]).**

where $T$ is temperature in °C. This relationship is plotted as a line in Figure 4.5. Relatively few additional values of $K_{NH}$ have been reported, and these compare poorly with the proposed expression (Figure 4.5). In light of the uncertainty regarding the magnitude of $K_{NH}$, use of a value of 1.0 mg/L at temperatures of 20°C and lower is recommended (Henze *et al.*, 1986).

**TOXICITY.** The autotrophic bacteria that perform nitrification are less robust than the heterotrophic bacteria that perform carbon removal and are particularly sensitive to a number of heavy metals and synthetic organic chemicals. A number of references are available that provide detailed information on the inhibitory effects of metals and organic. Data from several references are provided in Table 4.3.

It is important to recognize that the effects of the substances can be either inhibitory or fatal depending on the exact compound, its concentration, the duration of exposure, and the other environmental conditions in the nitrification reactor.

Special mention should be made of the inhibitory effect of gaseous (free or un-ionized) ammonia [$NH_3$(g)] and un-ionized nitrous acid ($HNO_2$). *Nitrosomonas* and *Nitrobacter* are inhibited by free ammonia and nitrous acid (Anthonisen *et al.*, 1976, and Turk and Mavinic, 1986). Table 4.4 from Randall *et al.* (1992) summarizes the range from Anthonisen where ammonium and nitrite concentrations may inhibit nitrification. *Nitrobacter* is more sensitive to free ammonia than *Nitrosomonas*. Turk and Mavinic (1986) found that nitrite oxidation was inhibited at concentrations of 0.1 to 1.0 mg/L $NH_3$(g)-N (free ammonia) and that ammonia oxidation was inhibited at concentrations of 5 to 20.0 mg/L $NH_3$(g)-N. Ford *et al.* (1980) found that nitrite oxidation was inhibited at a concentration of 10 to 150.0 mg/L $NH_3$(g)-N.

Ammonia oxidation at low pH is more sensitive to nitrite than is nitrite oxidation. Experiments indicate that inhibition by nitrite at low pH is caused by the presence of free nitrous acid (Beccari *et al.*, 1979). Turk and Mavinic (1986) found that nitrite-nitrogen levels as high as 100 mg/L caused no discernible inhibition to the treatment process. The conclusion again is that free nitrous acid, rather than nitrate, is inhibiting.

Alleman (1985) observed elevated nitrite concentrations of 27 mg/L in batch nitrification systems and concluded that *Nitrobacter* is more susceptible to environmental stresses than is *Nitrosomonas*. Alleman proposed that reduced temperature, limiting oxygen and carbon dioxide elevated pH, presence of free ammonia, and excess solid waste reduce *Nitrobacter* growth and nitrite oxidation. Alleman also reported that shock ammonium loadings and reduction of nitrate induce nitrite accumulation. Shock ammonium loads cause nitrite accumulation because *Nitrosomonas* can adapt its population more quickly than *Nitrobacter*. Mines (1983) observed nitrite concentrations ranging from 19 to 210 mg N/L at calculated free ammonia concentrations ranging from 9.5 to 73 mg/L in continuous-flow studies involving treatment of a high-strength nitrogenous wastewater. Tanaka and Dunn (1982) found

Table 4.3 **Organic compounds reported as inhibitory to nitrification (Blum and Speece, 1991; Christensen and Harremoës, 1977; Hockenbury and Grady, 1977; Painter, 1970; Payne, 1973; Richardson, 1985; and Sharma and Ahlert, 1977).**

| Name | mg/L |
|---|---|
| Acetone | 2 000 |
| Allyl alcohol | 19.5 |
| Allyl chloride | 180 |
| Allyl isothiocyanate | 1.9 |
| Allyl thiourea | 1.2 |
| AM (2-amino-4-chloro-6-methylprimidine) | 50 |
| Amino acids | 1-1 000 |
| Aminoethanol | 12.2 |
| Aminoguanidine | 74.0 |
| 2-Aminophenol | 0.27 |
| 4-Aminophinol | 0.07 |
| Aminopropiophenone | 43 |
| Aminotriazole | 70.0 |
| Ammonium | 1 000 |
| Aniline | 7.7 |
| 1-Arginine | 1.7 |
| Benzene | 13.00 |
| Benzidine dihydrochloride | 50.0 |
| Benzocaine | 100 |
| Benzothiazzole disulphide | 38.0 |
| Benzylamine | 100 |
| Benzyldimethyldodecylammonium chloride | 2.0 |
| Benzylthiuronium chloride | 40.0 |
| 2.2′ Bipyridine | 10.0 |
| Bisphenol A | 100 |
| Bromodichloropropane | 84.0 |
| 2-Bromophenol | 0.35 |
| 4-Bromophenol | 0.83 |
| *n*-Butanol | 8 200 |
| Cadmium | 14.3 |
| Carbamate | 2 |
| Carbon disulphide | 35.0 |
| Chlorine | 1 |
| Chlorobenzene | 0.71, 500 |
| Chloroform | 18.0 |
| 2-Chloronaphthol | 14.3 |
| 2-Chlorophenol | 2.70 |
| 3-Chlorophenol | 0.20 |
| 4-Chlorophenol | 0.73 |
| 5-Chloro 1-pentyne | 0.59 |

*(continued)*

**Table 4.3 Organic compounds reported as inhibitory to nitrification *(continued)*.**

| Name | mg/L |
|---|---|
| 2-Chloro-6-trichloromethyl-pyridine | 11.0 |
| Chromium (III) | 10 |
| Copper | 230 |
| *m*-Cresol | 01.-100 |
| *o*-Cresol | 11.4 |
| *p*-Cresol | 12.8 |
| Cyanide | 16.5 |
| Cyclohexylamine | 0.500 |
| Di-allyl ether | 100 |
| 1,2-Dibromoethane | 50.0 |
| Dibromethane | 60 |
| 1,2-Dichlorobenzene | 100 |
| 1,1-Dichloroethane | 0.91 |
| 2,4-Dichloroethane | 0.79 |
| 1,5-Dichloropentane | 13.00 |
| 2,3-Dichlorophenol | 0.42 |
| 2,3-Dichlorophenol | 0.61 |
| 2,6-Dichlorophenol | 8.10 |
| 3,5-Dichlorophenol | 3.00 |
| 1,3-Dichloropropene | 0.67 |
| 1,3-Dichloropropene | 0.48 |
| Dicyandiamide | 250 |
| Dicyclohexylcarbodiimide | 10.0 |
| Diethyl dithiothiosemicarbazide | 0.1 |
| Diguanide | 50.0 |
| Dimethylgloxime | 140 |
| Dimethylhydrazine | 19.2 |
| Dimethyl *p*-nitrosoaniline | 19.0 |
| Dimethyl *p*-nitrosoaniline | 30 |
| 2,4-Dinitrophenol | 37.0 |
| Diphenylthiocarbazone | 7.5 |
| Dithio-oxamide | 1 |
| Dodecylamine | <1 |
| Erythromycin | 50.0 |
| Ethanol | 2 400 |
| Ethanolamine | 100 |
| Ethyl acetate | 18 |
| Ethylenediamine | 100 |
| Ethyl urethane | 1 000 |
| Ethyl xanthate | 10 |
| Flavonoids | 0.01 |
| Guanidine | 4.7 |
| Hexamethylene diamine | 85 |

*(continued)*

**Table 4.3 Organic compounds reported as inhibitory to nitrification *(continued)*.**

| Name | mg/L |
|---|---|
| Histidine | 5 |
| Hydrazine | 58.0 |
| Hydrazine sulphate | 200 |
| Hydrogen sulfide | 50 |
| 8-Hydroxyquiniline mercaptobenzothiazole | 1 |
| Lauryl benzenesulphonate | 118 |
| Lead | 0.500 |
| 1-Lysine | 4.0 |
| Mercaptobenzothiazole | 3 |
| Methanol | 160 |
| Methionine | 9.0 |
| *n*-Methylaniline | <1 |
| Methylhydrazine | 12.3 |
| Methyl isothiocyanate | 0.800 |
| Methyl mercaptan | 300 |
| Methyl pyridines | 100 |
| 2-Methylpyridine | 100 |
| 4-Methylpyridine | 100 |
| Methylthiourea | 0.455 |
| Methyl thiuronium sulphate | 1 |
| Methylamine hydrochloride | 100 |
| Methylene blue | 30 |
| Monethanolamine | >200 |
| N-serve | 10 |
| Napthylethylenediamine dihydrochloride | 23 |
| Nickel | 5.0 |
| Ninhydrin | 10.0 |
| *p*-Nitroaniline | 10.0 |
| *p*-Nitrobenzaldehyde | 50.0 |
| Nitrobenzene | 50.0 |
| 4-Nitrophenol | 2.60 |
| 2-Nitrophenol | 11.00 |
| 2-Nitrophenol | 50.0 |
| Nitrourea | 1.0 |
| Panthothenic acid | 50 |
| Pentachloroethane | 7.90 |
| Perchloroethylene phenol | 5.6 |
| Phenolics (substituted) | 100 |
| Phenolic acids | 0.01 |
| *p*-Phenylazoaniline | 100 |
| Potassium chromate | 800 |
| Potassium chlorate | 2 500 |
| Potassium dichromate | 6.0 |

*(continued)*

**Table 4.3 Organic compounds reported as inhibitory to nitrification *(continued).***

| Name | mg/L |
|---|---|
| Potassium thiocyanate | 300 |
| *n*-Propanol | 20.0 |
| Purines | 50 |
| Pyridine | 10.0 |
| Primidines | 50 |
| Pyruvate | 400 |
| Resurcinol | 7.80 |
| Skatole | 7.0 |
| Sodium azide | 23.0 |
| Sodium azide | 20 |
| Sodium arsenite | 2 000 |
| Sodium chloride | 35 000 |
| Sodium cyanate | 100 |
| Sodium cyanide | 1 |
| Sodium dimethyl dithiocarbamate | 13.6 |
| Sodium chloride | 35 000 |
| Sodium cyanate | 100 |
| Sodium cyanide | 1 |
| Sodium dimethyl dithiocarbamate | 13.6 |
| Sodium methyl dithiocarbamare | 0.90 |
| Sodium pluoride | 1 218 |
| Sodium methyldithiocarbamate | 1 |
| ST (sulfathiazole) | 50 |
| Strychnine | 100 |
| Sulphides | 5.0 |
| Tannin | 0.01 |
| Tetrabromobisphenol | 100 |
| 1,2,3,4-Tetrachlorobenzene | 20.00 |
| 1,2,4,5-Tetrachlorobenzene | 9.80 |
| 1,1,1,2-Tetrachloroethane | 8.70 |
| 1,1,2,2-Tetrachloroethene | 1.40 |
| 12,3,5,6-Tetrachlorophenol | 1.30 |
| Tetramethylammonium chloride | 2 200 |
| Tetramethyl thiuram disulfide | 5 |
| Thiamine | 0.530 |
| Thioacetamide | 500 |
| Thiocyanates | 0.180 |
| Thiosemicarbazide (Aminothiourea) | 0.760 |
| Thiourea | 1 |
| Thiourea (substituted) | 3.6 |
| 1-Threonine | 5 |
| Threonine | 50.0 |
| 2,4,6-Tribromophenol | 7.70 |

*(continued)*

Table 4.3 **Organic compounds reported as inhibitory to nitrification *(continued).***

| Name | mg/L |
|---|---|
| 2,4,6-Tribromophenol | 50 |
| 2,4,6-Tribromophenol | 2.5 |
| 2,2,2-Trichloroethanol | 2.00 |
| 1,1,2-Trichloroethane | 1.90 |
| Trichloroethylene | 0.81 |
| Trichlorophenol | 100 |
| 2,3,5-Trichlorophenol | 3.90 |
| 2,3,6-Trichlorophenol | 0.42 |
| 2,4,6-Trichlorophenol | 7.90 |
| Triethylamine | 100 |
| Trimethylamine | 118 |
| 2,4,6-Trimethylphenol | 30.0 |
| 1-Valine | 1.8 |
| Vitamins riboflavin, A-lipolic acid, B-pyridoxine HCL | 50 |
| Zinc | 11.0 |

that oxygen concentrations approaching zero led to nitrite concentrations of 36 mg/L in a laboratory-scale fixed-film batch nitrification system. They found that when the DO concentrations were increased, the nitrite levels decreased to below 10.0 mg/L.

**ATTACHED-GROWTH SYSTEMS.** In addition to suspended-growth systems, there are a variety of alternative biological processes currently available to nitrify domestic and industrial wastewater. Collectively, they are

Table 4.4 **Ammonium-nitrogen and nitrate-nitrogen concentration range for *Nitrobacter* inhibition as function of pH (*T* = 20°C). (Reprinted from *Water Science and Technology*, Vol. 25, Randall, C.W., *et al.*, Nitrification Kinetics in Single-Sludge Biological Nutrient Removal Activated Sludge Systems, p. 195, Copyright 1992, with kind permission from Elsevier Science Ltd, The Boulevard, Langford Lane, Kidlington OX5 1GB, UK.)**

| pH | $NH_4^+$–N range, mg/L | $NO_2^-$–N range, mg/L |
|---|---|---|
| 6.0 | 210–2 100 | 30–330 |
| 6.5 | 70–700 | 88–1 050 |
| 7.0 | 20–210 | 260–3 320 |
| 7.5 | 7–70 | |
| 8.0 | 2–20 | |

referred to as attached-growth or fixed film systems. In these systems biomass is attached to solid support media contained within a reaction vessel. Although the kinetic (growth and oxidation) relationships used in the design of suspended-growth reactors are still valid, design of attached-growth systems is complicated by mass-transfer limitations of substrate into the biological films and definition of the biofilm surface itself (see Figure 4.6). Because of this, design of attached-growth reactors is primarily based on empirical results from pilot- and full-scale systems.

In fixed film nitrification reactors, the competition, or interactions, between heterotrophic and autotrophic bacteria are more important than in suspended-growth systems. The presence of significant organic substrate in a fixed film reactor will allow the heterotrophic biomass to overwhelm the autotrophs and effectively prevent their growth until the carbon substrate concentration in the bulk liquid is reduced to an acceptably low concentration. This concentration typically is accepted to be approximately 20 mg/L biochemical oxygen demand ($BOD_5$). Boller *et al.* (1994) provide an overview of the important parameters for nitrifying biofilms.

Mass transfer and competition for DO by heterotrophic bacteria must be considered in attached-growth systems. The concentration of substrates (such as ammonia-nitrogen or DO) within the biofilm can be significantly lower

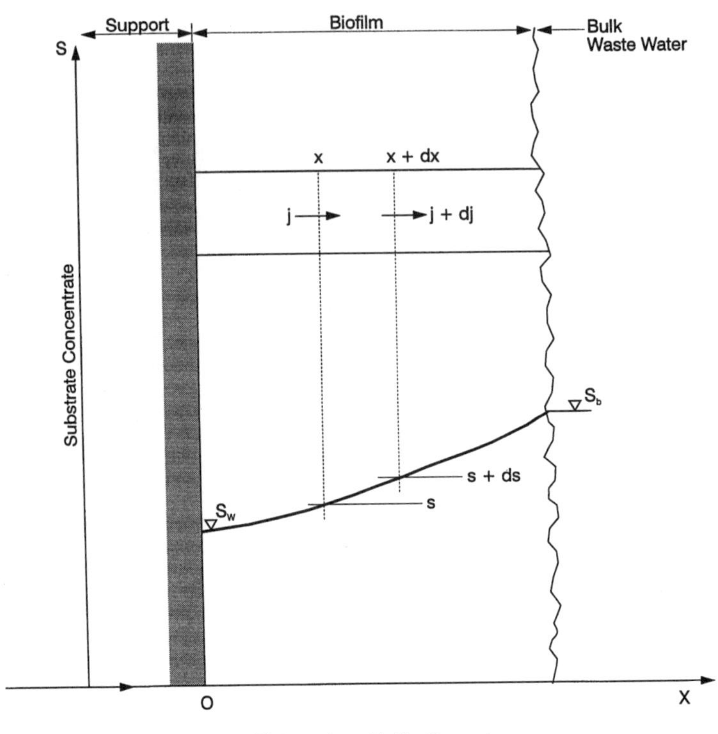


**Figure 4.6 Substrate concentration profiles in fixed film systems (Gujer and Boler, 1986).**

than in the bulk liquid because they must be transported through the biofilm. Lower concentrations within the biofilm will result in lower observed rates of nitrification. To accurately describe this, diffusion-reaction models that consider both external (liquid film) and internal (biofilm) mass-transfer resistances can be constructed. Several researchers have shown that external (liquid film) mass-transfer resistance is negligible in comparison to the internal resistance.

Examples of attached-growth systems include trickling filters, rotating biological contactors, submerged packed-bed reactors, and fluidized beds. Immobilized cells or high biomass processes also rely significantly on attached-growth biomass. Typically, attached-growth systems may tend to be more resistant to shock BOD loads and, when enclosed, are protected from excess losses in temperature. The short detention time in some fixed-film reactors, however, may result in breakthrough of ammonium-nitrogen at peak flows.

An additional advantage of attached-growth systems is that there is no need for recirculated sludge to maintain the necessary biomass to accomplish treatment because the biomass is attached to the solid support in the reactor. From a practical design viewpoint, this means process efficiency is not dependent on the settleability of the biomass, as is the case for suspended-growth systems.

As with suspended-growth systems, nitrification may be accomplished in a separate unit process or in combination with carbonaceous BOD removal. Both modes of operation will be presented.

# DENITRIFICATION

**BIOCHEMISTRY AND MICROBIOLOGY.** Most denitrifiers are facultative, meaning they can use either oxygen or nitrate as the terminal electron acceptor in respiration. The use of oxygen as the electron acceptor is termed *aerobic* respiration, whereas the term *anoxic* respiration refers to the use of nitrate or nitrite. These microorganisms use similar metabolic pathways. The only difference between aerobic respiration and anoxic respiration is the enzyme catalyzing the final electron transfer occurring in the electron transport chain. Oxygen must be excluded to promote denitrification. If both oxygen and nitrate are present, microorganisms will preferentially use oxygen as the terminal electron acceptor. This is because denitrification yields less energy than aerobic respiration.

Microorganisms require nitrogen for protein synthesis. The preferred source of nitrogen is ammonium because this form is used directly in synthesis. However, if sufficient ammonium is not available, some microorganisms can reduce nitrate to ammonium for this use (Gayle *et al.*, 1989). This process is referred to as *assimilatory nitrate reduction*, indicating that nitrogen is incorporated into the cell. It is therefore distinguished from dissimilatory nitrate reduction (denitrification), which is a respiratory process whereby the microorganism obtains energy.

Four steps are involved in biological denitrification, as shown schematically in Equation 4.9 (Grady and Lim, 1980):

$$NO_3^- \rightarrow NO_2^- \rightarrow NO(g) \rightarrow N_2O(g) \rightarrow N_2(g) \quad (4.9)$$

Nitrite ($NO_2^-$), nitric oxide (NO), and nitrous oxide ($N_2O$) are intermediates in the process. Each step involves a particular reductase enzyme that catalyzes the transfer of electrons to nitrogen. The electrons originate from the substrate, that is, the electron donor. Either inorganic (for example, hydrogen or sulfur) or organic waste compounds can serve as substrate for denitrification. As a result of denitrification, the electron donor is oxidized while nitrate is reduced.

At least 14 bacterial genera are known to contain denitrifying species (Gayle *et al.*, 1989). These include *Bacillus, Pseudomonas, Methanomonas, Paracoccus, Spirillum,* and *Thiobacillus.* Most of the denitrifying bacteria are heterotrophic, meaning that they obtain carbon for cell synthesis from organic compounds, often the same molecules that are oxidized for energy. There are relatively few species of autotrophic denitrifying bacteria. These bacteria obtain carbon for cell synthesis from organic compounds. One example is *Thiobacillus denitrificans.* This organism oxidizes elemental sulfur for energy and obtains carbon for cell biosynthesis from dissolved carbon dioxide (carbon dioxide) or bicarbonate ($HCO_3^-$).

Use of one-carbon organic compounds, such as methanol or methane, involves two types of bacteria. Methylotrophic bacteria convert a large fraction of assimilated carbon to soluble metabolites (for example, citrate/isocitrate), which then serve as electron donors for denitrifying bacteria (Rhee and Fuhs, 1978).

**STOICHIOMETRY.** The stoichiometric equations for denitrification depend on the carbon substrate and the source of nitrogen. The energy equations using methanol, acetic acid, methane, and wastewater as the carbon substrates and nitrate as the terminal electron acceptor can be written as in Equations 4.10 through 4.13:

$$6\ NO_3^- + 5\ CH_3OH \rightarrow 3\ N_2 + 5\ CO_2 + 7\ H_2O + 6\ OH^- \quad (4.10)$$

$$8\ NO_3^- + 5\ CH_3COOH \rightarrow 4\ N_2 + 10\ CO_2 + 6\ H_2O + 8\ OH^- \quad (4.11)$$

$$8\ NO_3^- + 5\ CH_4 \rightarrow 4\ N_2 + 5\ CO_2 + 6\ H_2O + 8\ OH^- \quad (4.12)$$

$$10\ NO_3^- + C_{10}H_{19}O_3N \rightarrow 5\ N_2 + 10\ CO_2 + 3\ H_2O + NH_3 + 10\ OH^- \quad (4.13)$$

The hydroxide ion formed during denitrification will react with carbon dioxide in the water to create bicarbonate ions according to the following stoichiometric equation:

$$OH^- + CO_2 \rightarrow HCO_3^- \tag{4.14}$$

As with nitrification, the inclusion of biosynthesis changes the stoichiometry. The overall result is an increase in the electron donor (carbon substrate) required per unit mass of nitrate reduced. Using methanol as the substrate the stoichiometry with biosynthesis can be presented as

$$\begin{aligned} &NO_3^- + 1.08\ CH_3OH + 0.24\ H_2CO_3 \\ &\rightarrow 0.056\ C_5H_7NO_2 + 0.47\ N_2 + 1.68\ H_2O + HCO_3^- \end{aligned} \tag{4.15}$$

$$\begin{aligned} &NO_2^- + 0.67\ CH_3OH + 0.53\ H_2CO_3 \\ &\rightarrow 0.040\ C_5H_7NO_2 + 0.48\ N_2 + 1.23\ H_2O + HCO_3^- \end{aligned} \tag{4.16}$$

$$\begin{aligned} &O_2 + 0.93\ CH_3OH + 0.056\ NO_3^- \\ &\rightarrow 0.056\ C_5H_7NO_2 + 1.04\ H_2O + 0.59\ H_2CO_3 + 0.56\ HCO_3^- \end{aligned} \tag{4.17}$$

The design significance of the stoichiometry is that nitrate is converted to nitrogen gas, the oxygen demand is reduced by 2.86 g oxygen per gram nitrate reduced, and alkalinity is produced by approximately 3.57 g as calcium carbonate per gram nitrate reduced if nitrate is the nitrogen source for biosynthesis. Where cells have ammonia available, alkalinity production can be lower, as suggested by reported values of 2.9 to 3.0 g alkalinity produced per gram nitrate-nitrogen reduced (Horstkotte *et al.*, 1974).

**SUBSTRATE REQUIREMENTS.** A number of factors influence substrate consumption in biological denitrification. The first factor is the concentrations of the electron acceptors present. These include nitrate, nitrite, DO, and sulfate. Most of the DO present must be reduced before denitrification can proceed. Nitrate and nitrite compete on approximately an equal basis for electrons from the substrate. Sulfate ($SO_4^{2-}$) is present in varying concentrations in wastewater. This anion can be reduced biologically, but only after almost all the DO, nitrate, and nitrite have been consumed. Hence, nearly complete denitrification can be obtained without appreciable sulfate reduction.

A second factor affecting electron donor requirements is the nature of the donor molecule. Organic compounds are used by bacteria as the source of electrons for energy metabolism, as well as the source of carbon for cell biosynthesis. Inorganic compounds such as molecular hydrogen and sulfur supply electrons for energy metabolism only.

A third factor is the extent of the denitrification reaction. A shortage of electron donor can cause the conversions depicted in Equation 4.10 to stop before nitrogen gas is produced, so that the quantity of nitrate-nitrogen removed exceeds the quantity of nitrogen gas produced. The electron donor requirement, expressed in terms of the mass of substrate consumed per unit mass of nitrate-nitrogen removed, will then vary directly with the percentage removal of nitrate, up to the point of complete conversion.

A fourth factor influencing electron donor requirement is the specific growth rate of denitrifying bacteria. Net cell yield is highest at high growth rates, decreasing as the specific growth rate declines because of decay. The electron donor requirement for high-rate systems will consequently exceed that for low-rate systems. The effect of the specific growth rate is most significant in systems using organic substrate.

The total concentration of substrate required to reduce the nitrate, nitrite, and DO present with no biosynthesis ($C_s$) is given by Equation 4.18:

$$C_s = 0.357\ W_e\ (1.0\ NO_3\text{-}N + 0.60\ NO_2\text{-}N + 0.35\ O_2) \qquad (4.18)$$

where $W_e$ is the electron equivalent of the substrate in grams and the concentrations of the terms are in grams per cubic metre (McCarty *et al.*, 1969). The first coefficient (0.357) converts from mass $NO_3^-$-N to electron equivalents. The term in parentheses on the right side of Equation 4.18 is termed the equivalent nitrate-nitrogen concentration ($NO_{3eq}$-N), and the coefficients preceding the concentrations in this term convert from nitrite and oxygen concentrations to equivalent nitrate-nitrogen concentration through the corresponding electron equivalent weights. Thus,

$$C_s = 0.357\ W_e\ NO_{3eq}\text{-}N \qquad (4.19)$$

Additional substrate will be consumed as a source of carbon for cell biosynthesis in heterotrophic systems. The total amount of substrate for denitrification ($C_s$) is then

$$C_s = 0.357\ C_r\ W_e\ NO_{3eq}\text{-}N \qquad (4.20)$$

where $C_r$ is the ratio between $C_s$ and $C_s$ and is termed the consumptive ratio (McCarty *et al.*, 1969). Values of the consumptive ratio are greater than 1.0 with higher values indicating greater substrate consumption for biosynthesis rather than for reduction of nitrate in respiration.

McCarty *et al.* (1969) investigated methanol, acetic acid, ethanol, acetone, and sucrose as substrates for denitrification of agricultural drainage waters. They found $C_r$ to vary from 1.2 to 1.7, with sugar giving the highest ratio and acetone the lowest. The average $C_r$ for methanol was 1.3. Consumptive ratios for methanol reported in other studies vary from 1.1 to 2.1.

If the substrate is quantified in terms of its chemical oxygen demand (COD), $W_e$ has a constant value of 8 g COD. In this case, the total substrate requirement is given by Equation 4.21:

$$\text{COD} = 2.86\ C_r\ \text{NO}_{3\text{eq}}\text{-N} \qquad (4.21)$$

where COD has units of mg COD/L. Monteith *et al.* (1980) tested 30 industrial organic wastes as substrates for denitrification. The $C_r$ of these wastes ranged from 0.8 to 3.6.

Another parameter for evaluating substrate requirements, the substrate consumption ratio (SCR), is calculated as follows:

$$\text{SCR} = \frac{\Delta C}{\Delta \text{NO}_{3\text{eq}}\text{-N}} \qquad (4.22)$$

where $\Delta NO_{3eq}$-N is the equivalent nitrate concentration consumed and $\Delta C$ is the corresponding change in substrate concentration expressed as COD. The SCRs of methanol are reported to vary from 3.2 to 6.0 g COD/g NO-N. Substrate consumption ratios reported by Monteith et al. (1980) for industrial organic wastes ranged from 2.2 to 10.2 g COD/g $NO_{3eq}$-N.

**ALKALINITY PRODUCTION.** The quantity of base produced by denitrification can be calculated from the following balanced reaction, as modified from McCarty *et al.* (1969):

$$\text{NO}_3^- + 1.08\,\text{CH}_3\text{OH}$$

$$\rightarrow 0.065\,\text{C}_5\text{H}_7\text{O}_2\text{N} + 0.47\,\text{N}_2 + 0.76\,\text{CO}_2 + 1.44\,\text{H}_2\text{O} + \text{OH}^- \qquad (4.23)$$

Therefore, 3.57 mg/L of alkalinity is produced per mg/L $NO_3^-$-N reduced when $NO_3^-$ is used by the denitrifying bacteria for cell synthesis. In systems where amonia is available, the alkalinity production can be less, as mentioned before.

**KINETICS.** The rate of denitrification has been found to vary depending on the type and concentration of compound used as the carbon substrate. The availability of soluble readily biodegradable substances results in higher denitrification rates. Denitrification is essentially zero-order with respect to nitrate concentration. The rate of denitrification is also strongly affected by the DO concentration, pH, temperature, and reactor configuration.

Researchers have developed several mathematical models for predicting denitrification rates. These include Monod kinetics and empirical rate expressions. Most empirical rate expressions are zero-order with respect to nitrate and organic substrate and first-order with respect to biomass concentration. Others vary with the organic loading or solids retention time. These different expressions for the denitrification rate ($r_v$,NO) in units of g $NO_3$-N reduced/m$^3$·d or the unit denitrification rate ($r_{v,x}$) in units of g $NO_3$-N reduced/g MLSS·d follow:

Monod (Henze *et al.*, 1986):

$$r_{v,\mathrm{NO}} = \left(\frac{1 - Y_H}{2.86\ Y_H}\right)\hat{\mu}_H\left(\frac{S_s}{K_s + S_s}\right)\left(\frac{S_{\mathrm{NO}}}{K_{\mathrm{NO}} + S_{\mathrm{NO}}}\right)\left(\frac{K_{O,A}}{S_o + K_{O,A}}\right)\eta_g X_{b,h} \tag{4.24}$$

Zero-order:

$$r_{v,\mathrm{NO}} = kX \tag{4.25}$$

Refling–Stensel, external substrate available (Refling and Stensel, 1978):

$$r_{x,\mathrm{NO}} = 0.03(\mathrm{F{:}M}) + 0.029 \tag{4.26}$$

No external substrate available:

$$r_{x,\mathrm{NO}} = 0.12\ \Theta_x^{-0.706} \tag{4.27}$$

Or

$$r_{x,\mathrm{NO}} = \frac{\eta_g}{2.86}\frac{A_n}{Y_n}\frac{1}{\Theta_x} \tag{4.28}$$

Where

$\hat{\mu}H$ = maximum specific growth rate for heterotrophs, d$^{-1}$;
$\eta_g$ = fraction heterotrophs using nitrate for electron acceptor;
$X_{b,h}$ = concentration of heterotrophs, mg/L COD;
$S_s$ = concentration of readily degradable organic substrate, mg/L COD;
$K_s$ = half-saturation coefficient for readily degradable substrate, mg/L COD;
$S_O$ = concentration of dissolved oxygen, mg $O_2$/L;
$K_{OH}$ = half-saturation coefficient for DO in heterotrophic growth, mg $O_2$/L;
$S_{\mathrm{NO}}$ = concentration of nitrate, mg N/L;
$K_{\mathrm{NO}}$ = half-saturation coefficient for nitrate, mg N/L;
$Y_H$ = heterotrophic yield, g biomass COD/g substrate COD;
$k$ = empirical rate constant, g $NO_3^-$-N/g MLSS·d;
$X$ = concentration of MLSS, g/m$^3$; and
F:M = food-to-microorganism ratio, g $BOD_5$ applied/g MLSS·d.

The values for the biokinetic coefficients for Monod-type expressions are not well defined. Table 4.5 summarizes recommended values from two references. Thus, typical practice has been to use empirical rate expressions that

**Table 4.5 Monod kinetic coefficients for denitrification (Baillod and Boyle, 1970, and Henze *et al.*, 1986).**

| Coefficient | Symbol | Typical range | Suggested |
|---|---|---|---|
| Maximum specific growth rate of heterotrophs | $\mu_{,H}$ | 3–13 | 4.0–6.0 |
| Heterotrophic biomass yield | $Y_H$ | 0.46–0.69 | 0.67 |
| Half-saturation coefficient organic substrate | $K_s$ | 10–180 | 10.20 |
| Half-saturation coefficient nitrate-nitrite | $K_{NO}$ | 0.06–0.5 | 0.2–0.5 |
| Correction factor for $\mu_H$ under anoxic conditions | $\eta_g$ | 0.5–1.0 | 0.8 |
| Half-saturation coefficient for dissolved oxygen for heterotrophic biomass | $K_{O,H}$ | 0.10–0.28 | 0.1–0.2 |
| Mass nitrogen per mass of COD[a] in biomass | $i_{X,B}$ | 0.06–0.12 | 0.06–0.086 |
| Decay coefficient for heterotrophic biomass | $b_H$ | | 0.05 |

[a] Chemical oxygen demand.

have been found to be appropriate for given types of substrate. These empirical rate expressions typically are zero-order with respect to nitrate and first-order with respect to biomass concentration according to Equation 4.29:

$$r_v = kX \tag{4.29}$$

Where

$r_v$ = rate of denitrification, days$^{-1}$;
$k$ = empirical rate constant, L/mg·d; and
$X$ = concentration of heterotrophs, mg/L.

The rate of denitirification is also strongly affected by the kinetic regime of the reactor. Plug-flow reactors and reactors in series will produce higher denitrification rates when the reaction order is greater than zero. This will typically happen when the availability of substrate limits the denitrification reaction.

Denitrifying bacteria grow well under the conditions typically experienced in wastewater, that is, pH between 7 and 8 and temperature between 5 and 25°C. The maximum rate of denitrification is temperature dependent, roughly doubling for every 10°C increase in temperature between 5 and 25°C. Reported values for the temperature coefficient are given in Table 4.6.

Soluble, readily degradable substrates support the highest rates of denitrification. Although methanol is the most typically used soluble substrate, it is not the best on a kinetic basis. Monteith *et al.* (1980) found that 22 of 30 industrial organic wastes tested (brewery and distillers' wastes) supported denitrification rates greater than those observed with methanol.

**Table 4.6 Denitrification temperature coefficients (Sutton *et al.*, 1975).**

| Θ Values | Substrate | Temperature range, °C | Type of system | Reference |
|---|---|---|---|---|
| 1.09 | Methanol | 6–25<br>6–16<br>10–20 | Suspended growth 6-day SRT[a] | Sutton *et al.*, 1975 |
| 1.07 | Methanol | 5–25<br>5–15<br>10–20 | Upflow packed column | Sutton *et al.*, 1975 |
| 1.094 | Wastewater (exogenous carbon) | 17–25 | Suspended growth | Barnard, 1975 |
| 1.20 | Endogenous (no external carbon) | 17–25 | Suspended growth | Barnard, 1975 |
| 1.12 | Methanol | 5–27 | Laboratory batch | Dawson and Murphy, 1972 |
| 1.10 | | 10–20 | | |
| 1.06 | | 15–25 | Batch activated sludge, SRT = 2 days | Stensel, 1970 |
| 1.13 | | 10–20 | Continuous activated sludge | Stensel, 1970 |
| 1.15 | Methanol | 10–20 | Suspended growth, SRT = 7.6 | Mulbarger, 1971 |
| | | 10–20 | Activated sludge | Johnson and Vania, 1971 |
| | Wastewater | 6–25 | Separate sludge | Murphy and Sutton, 1975 |
| | Wastewater | 6–25 | Single sludge | Murphy and Sutton, 1975 |
| 1.08 | Raw and settled wastewater | 12–24 | Suspended growth (first anoxic; SRT = 10–25 days) | Ekama *et al.*, 1984 |
| 1.03 | Raw and settled wastewater | 12–24 | Suspended growth (first anoxic; SRT = 10–25 days) | Ekama *et al.*, 1984 |
| 1.06 | — | — | | Dawson and Murphy, 1972 |

[a] Solids retention time.

**TOXICITY.** The heterotrophic bacteria that perform denitrification typically are less sensitive to inhibition from toxic chemicals; however, it is still a concern. Oxygen has been found to inhibit nitrite reductase, slowing the rate of nitrite reduction. Oxygen has been found to inhibit nitrite reductase to an even greater extent than nitrate reductase. Hernandez and Rowe (1987) found that nitrite began to accumulate when oxygen was added to a batch denitrifying system and stopped accumulating when the oxygen supply was terminated and the system was flushed with argon gas.

Hochstein *et al.* (1984) found that DO concentrations of 5% saturation reduced nitrous oxide production rates, that nitrite reductase was sensitive to DO concentrations of 13% of saturation, and that nitrate reductase was least sensitive to oxygen and was completely inhibited only at oxygen concentrations greater than 13% saturation. These data suggest that nitrate reduction continues at low DO, whereas nitrite reduction is inhibited.

Nitrite can suppress rates and decrease its own use. Rowe *et al.* (1979) reported that $NO_2$-N concentrations above 14.0 mg/L at a pH of 7.0 inhibited active transport of carbohydrates and amino acid in *Pseudomonas aeruginosa,* and that concentrations above 350 mg/L completely inhibited active transport by microorganisms in both an anoxic environment with $NO_3^-$ as the terminal electron acceptor and an oxic environment with oxygen as the terminal electron acceptor. Beccari *et al.* (1979) suggested that inhibition by nitrite is caused by free nitrous acid ($HNO_s$). They found that nitrite reduction rates drop sharply for pH values less than 7.5 in contrast to nitrate reduction rates.

# *REFERENCES*


Alleman, J.E. (1985) Elevated Nitrite Occurrence in Biological Wastewater Treatment Systems. *Water Sci. Technol.*, **17**, 409.

Anthonisen, A.C., *et al.* (1976) Inhibition of Nitrification by Ammonia and Nitrous Acid. *J. Water Pollut. Control Fed.*, **48**, 835.

Antoniou, P. (1989) Determination of Biokinetic Coefficients for Nitrification in the Activated Sludge Process. M.E. thes., Dep. Environ. Eng. Sci., Univ. Fla., Gainesville.

Antoniou, P., *et al.* (1990) Effect of Temperature and pH on the Effective Maximum Specific Growth Rate of Nitrifying Bacteria. *Water Res.* (G.B.), **24**, 97.

Atkinson, B., and Rahman, F.U. (1979) Effect of Diffusion Limitations and Floc Size Distributions on Fermentor Performance and the Interpretation of Experimental Data. *Biotechnol. Bioeng.* **21,** 221.

Bailey, J.E., and Ollis, D.F. (1986) *Biochemical Engineering Fundamentals.* McGraw-Hill, Inc., New York, N.Y.

Baillod, C.R., and Boyle, W.C. (1970) Mass Transfer Limitations in Substrate Removals. *J. Sanit. Eng. Div., Proc. Am. Soc. Civ. Eng.*, **96**.

Bakti, N.A.K., and Dick, R.I. (1992) A Model for a Nitrifying Suspended-Growth Reactor Incorporating Intraparticle Diffusional Limitation. *Water Res.* (G.B.), **26**, 1681.

Barnard, J.L. (1975) Biological Nutrient Removal Without the Addition of Chemicals. *Water Res.* (G.B.), **9**, 485.

Barritt, N.W. (1933) The Nitrification Process in Soils and Biological Filters. *Ann. Appl. Biol.* (G.B.), **20**, 165.

Beccari, M., *et al.* (1979) A Critical Analysis of Nitrification Alternatives. *Water Res.* (G.B.), **13**, 185.

Benefield, L.D., and Randall, C.W. (1980) *Biological Process Design for Wastewater Treatment.* Prentice-Hall, Inc., Englewood Cliffs, N.J.

Blum, D.J., and Speece, R.E. (1991) A Database of Chemical Toxicity to Environmental Bacteria and Its Use in Interspecies Comparison and Correlations. *Water Environ. Res.*, **63**, 198.

Boller, M., *et al.* (1994) Parameters Affecting Nitrifying Biofilm Reactors. *Water Sci. Technol.*, **29,** 10/11, 1.

Christensen, M.H., and Harremoës, P. (1977) Biological Denitrification of Sewage: A Literature Review. *Prog. Water Technol.*, **8**, 509.

Clayfield, G.W. (1974) Respiration and Denitrification Studies on Laboratory and Works Activated Sludges. *J. Water Pollut. Control Fed.*, **73**, 51.

Dawson, R.N., and Murphy, K.L. (1972) The Temperature Dependency of Biological Denitrification. *Water Res.* (G.B.), **6**, 71.

Downing, A.L., and Hopwood, A.P. (1964) Some Observations on the Kinetics of Nitrifying Activated Sludge Plants. *Schweizerische Aeitschrift fur Hyddrologie,* **26**, 271.

Downing, A.L., *et al.* (1964) Nitrification in the Activated Sludge Process. *J. Inst. Sew. Purification*, 130.

Ekama, G.A., *et al.* (1984) Biological Excess Phosphorus Removal. In *Theory, Design, and Operation of Nutrient Removal Activated Sludge Processes.* Water Resour. Comm., Pretoria, S. Afr.

Engel, M.S., and Alexander, M. (1958) Growth and Autotrophic Metabolism of Nitrosomonas Europaea. *J. Bacteriol.*, **76**, 217.

Focht, D.D., and Chang, A.C. (1975) Nitrification and Denitrification Processes Related to Wastewater Treatment. *Adv. Appl. Microbiol.*, **20**, 153.

Ford, D.L., *et al.* (1980) Comprehensive Analysis of Nitrification of Chemical Processing Wastewaters. *J. Water Pollut. Control. Fed.*, **52**, 2726.

Forster, J.R.M. (1974) Studies on Nitrification in Marine Biological Filters. *Aquaculture,* **4**, 387.

Gayle, B.P., *et al.* (1989) Biological Denitirification of Water. *J. Environ. Eng.*, **115**, 930.

Grady, C.P.L., and Lim, H.C. (1980) *Biological Wastewater Treatment: Theory and Applications.* Marcel Dekker, New York, N.Y.

Gujer, W., and Boler, M. (1986) Design of a Nitrifying Tertiary Trickling Filter Based on Theoretical Concepts. *Water Res.* (G.B.), **20**, 1353.

Gujer, W., and Jenkins, D. (1974) The Contact Stabilization Process-Oxygen and Nitrogen Mass Balances. Report No. 74-2, Sanit. Eng. Res. Lab., Univ. Calif., Berkeley.

Hall, I.R., and Murphy, K.L. (1980) Estimation of Nitrifying Biomass and Kinetics in Wastewater. *Water Res.* (G.B.), **14**, 297.

Hanaki, K., *et al.* (1990) Nitrification at Low Levels of Dissolved Oxygen With and Without Organic Loading in a Suspended-Growth Reactor. *After Res.*, **24**, 297.

Haug, R.T., and McCarty, P.L. (1971) Nitrification with the Submerged Filter. Technical Report No. 149, Dep. Civ. Eng., Stanford Univ. Environ. Protection Agency, Research Grant No. 107010 EPM, Palo Alto, Calif.

Henze, M., *et al.* (1986) Activated Sludge Model No. 1. Scientific and Technical Reports No. 1, IAWPRC, London, U.K.

Hernandez, D., and Rowe, J.J. (1987) Oxygen Regulation of Nitrate Uptake in Denitrifying *Pseudomonas aeruginosa. Appl. Environ. Microbiol.,* **53,** 745.

Hochstein, L.I., *et al.* (1984) The Effect of Oxygen on Denitrification During Steady-State Growth of Paracoccus Halodenitrificans. *Arch. Microbiol.*, **137**, 74.

Hockenbury, M.R., and Grady, C.P.L. (1977) Inhibition of Nitrification-Effects of Selected Organic Compounds. *J. Water Pollut. Control Fed.*, **49,** 768.

Horstkotte, G.A., *et al.* (1974) Full-Scale Testing of a Water Reclamation System. *J. Water Pollut. Control Fed.,* **45,** 181.

Hultman, B. (1971) Kinetics of Biological Nitrogen Removal. Inst. Vattenforsorjmingsoch Avloppsteknik samt Vattenkemi, Stockholm: KTH Pub. 71: 5.

Johnson, W.K., and Vania, G.B. (1971) Nitrification and Denitrification of Wastewater. Sanit. Eng. Rep. No. 1755, Univ. Minn., St. Paul.

Kawai, A., *et al.* (1965) Biochemical Studies on the Bacteria in the Aquarium with a Circulation System. II. Nitrifying Activity of the Filter Sand. *Bull. Jpn. Soc. Sci. Fish.*, **31**, 1, 65.

Knowles, G., *et al.* (1965) Determination of Kinetic Constants for Nitrifying Bacteria in Mixed Culture, with the Aid of an Electronic Computer. *J. Gen. Microbiol.*, **38**, 263.

LaMotta, E.J., and Shieh, W.K. (1978) Diffusion and Reaction in Biological Nitrification. *J. Environ. Eng.*, **105,** 655.

Laudelot, H., *et al.* (1974) Effect of Temperature on the Velocity of Oxidation of Ammonia to Nitrate in Mixed Nitrifier Culture. *Ann. Microbiol.*, **125B**, 75.

Laudelot, H., *et al.* (1976) Effect of pH and Partial Pressure of Oxygen on Nitrification. *Ann. Microbiol.*, **12**, 367.

Lawrence, A.W., and Brown, C.G. (1976) Design and Control of Nitrifying Activated Sludge Systems. *J. Water Pollut. Control Fed.*, **48,** 1779.

Loveless, J.E., and Painter, H.A. (1968) The Influence of Metal Ion Concentrations and pH Value on the Growth of Nitrosomonas Strain Isolated from Activated Sludge. *J. Gen. Microbiol.*, **52**, 1.

McCarty, P.L., *et al.* (1969) Biological Denitrification of Wastewater by Addition of Organic Materials. *Proc. 24th Ind. Waste Conf., Purdue Univ.,* West Lafayette, Ind.

Mines, R.O. (1983) Oxygen Transfer Studies in the Completely Mixed Activated Sludge Process. Ph.D. diss., Virginia Tech, Blacksburg, Va.

Monod, J. (1949) The Growth of Bacterial Cultures. *Annu. Rev. Microbiol.*, **3**, 371.

Monteith, H.D., *et al.* (1980) Industrial Waste Carbon Source for Biological Denitrification. *Prog. Water Technol.*, **12**, 127.

Mueller, J.A., *et al.* (1968) Oxygen Diffusion Through Zoogloeal Flocks. *Biotechnol. Bioeng.*, **10,** 331.

Mulbarger, M.C. (1971) Nitrification and Denitrification in Activated Sludge Systems. *J. Water Pollut. Control Fed.*, **43**, 2059.

Murphy, K.L., and Sutton, P.M. (1975) Pilot Scale Studies on Biological Denitrification. *Prog. Water Technol.*, **7**, 317.

Nagel, C.A., and Haworth, J.G. (1969) Operation Factors Affecting Nitrification in the Activated Sludge Process. Paper presented at 42nd Annu. Conf. Water Pollut. Control Fed., Dallas, Tex.

Painter, H.A. (1970) A Review of Literature on Inorganic Nitrogen Metabolism in Microorganism. *Water Res.* (G.B.), **4**, 393.

Painter, H.A. (1977) Microbial Transformations of Inorganic Nitrogen. *Prog. Water Technol.*, **8**, 4/5, 3.

Painter, H.A., and Loveless, J.E. (1983) Effect of Temperature and pH Value on the Growth-Rate Constants of Nitrifying Bacteria in the Activated Sludge Process. *Water Res.* (G.B.), **7**, 238.

Parker, D.S., *et al.* (1975) Process Design Manual for Nitrogen Control. EPA-625/1-75-007, U.S. EPA, Washington, D.C.

Payne, W.J. (1973) Reduction of Nitrogenous Oxides by Microorganisms. *Bacteriol. Rev.*, **37**, 409.

Peeters, T.L., *et al.* (1969) Kinetic Study of Oxygen Limited Respiration in Nitrifying Bacteria. *Bacteriol. Proc.,* 141.

Poduska, R.A., and Andrews, J.F. (1974) Dynamics of Nitrification in the Activated Sludge Process. *Proc. 29th Ind. Waste Conf., Purdue Univ. Ext. Ser. 145,* West Lafayette, Ind., 1005.

Poduska, R.A., and Andrews, J.F. (1975) Dynamics of Nitrification in the Activated Sludge Process. *J. Water Pollut. Control Fed.*, **47**, 2599.

Randall, C.W., *et al.* (1992) Nitrification Kinetics in Single-Sludge Biological Nutrient Removal Activated Sludge Systems. *Water Sci. Technol.*, **25**, 195.

Refling, D.R., and Stensel, H.D. (1978) A Rational Approach to Biological Nitrogen Removal. Paper presented at 51st Annu. Conf. Water Pollut. Control Fed., Anaheim, Calif.

Richardson, M. (1985) *Nitrification Inhibition in the Treatment of Sewage*. The Royal Society of Chemistry. Burlington House: Thames Water, Reading, Pa.

Rhee, G-Y., and Fuhs, G.W. (1978) Wastewater Denitrification with One-Carbon Compounds as Energy Source. *J. Water Pollut. Control Fed.*, **50,** 1978.

Rowe, J.J., *et al.* (1979) Nitrite Inhibition of Aerobic Bacteria. *Curr. Microbiol.,* **2,** 51.

Sharma, B., and Ahlert, R.C. (1977) Nitrification and Nitrogen Removal. *Water Res.* (G.B.), **11**, 897.

Stensel, H.D. (1970) Biological Kinetics of Suspended Growth Denitrification Process. Ph.D. thes., Cornell Univ., Ithaca, N.Y.

Stenstrom, M.K., and Poduska, R.A. (1980) The Effect of Dissolved Oxygen Concentration on Nitrification. *Water Res.* (G.B.), **14**, 643.

Stenstrom, M.K., and Song, S.S. (1991) Effects of Oxygen Transport Limitation on Nitrification in the Activated Sludge Process. *Res. J. Water Pollut. Control Fed.*, **63**, 208.

Stratta, J.M., *et al.* (1982) Improvement of Nitrification in Rotating Biological Contactors by Means of Alkaline Chemical Addition. *Proc. 1st Int. Conf. Fixed-Film Biol. Proc.,* Y.C. Wu (Ed.).

Stratton, F.E., and McCarty, P.L. (1967) Prediction of Nitrification Effects on the Dissolved Oxygen Balance of Streams. *Environ. Sci. Technol.,* **1,** 5, 405.

Sutton, P.M., *et al.* (1975) Low-Temperature Biological Denitrification of Wastewater. *J. Water Pollut. Control Fed.*, **47**, 122.

Tanaka, H., and Dunn, I.J. (1982) Kinetics of Biofilm Nitrification. *Biotechnol. Bioeng.*, **24,** 669.

Turk, O., and Mavinic, D.S. (1986) Preliminary Assessment of a Shortcut in Nitrogen Removal from Wastewater. *Can. J. Civ. Eng.*, **13**, 600.

U.S. Environmental Protection Agency (1975) *Process Design Manual for Nitrogen Control.* Off. Technol. Transfer, Washington, D.C.

Wild, H.E., *et al.* (1971) Factors Affecting Nitrification Kinetics. *J. Water Pollut. Control Fed.*, **43**, 1845.

# Chapter 5
# Principles of Biological Phosphorus Removal



Phosphorus is an essential nutrient for biological growth. Conventional secondary biological treatment systems accomplish phosphorus removal by using phosphorus for biomass synthesis during biochemical oxygen demand (BOD) removal. Phosphorus is an important element in microorganisms for energy transfer and for such cell components as phospholipids, nucleotides, and nucleic acids. In typical growth conditions, microbial solids contain 1.5 to 2.5% phosphorus on a dry weight basis. The only mechanism for removal of phosphorus in the activated-sludge process is through the phosphorus contained in the waste activated sludge (WAS). The amount of phosphorus removed is therefore directly related to the phosphorus content of the WAS. Waste activated sludge with a phosphorus content of 1.5 to 2.5% can result in a total phosphorus removal of 10 to 30%, depending on the BOD-to-phosphorus ratio and other operating conditions.

# ENHANCED BIOLOGICAL PHOSPHORUS REMOVAL

Additional removal of phosphorus in the activated-sludge process can be achieved if the phosphorus content of the WAS is increased beyond the typical 1.5 to 2.5% level. Early developments pointed in this direction. In 1955, Greenburg *et al.* proposed that activated sludge could take up phosphate at a level beyond its normal microbial growth requirements (Greenburg *et al.,* 1955). Others soon investigated the excess phosphorus uptake, and by the 1970s, several wastewater treatment plants (WWTPs) reported total phosphorus removals of 85 to 95%, with the phosphorus content of the waste sludge between 2 and 7% on a dry weight basis.

The following operating characteristics emerged with respect to enhanced biological phosphorus removal systems (EBPRs) (Barnard, 1976):

- The return sludge and the mixed liquor should be introduced at the inlet end of the aeration basin.
- Aeration should provide a high dissolved oxygen (DO) content at the outlet of the basin to prevent anaerobic conditions from developing in the secondary clarifier.
- The degree of nitrification should be kept to a minimum.
- Solids should be rapidly removed from the secondary clarifier and returned to the aeration basin. If the wastewater is allowed to become anaerobic, phosphorus release will lead to an increase in effluent phosphorus concentrations.
- Liquid should not be recycled to the activated-sludge process from a subsequent sludge treatment step that may release phosphates.

The importance of an anaerobic zone in the activated-sludge system to remove phosphorus and nitrogen was clearly established. Because the mechanism (microbiology and biochemistry) was still in dispute, the design of systems remained largely empirical. Several definitions of various zones within the activated-sludge system emerged:

- *Anaerobic* zone referred to a zone without aeration, devoid of DO, nitrite, or nitrate. However, the redox potential in the zone remained above that for reducing sulfate to hydrogen sulfide but below the nitrate-nitrogen gas level.
- *Anoxic* conditions referred to zones where DO was absent but nitrate was present. Anoxic zones are used for denitrification.

- *Aerobic* conditions referred to zones with DO in the liquid. Nitrification only occurs under aerobic conditions.

# MICROBIOLOGY

The microbiology of the activated-sludge system is complex, with environmental conditions controlling the "selection" of certain organisms. The survival of certain microorganisms in the process is not only a function of environmental conditions such as temperature, pH, and DO, but also depends on the substrate composition, such as the availability of soluble, readily biodegradable, or slowly biodegradable organic compounds. In addition, conditions in an activated-sludge process operated in an enhanced phosphorus removal mode change as the microorganisms are subjected to aerobic, anaerobic, and anoxic conditions.

In an effort to identify the microorganisms responsible for EBPR, Fuhs and Chen (1975) examined activated sludge from the Baltimore Back River and the Seneca Falls, New York, WWTPs when the plants were exhibiting high levels of phosphorus removal. They concluded that the organism associated with phosphorus removal belonged to the *Acinetobacter* genus. They also subjected a pure culture of *Acinetobacter* to batch anaerobic–aerobic cycles and noted excess phosphorus removal when acetate was fed to the system. In subsequent years, many other investigators confirmed the abundance of *Acinetobacter* in EBPR systems (Brodisch, 1985; Brodisch and Joyner, 1983; Buchnan, 1983; Florenz and Hartemann, 1984; and Suresch *et al.*, 1985). These investigations clearly established the dominance of *Acinetobacter* in EBPR systems.

In addition, investigators found microorganisms other than *Acinetobacter* in EBPR plants. Brodish and Joyner (1983) found only low *Acinetobacter* populations (1 to 10%) in an EBPR system and concluded that other microorganisms may be responsible for EBPR. Lotter (1985) found significant levels of *Aeromonas* and *Pseudomonas* capable of storing polyphosphates.

Pure culture studies to determine *Acinetobacter* growth kinetic coefficients on aerobically grown cultures, using substrates such as acetate and ethanol, demonstrated maximum growth rates ($\mu_{\max}$) in the range 4 to 30/day, dry cell yields ($Y$) of approximately 0.4 g/g chemical oxygen demand (COD), and endogenous decay rates. Rates of ($k_d$) of 1 to 5 g/g·d have been obtained (Abbott *et al.,* 1973 and 1974; Deinema *et al.*, 1985; Ensley and Finnerty, 1980; and Hao and Chang, 1987). Cells grown under such conditions have high phosphorus contents (approximately 4 to 7% [Deinema *et al.*, 1985], maximum of 4 to 8% [Hao and Chang, 1987]). Values of $\mu_{max}$ and $k_d$ obtained in these experiments are far higher than expected for a polyphosphate- (poly-P-) accumulating organism from the observed performance of EBPR activated-sludge systems. For example, Wentzel *et al.* (1987 and

1988) determined the kinetic growth coefficients on an *Acinetobacter* culture "weaned" from an EBPR activated-sludge system by feeding the system an acetate–yeast extract–mineral salts–micronutrients mixture. Batch tests with this enriched culture gave $\mu_{max}$ = 0.75–0.95/day without phosphorus limitation, 0.35/day with phosphorus limitation, and $k_d$ = 0.03–0.04/day. The culture showed all of the attributes of EBPR activated sludge.

Tandoi *et al.* (1987) grew *A. calcoaceticuslwoffi* AC-7 (isolated from anaerobic/aerobic activated sludge) on an acetate mineral salt medium containing yeast extract in a two-stage chemostat under both completely aerobic and alternating anaerobic/aerobic conditions. Values of $\mu_{max}$ = 0.84/d and $Y$ = 0.13 g volatile suspended solids (VSS)/g COD were obtained—consistent with Wentzel *et al.* (1987). Aerobic chemostat operation gave significantly higher values of $\mu_{max}$ = 4.8/day and $Y$ = 0.22 g VSS/g COD.

For simplicity, the microorganisms that grow and participate in phosphorus uptake–release–storage typically are referred to as biological phosphorus removing microbes (bio-P). While *Acinetobacter* is the dominant bio-P organism, there are several others. These organisms flourish under the cyclic anaerobic–aerobic activated-sludge operation and incorporate a storage–uptake mechanism for soluble organic compounds.

# *BIOCHEMISTRY*

The biochemistry of EBPR is difficult to study because of the complexity of the activated-sludge system. Jenkins and Tandoi (1991) pointed out in their review that all the biochemical models and conclusions regarding their applications have been based on experiments with activated-sludge systems operated under aerobic–anaerobic cycles. In these mixed cultures it is impossible to isolate the effects of the phosphate-removing bacteria and their biochemistry. Detailed biochemical mechanisms have been developed by several investigators, based on a single bio-P organism (such as *Acinetobacter*) responsible for the uptake–release–storage of phosphate and organics in anaerobic conditions.

The role of the anaerobic zone can be summarized from various observations reported in the literature.

- Addition of wastewater or a simple carbon source (acetate) results in phosphorus release and simultaneous carbon uptake and storage (Fukase *et al.*, 1982). Phosphate release is also triggered by other simple carbon compounds such as acetate, propionate, butyrate, lactate (Gerber *et al.*, 1986), propionate and glucose (Potgieter and Evans, 1983), fermented sludge (Barnard, 1984, and Oldham, 1985), and septic wastewater (Paepcke, 1983).

- The phosphate released from bio-P organisms in the anaerobic zone originates from polyphosphate reserves (Arvin, 1985, and Marais *et al.*, 1983).
- Carbon-rich intracellular granules form in the anaerobic zone. Deinema *et al.* (1980), Fukase *et al.* (1982), and Timmerman (1979) reported accumulation of poly-ß-hydroxybuterate in bio-P bacteria. Other storage products have been identified: polyhydroxyvalerate (Comeau *et al.*, 1987) and glycogen (Fukase *et al.*, 1982). Organic storage products decrease in the aerobic stage.
- Addition of nitrate or DO will prevent and reverse the uptake–release reactions in the anaerobic zone (Barnard, 1976; Comeau *et al.*, 1986; and Jenkins and Tandoi, 1991).
- The ratio of organic substrate uptake and phosphorus release has not been clearly established. With acetate as the carbon source, the acetate uptake:phosphorus release molar ratio ranges from 0.6 (Rabinowitz, 1985) to 0.7 (Arvin, 1985), 1.9 (Fukase *et al.*, 1982, and Wentzel *et al.*, 1984), and up to 3.4 (Mino *et al.*, 1987). Rabinowitz (1985) also found that the ratio changes with different types of substrate: in decreasing order were sodium acetate, propionic acid, glucose, acetic acid, and butyric acid. Jones *et al.* (1985) found a greater phosphorus release with, in decreasing order, butyric acid, ethanol, acetic acid, methanol, and sodium acetate.

Various models have been proposed to describe the biochemical pathways for EBPR (Arun *et al.*, 1988; Comeau *et al.*, 1986; and Mino *et al.*, 1987). The exact mechanism for the biochemistry of phosphorus and carbon uptake–release is still elusive and a topic of debate. However, many of the key aspects and phenomena required for the successful growth of bio-P bacteria and design of EBPR systems have been established.

# TEMPERATURE AND PH

The EBPR processes are relatively insensitive to temperature changes, compared to other biological processes. Sell *et al.* (1981) found no difference in EBPR efficiency at temperatures of 10 and 15°C but improved performance when temperature was lowered to 5°C. Ekama *et al.* (1984) reported better removal at 14°C than at 22°C. However, Waltrip (1991) reported a decrease in phosphorus removal with decreasing temperatures between 27 and 13°C. These somewhat contradictory results indicate that the effect of temperature on EBPR systems is probably attenuated by other environmental factors, such as changes in wastewater composition and other uncontrollable factors.

Tracy and Flammino (1985) investigated the effects of pH on EBPR and showed that pH affects the phosphorus uptake rate. The uptake rate remained at its optimum between pH 6.6 and 7.4 but decreased rapidly as the pH dropped below 6.2.

# SETTLEABILITY OF ENHANCED BIOLOGICAL PHOSPHORUS REMOVAL ACTIVATED SLUDGES

The use of "selectors" has been widely implemented to control activated-sludge bulking. Selectors have been aerobic or aerated (Albertson, 1987; Casey *et al.*, 1975; Chudoba *et al.*, 1973; Lee *et al.*, 1982; Tomlinson, 1976; and van Niekerk and Jenkins, 1986), anoxic (Shao and Jenkins, 1989, and Warner and Grau, 1987), or anaerobic (Barnard, 1976; Heide and Pasveer, 1974; and Spector, 1977). A selector refers to the addition of one or more small compartments at the head end of the aeration basin. The initial compartment could be aerated or not aerated, as in the case of nutrient removal plants. The reason for the nonbulking sludge is attributed to the fact that the operating regime (semi–plug-flow conditions with a substrate gradient) favors the growth of nonfilamentous organisms. It is fortunate that the basin configuration for EBPR systems has many similarities to the selector system and tends to produce a sludge that settles well.

# OXYGEN CONSUMPTION

Some EBPR systems have reported a reduction in oxygen requirements (Bordacs and Tracy, 1988; Lan *et al.*, 1983; and Randall, 1984). A detailed study to confirm the reduction in oxygen requirements in EBPR systems is reported by Randall *et al.* (1992b). A COD mass balance was developed to account for all COD and nitrogen fractions. They assumed that a fraction of the COD is stabilized in the anaerobic zone and determined the mass balance as

$$\mathrm{AnS} = \mathrm{COD_u} + \mathrm{NOD} - \mathrm{AxS} - \mathrm{COD_w} - \mathrm{TOR}$$

$$\begin{aligned}\mathrm{AnS} = {} & (\mathrm{COD_{influent}} - \mathrm{COD_{effluent}}) + 4.57\,(\mathrm{NO_3\text{-}N})_p \\ & + 3.43\,(\mathrm{NO_2\text{-}N})_p - 2.86\,(\mathrm{NO_3\text{-}N})_u \\ & - 1.71\,(\mathrm{NO_2\text{-}N})_u - (\mathrm{COD/VSS})\,(\mathrm{MX})_w \\ & - (\mathrm{COD/VSS})\,(\mathrm{MX})_e - \mathrm{TOR}\end{aligned} \tag{5.1}$$

Where

| | | |
|---|---|---|
| AnS | = | anaerobically stabilized COD, |
| $(NO_2\text{-}N)_p$ | = | $NO_2$-N produced, |
| $(NO_2\text{-}N)_u$ | = | used, |
| $MX_w$ | = | biomass wasted, |
| $MX_e$ | = | effluent biomass, |
| TOR | = | total oxygen used in the system, |
| NOD | = | nitrogenous oxygen demand, and |
| AxS | = | COD stabilized in anoxic reactor (nitrate + nitrite reduction). |

A series of experiments was conducted to test a number of hypotheses that could explain the source for AnS. The results of these tests indicated the following conclusions.

- In a nonnitrifying EBPR system, between 13 and 48% of the potential oxygen requirements are satisfied by AnS.
- In a nitrifying EBPR system, between 10 and 27% of the potential oxygen requirements are satisfied by AnS.
- In parallel tests, a University of Cape Town EBPR system consistently required less oxygen than a conventional system (because of denitrification credit). In addition, measured oxygen requirements indicated that anaerobic stabilization of COD ranged from 8 to 18% and 12 to 27% at sludge ages of 5 and 15 days, respectively.

The search for a mechanism to explain these observations has proven difficult and inconclusive. Measurements of the COD:VSS ratio were comparable to conventional values (1.42 and 1.44 g COD/g VSS, and 1.34 and 1.30 for the side-by-side EBPR and conventional systems, respectively) discounting a suggestion that the WAS accounts for AnS. Sulfate reduction was insignificant. They suggested a mechanism whereby organic compounds are fermented to produce hydrogen and carbon dioxide as products. No experimental data supporting this mechanism have been presented yet.

In subsequent work, Wable and Randall (1994) found that

- Oxygen/water, COD/$CO_2$, and sulfate/sulfide redox pairs were the main components of an EBPR system.
- Hydrogen and methane production explained less than 1% of AnS, except when formate was present in the feed, in which case methane production explained 19% of AnS.
- Aeration-induced stripping of reduced volatiles explained less than 6% of AnS.
- Chemical oxygen demand test limitations (digestion time) were insignificant, but production of oxidation-resistant species has potential for explaining AnS.

In conclusion, there appears to be a reduction in oxygen requirements exerted by EBPR systems. This stabilization occurs in the anaerobic zone. However, the mechanism of this phenomenon has not been clearly established, making it difficult to quantify or account for anaerobic stabilization for design purposes.

# References


Abbott, B.J., *et al.* (1973) Growth of *Acinetobacter Calcoaceticus* on Ethanol. *Appl. Microbiol.*, **25**, 787.

Abbott, B.J., *et al.* (1974) Effect of Growth Rate and Nutrient Limitation on the Composition and Biomass Yield of *Acinetobacter Calcoaceticus. Appl. Microbiol.*, **28**, 58.

Albertson, O. (1987) The Control of Bulking Sludges: From the Early Innovators to Current Practice. *J. Water Pollut. Control Fed.*, **59**, 172.

Arun, V., *et al.* (1988) Biological Mechanism of Acetate Uptake Mediated by Carbohydrate Consumption in Excess Phosphorus Removal Systems. *Water Res.* (G.B.), **22**, 565.

Arvin, E. (1985) Biological Removal of Phosphorus from Wastewater. *Crit. Rev. Environ. Control,* **15,** 25.

Barnard, J.L. (1976) A Review of Biological Phosphorus Removal in the Activated Sludge Process. *Water SA*, **2**, 3, 136.

Barnard, J.L. (1984) Activated Primary Tanks for Phosphate Removal. *Water SA*, **10,** 3, 121.

Bordacs, K., and Tracy, K.D. (1988) Retention of Organic Storage Products in Anaerobic-Aerobic Activated Sludge. Paper presented at 61st Annu. Conf. Water Pollut. Control Fed., Dallas, Tex.

Brodisch, K.E.U. (1985) Interactions of Different Groups of Microorganisms in Biological Phosphate Removal. *Water Sci. Technol.*, **17**, 89.

Brodisch, K.E.U., and Joyner, S.J. (1983) The Role of Microorganisms Other than *Acinetobacter* in Biological Phosphate Removal in Activated Sludge Processes. *Water Sci. Technol.*, **15**, 3/4, 117.

Buchnan, L. (1983) Possible Biological Mechanism of Phosphorus Removal. *Water Sci. Technol.*, **15**, 3/4, 87.

Casey, J., *et al.* (1975) Non-Bulking Activated Sludge Process. U.S. Patent No. 3,864,246.

Chudoba, J., *et al.* (1973) Control of Activated Sludge Filamentous Bulking—I and II, Effect of Hydraulic Regime of Degree Mixing in an Aeration Tank. *Water Res.* (G.B.), **7**, 1163.

Comeau, Y., *et al.* (1986) Biochemical Model of Enhanced Biological Phosphorus Removal. *Water Res.* (G.B.), **20**, 1511.

Comeau, Y., *et al.* (1987) Dynamics of Carbon Reserves in Biological Dephosphation of Wastewaters. In *Advances in Water Pollution Control: Biological Phosphate Removal from Wastewaters*. R. Ramadori (Ed.), Pergamon Press, Oxford, Eng.

Deinema, M.H., *et al.* (1980) The Accumulation of Polyphosphate in *Acinetobacter* spp. *FEMS Microbiol. Lett.*, **9**, 275.

Deinema, M.H., *et al.* (1985) Some Physiological Characteristics of *Acinetobacer* spp. Accumulating Large Amounts of Phosphate. *Water Sci. Technol.*, **17**, 119.

Ekama, G.A., *et al.* (1984) Biological Excess Phosphorus Removal. In *Theory, Design, and Operation of Nutrient Removal Activated Sludge Processes*. Water Resour. Comm., Pretoria, S. Afr.

Ensley, B.D., and Finnerty, W.R. (1980) Influences of Growth Substrates and Oxygen on the Electron Transport System in *Acinetobacter* sp. HO1-N. *J. Bacteriol.*, **142**, 859.

Florenz, M., and Hartemann, P. (1984) Screening for Phosphate Accumulating Bacteria Isolated from Activated Sludge. *Environ. Technol. Lett.*, **5**, 457.

Fuhs, G.W., and Chen, M. (1975) Microbial Basis for Phosphate Removal in the Activated Sludge Process for the Treatment of Wastewater. *Microbiol. Ecol.*, **2**, 119.

Fukase, T., *et al.* (1982) Studies on the Mechanism of Biological Phosphorus Removal. *Jpn. J. Water Pollut. Res.*, **5**, 309.

Gerber, A., *et al.* (1986) The Effect of Acetate and Carbon Compounds on the Kinetics of Biological Nutrient Removal. *Water SA*, **12**, 7.

Greenburg, A.E., *et al.* (1955) Effect of Phosphorus Removal on the Activated Sludge Process. *Sew. Ind. Wastes*, **27**, 227.

Hao, O.J., and Chang, C.H. (1987) Kinetics of Growth and Phosphate Uptake in Pure Culture Studies of *Acinetobacter* Species. *Biotechnol. Bioeng.*, **29**, 819.

Heide, B.A., and Pasveer, A. (1974) Oxidation Ditch-Prevention and Control of Filamentous Sludge. *H2O*, **18**, 373.

Jenkins, D., and Tandoi, V. (1991) The Applied Microbiology of Enhanced Biological Phosphate Removal-Accomplishments and Needs. *Water Res.* (G.B.), **25,** 1471.

Jones, P.H., *et al.* (1985) Studies in the Enhanced Uptake of Phosphorus by Activated Sludge: Effect of Substrate Addition. *Proc. New Dir. Res. Waste Treat. Residuals Manage.,* Univ. Brit. Colombia, Vancouver, Can.

Lan, J.C. *et al.* (1983) Phosphorus Removal in the Activated Sludge Process. *Water Res.* (G.B.), **17**, 1193.

Lee, S.E., *et al.* (1982) The Effect of Aeration Basin Configuration on Activated Sludge Bulking at Low Organic Loading. *Water Sci. Technol.*, **14**, 407.

Lotter, L.H. (1985) The Role of Bacterial Phosphate Metabolism in Enhanced Phosphorus Removal from the Activated Sludge Process. *Water Sci. Technol.*, **17**, 127.

Marais, G.v.R., *et al.* (1983) Observations Supporting Phosphate Removal by Biological Excess Uptake—A Review. *Water Sci. Technol.*, **15**, 3/4, 15.

Mino, T., *et al.* (1987) Effect of Phosphorus Accumulation on Acetate Metabolism in the Biological Phosphorus Removal Process. In *Advances in Water Pollution Control: Biological Phosphate Removal from Wastewaters.* R. Ramadori (Ed.), Pergamon Press, Oxford, Eng.

Oldham, W.K. (1985) Full Scale Optimization of Biological Phosphorus Removal at Kelowana. *Water Sci. Technol.*, **15**, 3/4, 105.

Paepcke, B.H. (1983) Performance and Operational Aspects of Biological Phosphate Removal Plants in South Africa. *Water Sci. Technol.*, **15**, 3/4, 219.

Potgieter, D.J.J., and Evans, B.W. (1983) Biochemical Changes Associated with Luxury Phosphate Uptake in a Modified Phoredox Activated Sludge System. *Water Sci. Technol.*, **15**, 3/4, 105.

Rabinowitz, B. (1985) The Role of Specific Substrates in Excess Biological Phosphorus Removal. Ph.D. thes., Univ. Brit. Columbia, Vancouver, Can.

Randall, C.W. (1984) The Utilization of Biological Phosphorus Removal Processes to Reduce the Aeration Requirements of Activated Sludge Systems. *Proc. IAWPRC Post-Conf. Sem. Enhanced Biol. Phosphorus Removal Wastewater,* Paris, Fr.

Randall, C.W., *et al.* (1992a) Design of Activated Sludge Biological Nutrient Removal Plants. In *Design and Retrofit of Wastewater Treatment Plants for Biological Nutrient Removal.* C.W. Randall, J.L. Barnard, and H.D. Stensel (Eds.), Technomic Publishing Co., Lancaster, Pa.

Randall, C.W., *et al.* (1992b) The Case for Anaerobic Reduction of Oxygen Requirements in Biological Phosphorus Removal Systems. *Water Environ. Res.*, **64**, 824.

Sell, R.L., *et al.* (1981) Low Temperature Biological Phosphorus Removal. Paper presented at 54th Annu. Conf. Water Pollut. Control Fed. Detroit, Mich.

Shao, Y., and Jenkins, D. (1989) The Use of Anoxic Selectors for the Control of Low F/M Activated Sludge Bulking. *Water Sci. Technol.*, **21**, 609.

Spector, M.L. (1977) Production of Non-Bulking Activated Sludge. U.S. Patent No. 4,056,465 and the supporting documents in the PTO file wrapper.

Suresch, N., *et al.* (1985) New Strategies for the Isolation of Microorganisms Responsible for Phosphate Accumulation. *Water Sci. Technol.*, **17**, 99.

Tandoi, V., *et al.* (1987) *Acinetobacter* sp Growth in Alternate Anaerobic-Aerobic Conditions. In *Advances in Water Pollution Control: Biological Phosphate Removal from Wastewaters.* R. Ramadori (Ed.), Pergamon Press, Oxford, Eng.

Timmerman, M.W. (1979) Biological Phosphate Removal from Domestic Wastewater Using Anaerobic/Aerobic Treatment. In *Developments in Industrial Microbiology.* Soc. Ind. Microbiol., Washington, D.C., 285.

Tomlinson, E. (1976) Bulking—A Survey of Activated Sludge Plants. Technical Reports TR35, Water Res. Cent., 1.

Tracy, K.D., and Flammino, A. (1985) Kinetics of Biological Phosphorus

Removal. Paper presented at 58th Annu. Conf. Water Pollut. Control Fed., Kansas City, Mo.

Van Niekerk, A.M., and Jenkins, D. (1986) The Competition of Type 021N and Floc-Forming Bacteria as a Model for Activated Sludge Bulking at Low F/M. Paper presented at 59th Annu. Conf. Water Pollut. Control Fed., Los Angeles, Calif.

Wable, M.V., and Randall, C.W. (1994) Investigation of Hypothesized Anaerobic Stabilization Mechanisms in Biological Nutrient Removal Systems. *Water Environ. Res.*, **66**, 161.

Waltrip, G.D. (1991) Full-Scale Nutrient Removal Demonstration Project at the York River Treatment Plant with the VIP Process. Paper presented at 45th Annu. Conf. Virginia Water Pollut. Control Assoc., Williamsburg, Va.

Warner, J., and Grau, P. (1987) Filamentous Bulking in Nutrient Removal Activated Sludge Systems. *Proc. IAWPRC Workshop Des. Oper. Large Wastewater Treat. Plants,* Budapest, Hung.

Wentzel, M.C., *et al.* (1984) Kinetics of Biological Phosphorus Release. Enhanced Biological Phosphorus Removal from Wastewater. *IAWPRC Post Conf. Semin.*, **1.**

Wentzel, M.C., *et al.* (1987) Experiments Toward Establishing the Kinetics of Biological Excess Phosphorus Removal. In *Advances in Water Pollution Control: Biological Phosphate Removal from Wastewaters.* R. Ramadori (Ed.), Pergamon Press, Oxford, Eng.

Wentzel, M.C., *et al.* (1988) Enhanced Polyphosphate Organism Cultures in Activated Sludge Systems. Part 1: Enhanced Culture Development. *Water SA*, **14**, 81.

# Chapter 6 Nitrification Design



## OVERVIEW AND PROCESS OPTIONS

Nitrification designs are used to convert unoxidized nitrogen (total Kjeldahl nitrogen, or TKN) to oxidized nitrogen (typically to nitrate, $NO_3^-$). In practice, this is virtually always done by some type of biological nitrification process.

Nitrification typically is used in plants that have to meet an ammonia-nitrogen ($NH_4$-N) standard. In these cases, it typically is the last biological process in the process train; most nitrification systems today are combined in a single stage (that is, integrated) with the biological process.

In some cases, nitrification is part of system that removes nitrogen by biological denitrification. Nitrification is a form of nitrogen control and does not need to be accompanied by denitrification. Chapter 9 covers the design of systems that provide the bulk of the denitrification in the same stage.

A basic assumption in the designs discussed in this chapter is that neither dissolved oxygen (DO) nor alkalinity is limiting. Alkalinity and oxygen requirements are discussed elsewhere in this manual. Methods of aeration are discussed in this chapter.

The most typical biological process used at a number of wastewater treatment plants (WWTPs) throughout the world for nitrification is the suspended-growth (activated-sludge) process. One of the oldest methods of nitrification uses trickling filters and is called an attached-growth process. While less common than activated sludge, this process is also used frequently.

Effluents with an ammonia-nitrogen ($NH_4$-N) content less than 1.0 mg/L typically are achievable with suspended-growth nitrification processes. Effluent from the attached-growth nitrification processes, conversely, typically contains approximately 1.0 to 3.0 mg $NH_4$-N/L (Metcalf and Eddy, 1991) and typically is not reliable below 2 mg/L (U.S. EPA, 1992). In recent years a variety of new attached-growth systems, in addition to trickling filters, has been introduced. These include packed-bed reactors, rotating biological contactors (RBCs), biological fluidized-bed reactors (BFBs), and biological aerated filters (BAFs). The design of each of these technologies is discussed in this chapter. Newer, less typical technologies that combine suspended-growth and attached-growth (fixed film) systems using various surfaces also exist but are beyond the scope of this chapter. Information on these may be found elsewhere (Lessel, 1993, and Sen *et al.*, 1993 and 1994). Effluent criteria and wastewater characterization are discussed in Chapter 2.

# NITRIFICATION

**SUSPENDED-GROWTH SYSTEMS.** Biological nitrification can be accomplished using suspended-growth (activated-sludge) systems in the following two basic ways (see Figure 6.1):

- Single stage—nitrification and carbonaceous biochemical oxygen demand ($BOD_5$) removal are done in a single activated-sludge unit, with one set of aeration tanks and one set of clarifiers.
- Two stage—nitrification and carbonaceous $BOD_5$ removal are done in separate activated-sludge units, each with its own set of aeration tanks and clarifiers.

Aeration tank volume for both approaches is calculated based on solids retention time (SRT) as follows:

$$\text{SRT} = X_a \theta / X_w \tag{6.1}$$

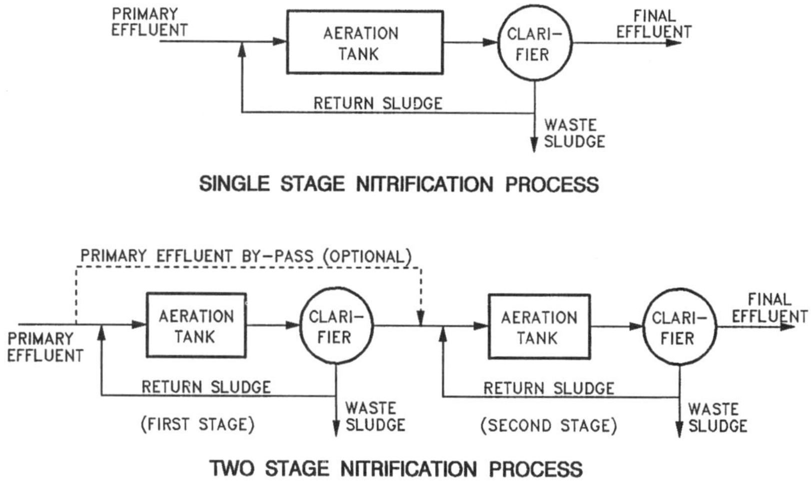


**Figure 6.1 Suspended-growth nitrification systems.**

Where

SRT = solids retention time, days.
$X_a$ = average mixed liquor suspended solids [MLSS] in aeration tank, mg/L;
$\theta$ = aeration tank hydraulic detention, days; and
$X_w$ = biomass production, mg/L.

Thus, aeration tank volume can be calculated as shown in Equation 6.2:

$$V = \text{SRT} \times X_w \times Q/X_a \tag{6.2}$$

Where

$V$ = aeration tank volume, m$^3$; and
$Q$ = design flow, m$^3$/d.

**CHOICE OF DESIGN SOLIDS RETENTION TIME.** The most important design parameter in suspended-growth nitrification systems may be the selection of SRT, also known as solids residence time or mean cell residence time (MCRT). In comparison with simple secondary treatment designed for BOD removal only, a longer SRT typically is needed to achieve nitrification. The design SRT primarily is a function of temperature. An effective design provides at least a 2.0 mg/L DO value throughout the aeration basins at all times, and alkalinity supplements, if needed, are used to maintain a pH above 6.5 so that DO and pH effects are reduced. Designs in the 1970s typically used a safety factor of 2.0 or higher on calculated SRTs. Twenty years of operating experience in numerous locations has shown that, typically, as long as the design minimum described by Equation 6.2 (U.S. EPA, 1992) is met, adequate nitrification can be achieved.

$$SRT_{min} = (0.47)\exp^{0.098(t-15)} \qquad (6.3)$$

Alternative equations can be considered. The actual SRT selected will depend on wastewater characteristics, the presence of inhibitors, and environmental factors. Accordingly, using this SRT at the minimum design monthly temperature typically is adequate. This equation corresponds to values ranging from approximately 2.5 times the theoretical minimum [$\theta_{min} = 2.13 \exp (0.098)(15 - T)$] at 20°C, to approximately 3 times it at temperatures below 10°C. It is interesting to note that, in Germany, the standard is to design small plants (less than 20 000 population equivalent, or PE) to a 10-day SRT and large plants (more than 100 000 PE) to an 8-day SRT. For unusual conditions, they use a temperature-dependent SRT calculation. A comparison of models from a U.S. Environmental Protection Agency (U.S. EPA) manual and the Abwassertechnischen Vereinigung e.V., Germany (ATV), criteria is given in Figure 6.2.

**SINGLE-STAGE NITRIFICATION.** The design of single-stage nitrification (also known as single-sludge nitrification) is straightforward. The tank sizing is done the same way as it is for secondary treatment, but maintaining an adequate aerobic SRT under all design conditions is emphasized. (Aerobic SRT is defined as the SRT of the aerated portion of the reactor. It does not include solids in anaerobic zones or, for most designers, those in anoxic zones.)

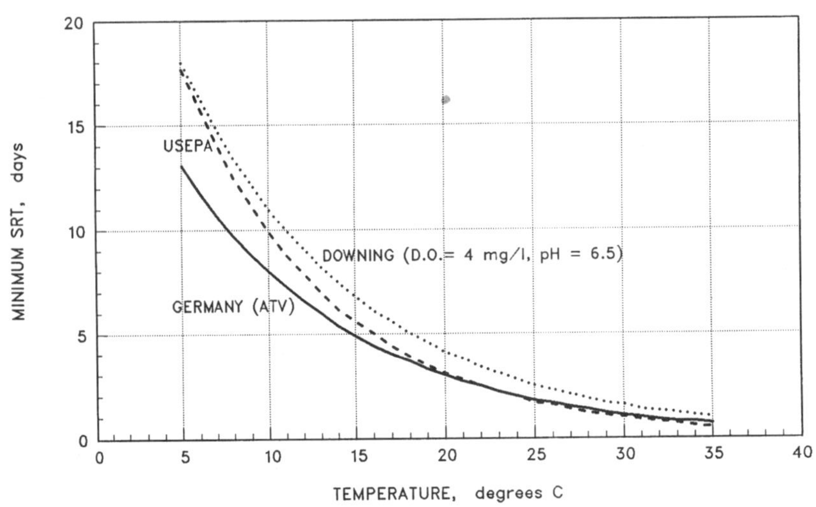


**Figure 6.2 Activated-sludge nitrification—comparison of minimum design SRTs.**

Typically, it is advantageous to precede single-stage nitrification systems with a plug-flow anoxic section. As discussed in Chapter 9, such a section offers many advantages, including the following:

- Improved settleability by reducing many types of filamentous organisms;
- Alkalinity recovery, with approximately 3 mg/mg N denitrified; and
- Denitrification, with more than 1 mg N/5 mg soluble $BOD_5$ ($SBOD_5$) applied (assuming adequate recycle of nitrate).

This anoxic zone, typically known as an anoxic selector, can be a part of a BNR system. A minimal anoxic section, to achieve improved settleability and limited other benefits, should be sized at a 15- to 30-minute detention time at average design flow (Dale Mabry, Florida; Brownsburg, Indiana) and compartmentalized into at least three compartments to approach a plug-flow condition. Each compartment should be completely mixed; typical mixing designs allow 0.3 hp/1 000 cu ft tankage (complete turnover of the zone contents every 20 minutes). This minimal system will use the easily biodegradable portion of the wastewater at a very high rate (see Chapter 9 for a discussion of denitrification rates).

The interaction between clarifier and reactor sizing can be optimized. Optimization can be done on a cost basis or, for limited sites, on a volume or area basis. Figure 6.3 is a typical example of such analysis, using Wilson's design approach (Wilson, 1992). Note that higher wastewater strength and design SRTs favor larger clarifiers (smaller surface overflow rates, or SORs) and that single-stage nitrification systems appear to be optimized when using relatively large clarifiers with SORs between 12 and 24 m/d (300 and 600 gpd/sq ft) (see 10-day SRT curves).

Return sludge systems typically should be designed to provide a firm capacity of at least 100% design plant flow and to return as little as 40 to 50% of expected startup flow.

Waste activated sludge (WAS) systems should be designed to continuously handle at least two times the average wastage requirement—four times for step-feed systems (which can produce very dilute sludges in some modes) and more in small plants where sludge might be only intermittently wasted (although continuous wasting is recommended for all sizes).

Aeration systems should be designed to provide at least peak day oxygen demands. In typical medium to large municipal plants, if peaks are not known, 1.5 to 2 times the average daily requirements should be provided.

**TWO-STAGE NITRIFICATION.** The design of the first, carbonaceous, stage of a two-stage system is as a high-rate activated-sludge process. Some first stages are designed, or required by local standards to be designed, at conventional secondary design loadings. However, experiences at both Tampa, Florida, and Blue Plains, Washington, D.C., have shown that a high-rate first stage works well.

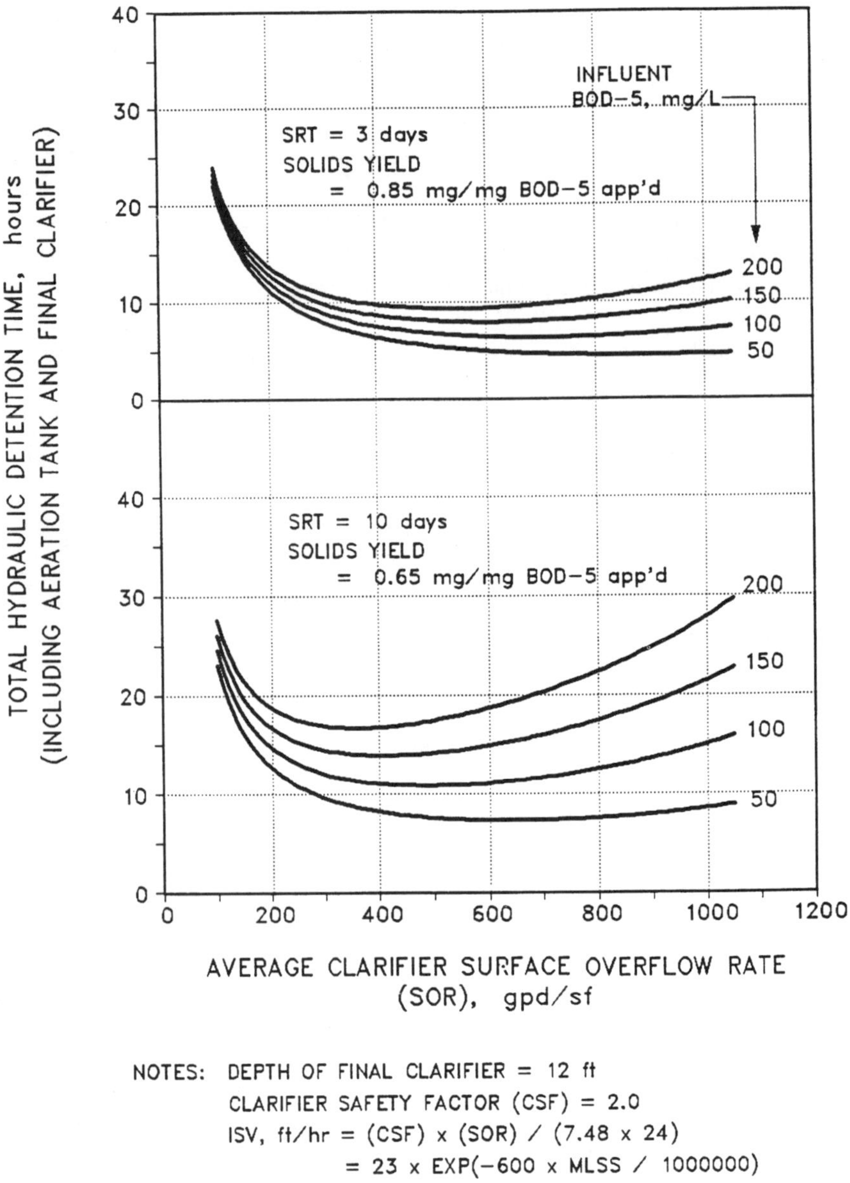


**Figure 6.3** **Tankage optimization for activated-sludge process (gpd/sq ft × [4.074 × $10^{-2}$] = $m^3/m^2 \cdot d$; ft × 0.304 8 = m; ft/hr × [8.467 × $10^{-5}$] = m/s).**

The object of the first-stage design is to produce less than a 30 mg/L-to-30 mg/L ($BOD_5$-to-total suspended solids [TSS]) effluent consistently, using as little tankage as necessary. The following approach works well:

1. Design at an SRT of 0.5 (warm waters) to 1.5 (cool waters) days.
2. Load aeration tanks at as high a loading as possible, while being able

to meet SRT, and transfer adequate oxygen. For air systems, 1.6 kg $BOD_5/m^3{\cdot}d$ (100 lb $BOD_5$/d/1 000 cu ft) (average) has worked well, and for oxygen systems, 4.0 kg $BOD_5/m^3{\cdot}d$ (250 lb $BOD_5$/d/1 000 cu ft) (average) has worked well.

3. Provide 1.0 kg $O_2$/kg $BOD_5$ applied and size so this can be met at maximum daily loading. Assume at least 1 mg/L at all times and 4 mg/L DO at average load in design calculations. Do not assume that such a low SRT system needs less oxygen than a conventional system. Actual usage may be close to 0.8 kg/kg $BOD_5$, and this is suggested for operational cost estimates, but it is imperative that enough oxygen be supplied at all times, or else a bulky sludge will result.
4. Size clarifier at 24 to 33 $m^3/m^2{\cdot}d$ (600 to 800 gpd/sq ft) at average flow and 49 to 65 $m^3/m^2{\cdot}d$ (1 200 to 1 600 gpd/sq ft) at peak flow.
5. Design MLSS concentration should be a function of clarifier size and design settling characteristics. A good choice for a 24-$m^3/m^2{\cdot}d$ (600-gpd/sq ft) (average) clarifier would be 2 000 to 2 500 mg/L.

The second, nitrifying, stage of two-stage nitrification is designed identically to a single-stage nitrification system.

The choice of SRT is identical to that of a single-stage system. Alkalinity requirements are calculated the same way but may be less than for a single-stage system. The reason for this is that some of the nitrogen is removed from the system as WAS, which is a part of solids created in the first carbonaceous stage. If this nitrogen is not recycled to the system, less nitrogen is fed to the second stage and the alkalinity requirements are correspondingly lower. If, however, anaerobic digestion is practiced, most of this nitrogen will be recycled, thereby increasing the nitrogen loading. Thus, alkalinity requirements will be similar to those for a single-stage system. When Tampa, Florida, for example, switched from aerobic to anaerobic digestion to treat its WAS, nitrogen to the second stage increased by approximately 5 mg/L.

A significant design parameter to be chosen is the biomass production, $X_{NW}$, to be used (see Wilson, 1992). To aid in understanding this, SRT is calculated as follows:

$$\text{SRT} = \frac{X_a \theta}{X_{NW}} \tag{6.4}$$

Where

$X_a$ = average MLSS concentration in nitrification aeration tank, mg/L;
$\theta$ = nitrification aeration tank hydraulic detention time, days; and
$X_{NW}$ = biomass production in nitrification stage, mg/L.

An important design feature for a two-stage system is providing a partial bypass of the first stage. The reason for this is that, if the first stage performs too well, for example producing 5 to 10 mg/L $BOD_5$ and suspended solids, then the mixed liquor solids of the second stage can start to drop, resulting in too low of an SRT and eventually loss of nitrification. This happens when the effluent suspended solids content from the nitrifying stage is higher than from the first stage. These bypass lines typically are sized to maintain 50 mg/L $BOD_5$ and TSS to the second stage.

Some designers prefer to achieve a similar effect by using WAS from the first stage. While this does not reduce the load to the first stage, it can be a good alternative method in WWTPs where a good quality (that is, not poor settling) first-stage sludge can be ensured.

Another important design feature in WWTPs with effluent filters is provision for the filter backwash water to be returned to the head of the second stage. This provision allows any mixed liquor that is washed out from the second stage to be recaptured and effectively sets the minimum $X_{NW}$ required to maintain an adequate SRT at the filter effluent suspended solids concentration, not the clarifier effluent suspended solids concentration. Plants with filters and this type of recycle typically have less or no need for the first-stage bypass. For SRT and mass-balance calculations, these recycled solids should be treated the same as the second-stage return sludge solids.

**MID-STEP NITRIFICATION.** One way to profit from the advantages of both systems is to design the nitrification plant to be operated as either a one- or two-stage system. This type of design has been used at Tampa, Florida; Gurnee, Illinois; and Waukegan, Illinois. As shown in Figure 6.4, these systems can be operated as one- or two-stage systems, distributing the aeration and clarifier tankages in the two-step mode in any way the WWTP operator may choose.

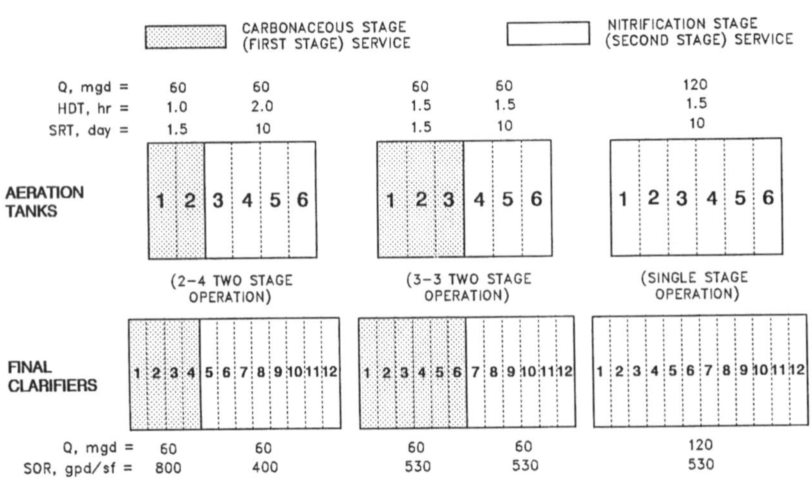


**Figure 6.4 Sample mid-step nitrification process (mgd × [3.785 × $10^3$] = $m^3/d$; gpd/sq ft × [4.074 × $10^{-2}$] = $m^3/m^2 \cdot d$).**

Some of the advantages of this design follow:

- Clarifier area—when in two-stage mode, clarifier area can rapidly be increased by switching to single-stage mode. At the Tampa facility, this is done routinely in rainy seasons to accommodate peak hydraulic loads.
- Tankage minimization—by operating in single-stage mode in summer and two-stage mode in colder months, less tankage is required. This also allows one to take advantage of anoxic zones in summer.

**SELECTION OF CONFIGURATION.** As discussed previously, a suspended-growth nitrification system can be either single stage or two stage. In addition, any single- or two-stage aeration tank can be designed to include step-feed options. More detailed discussion on the advantages of using step-feed operation can be found elsewhere (Buhr *et al.*, 1984, and Carrio *et al.*, 1992). All the aeration tanks can be designed as a plug-flow reactor, a completely mixed activated-sludge system, an oxidation ditch, or a sequencing batch reactor. Aeration tanks can use diffused aeration, mechanical aeration, or pure-oxygen systems.

In principle, two-stage nitrification plants are more flexible to operate. The BOD removal and nitrification are to be accomplished in two different tanks. Because much of the BOD is removed in the first stage, the influent to the second stage has a lower BOD:TKN ratio. It has been found that to achieve a 10% or higher population density of nitrifiers, that is, *Nitrosomonas* and *Nitrobacter*, an influent $BOD_5$:TKN ratio of less than 1 and $SBOD_5$ less than 15 mg/L are recommended. There are other considerations involved in the selection of the system configuration; the major items are summarized as follows:

- Tankage requirement—two-stage systems, designed according to the principles set forth in this chapter, typically take less tankage than single-stage systems. The stronger the wastewater, the greater the advantage will be (see Figure 6.5).
- Site constrictions—because two-stage systems typically require less tankage, they also require less area for the site.
- Anoxic zones—two-stage systems inherently are compatible with anoxic zones because it may not be convenient to return nitrate to where the carbon source is (to the first stage) or return carbon to where the nitrate is (the second stage).
- Enhanced biological phosphorus removal—this typically is more easily achieved in the first stage of a two-stage system than in a single-sludge system. This is because there is no interfering nitrate in the return sludge and because the easily degradable $BOD_5$ does not have to be shared with a denitrification process.

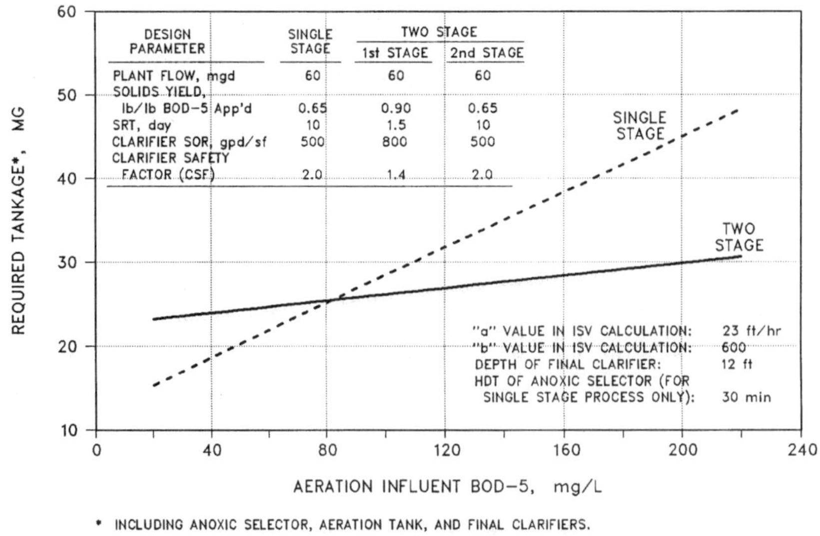


**Figure 6.5 Sample required tankage for nitrification process (HDT = hydraulic detention time) (mgd × [3.785 × 10$^3$] = m$^3$/d; gpd/sq ft × [4.074 × 10$^{-2}$] = m$^3$/m$^2$·d; ft × 0.304 8 = m; ft/hr × [8.467 ×10$^{-5}$] = m/s).**

- Energy considerations—two-sludge systems typically require less oxygen (because less solids are aerobically digested and more nitrogen is removed in the biomass instead of being nitrified) but more pumping energy (to get through the second-stage tankage). (Depending on the aeration system, this may or may not correspond to less energy. Fine-pore systems typically are less efficient in the first stage of a two-stage system. For other types of aeration, aeration energy savings should occur for two-stage systems.)
- Denitrification requirements—if high levels (to less than 3 mg/L effluent total nitrogen) of denitrification are required in conjunction with low effluent suspended solids, the most economical solution may be to use deep-bed filters for denitrification. Under these circumstances, it may be more economical to use a two-stage system ahead of deep-bed filters. Two large plants in Munich, Germany, are currently being designed this way (Eichinger, 1992); the existing Tampa, Florida, Hookers Point advanced wastewater treatment plant also makes use of this.
- Sludge disposal and handling—if anaerobic digestion is not used, operating costs for two-stage systems can be somewhat less because less alkalinity and oxygen are required. These advantages typically are lost when sludges are anaerobically digested. Because two-stage systems tend to produce more sludge than single-stage systems, they would be favored if sludge disposal costs are low.

- Plant size—typically, the complexity of operating two separate stages is not warranted in small WWTPs. Most two-stage plants are larger than 430 L/s (10 mgd). Similarly, lower tankage, site, and energy considerations typically become more pronounced and important in larger plants; therefore, two-stage systems tend to be the better choice in larger plants.
- Wastewater strength—the stronger the wastewater, the more attractive a two-stage system becomes (see Figure 6.5).
- Site constraints—an existing site might not lead itself to one or the other type of system. Preliminary layouts should be made to determine this before a decision is made.
- Protection—two-stage systems inherently protect the sensitive nitrifying organisms by removing significant amounts of toxic in the first stage.
- Stability—two-stage systems are more likely to absorb wide ranges in organic loads than single-stage systems, particularly if the first stage uses pure oxygen. The Howard F. Curren (Hookers Point) advanced wastewater treatment plant in Tampa, Florida, has absorbed swings up to 8.0 kg $BOD_5/m^3\cdot d$ (500 lb $BOD_5$/d/1 000 cu ft) for extended periods, without upset. The mid-step configuration allows the operator to quickly reduce clarifier loadings at high flows.

**ATTACHED-GROWTH SYSTEMS.** Attached-growth systems in wastewater treatment typically consist of trickling filters, RBCs, packed-bed reactors, biological fluidized-bed reactors, and BAFs. Trickling filters can be either rock media or plastic media; plastic media trickling filters are also known as biotowers. Plastic media can be either in sheets (corrugated bundles) or in randomly packed modules (WEF, 1992).

As with suspended-growth systems, nitrification can be accomplished in either a single-stage or a two-stage attached-growth system. In two-stage systems, the first stage is designed for the removal of carbonaceous organic substances (that is, influent BOD), and the second stage is for nitrification. Organic nitrogen is either assimilated to biomass or broken down (hydrolyzed) to ammonia-nitrogen in the first-stage reactions. Considerations for selecting the configurations are the same as those discussed above for the suspended-growth systems. Enough alkalinity should also be present or added for successful nitrification.

**TRICKLING FILTERS. Single-Stage Trickling Filter.** The design of a single-stage trickling filter to include nitrification is basically the same as that for BOD removal, except that the loading rate, in terms of kilograms $BOD_5$ per cubic metre per day (pounds $BOD_5$ per day per 1 000 cubic feet), is much lower. Lower organic loading to a trickling filter is equivalent to the increase of SRT in the suspended-growth system (Mulbarger, 1991). The accepted loading rates for achieving a 75 to 85% nitrification are as follows (Metcalf and Eddy, 1991):

- Rock media trickling filter: 0.8 to 1.3 kg $BOD_5/m^3 \cdot d$ (6 to 10 lb $BOD_5$/d/1 000 cu ft).
- Plastic media trickling filter: 1.5 to 2.3 kg $BOD_5/m^3 \cdot d$ (12 to 18 lb $BOD_5$/d/1 000 cu ft).

The loading rates should be further reduced to 0.4 to 0.8 kg $BOD_5/m^3 \cdot d$ and 0.8 to 1.5 kg $BOD_5/m^3 \cdot d$, respectively, for a 85% and 95% required nitrification. The corresponding SRT for trickling filters, according to the graphs developed by Mulbarger (1991), would be approximately 10 days for 75 to 85% nitrification and 20 days for 85 to 95% nitrification. These SRTs are, in general, adequate for achieving nitrification at approximately 10°C in the suspended-growth systems.

The minimum hydraulic loading for nitrification trickling filters, while varying with media used, is also lower but should be maintained in the range of 30 to 60 $m^3/m^2 \cdot d$ (0.5 to 1.0 gpm/sq ft). This loading is needed to keep the media adequately wetted. This is achieved by providing enough recirculation of filter or clarifier effluent to ensure this minimum under all flow conditions.

Depth of rock media filters in the U.S. typically is 1 to 2 m (3 to 6 ft) and, as stated in the Ten State Standards (Great Lakes, 1990), should not exceed 3 m (10 ft). This is because of the limited ventilation in the rock media. Weight of the media, and thus the required supporting structure, is another concern. Plastic media weigh much less and have higher surface areas as well as a better ventilation property, and thus can be much taller but generally do not exceed 12 m (40 ft). Higher depth will result in greater cost to pump the water, both raw and recycle flows, during plant operation. The advantage of using a higher depth is reduction of construction site requirements and also reduction of recycle flow for keeping the media wetted. It may, however, require a forced ventilation because of the increased loading per unit floor area. Modern nitrification trickling filters typically are plastic media and often equipped with ventilation blowers.

**Two-Stage Trickling Filter.** For separate-stage trickling filters, the required media volume for nitrification stage (the second filter) can be determined by using the curves shown on Figure 6.6 (U.S. EPA, 1992). The $BOD_5$ of the nitrification filter influent, after treatment through the first stage, was 7 to 20 mg/L. It should be noted that the surface areas shown are for the media, not the floor area of the filter. The required media volume may be determined by using the specific surface area of the filter media—typically 46 $m^2/m^3$ (15 sq ft/cu ft) for rock media filters and 122 $m^2/m^3$ (40 sq ft/cu ft) for plastic media filters.

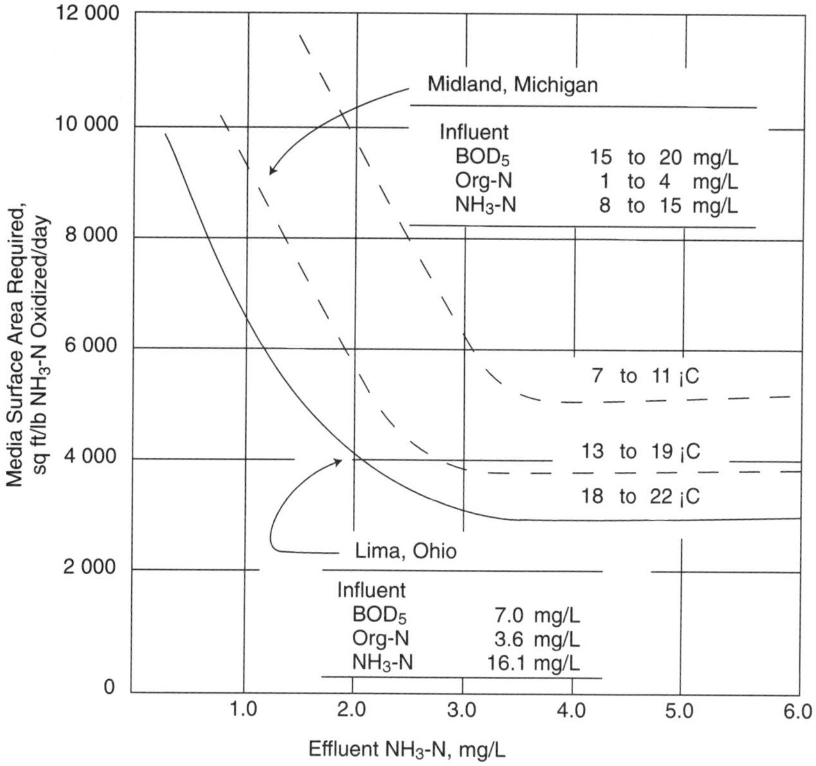


**Figure 6.6 Performance of nitrification stage for separate-stage trickling filters (pilot study results) (U.S. EPA, 1992).**

The typical design value for obtaining an effluent $NH_4$-N of 3.0 mg/L is, as shown on Figure 6.6, 5 000 sq ft/lb $NH_4$-N oxidized/d, which is equivalent to a nitrification rate of 1.0 $g/m^3 \cdot d$ (0.2 lb/d/1 000 sq ft). Various studies conducted in the past years have shown that this is still the adequate design value and this removal rate could be obtained with an effluent $NH_4$-N lower than 3.0 mg/L within the temperature range of 10 to 30°C. The nitrification trickling filter should, however, be

- Equipped with plastic media having a depth of 6 to 12 m (20 to 40 ft);
- Operated with a hydraulic loading (including recycle flow) of no lower than 47 $m^3/m^2 \cdot d$ (0.8 gpm/sq ft); and
- Provided with forced ventilation.

The reason for using a deeper media is to reduce recirculation flow that would otherwise decrease the pH of the filter influent and thus affect the rate of nitrification. The recommended air flow rate for the forced ventilation is 50 kg $O_2$/kg oxygen demand, which is equivalent to 2% oxygen transfer efficiency.

**ROTATING BIOLOGICAL CONTACTORS. Single-Stage Rotating Biological Contactor.** In a single-stage RBC system, ammonia is converted to nitrate following the removal of BOD and organic nitrogen in a single-series RBC train before final clarification. The nature of the nitrification RBC system is such that BOD is removed in the first several passes (or the front portion of a series of RBC shafts) and nitrification occurs in the following passes. This is because the autotrophic nitrifiers do not develop in sufficient quantity for nitrification until the soluble $BOD_5$ ($SBOD_5$) of the processing wastewater is removed to less than 15 mg/L. The process function, thus the design, of the RBC system is therefore sequential. Sufficient disk area must be provided to satisfy the needs for both the removal of BOD and the conversion of TKN to ammonia and subsequently to $NO_3$-N.

It has been found (and has also become a typical procedure in RBC design) that to achieve an effluent $SBOD_5$ lower than 15 mg/L, the whole series of RBC shafts should be separated into at least four passes or "stages." To distinguish this from the stages designated for the nitrification process, the word "pass" has been selected here to indicate each of the compartments divided from a single RBC tank. Each compartment (that is, each pass) may contain different numbers of RBC shafts determined from the design loading to be addressed below. Flow to the RBC shafts in each pass is to be distributed as evenly as possible.

Figures 6.7 and 6.8 show design curves for determining the area requirements for the two functions of a nitrifying RBC system at a water tempera-

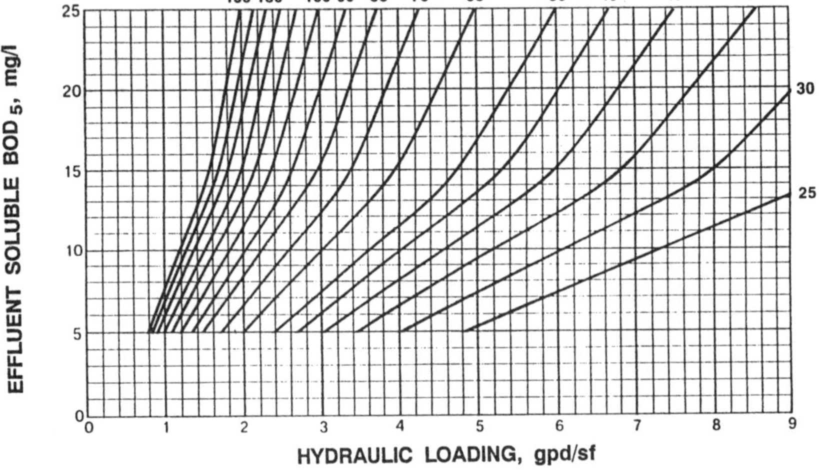


**Figure 6.7** **Rotating biological contactor design curves for BOD removal (gpd/sq ft $\times$ [$4.074 \times 10^{-2}$] = $m^3/m^2 \cdot d$; 55°F = 13°C).**

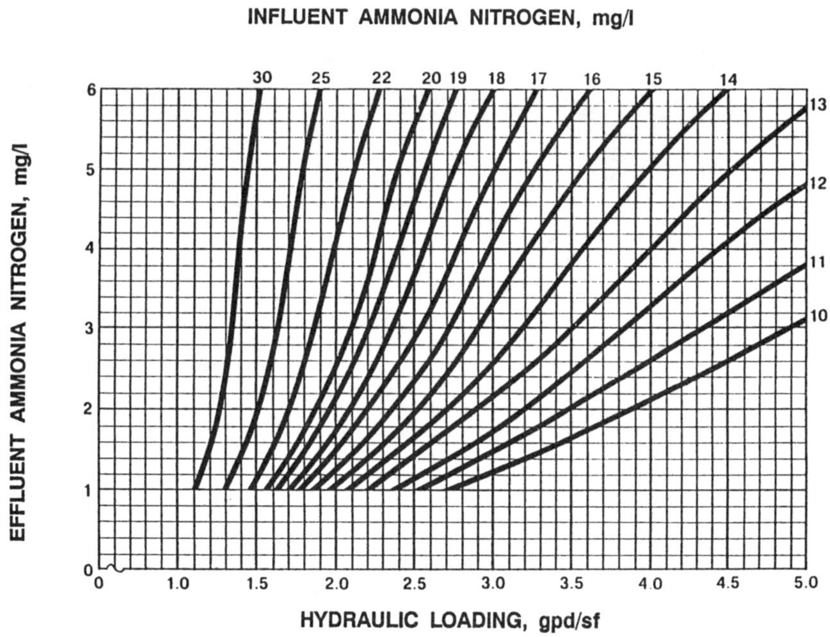


**Figure 6.8 Rotating biological contactor design curves for nitrification (gpd/sq ft × [$4.074 \times 10^{-2}$] = $m^3/m^2 \cdot d$; 55°F = 13°C).**

ture of more than 12°C (55°F) (Envirex, 1983). All the curves suggest that at a fixed hydraulic loading, the rates of both SBOD removal (grams $SBOD_5$ removed per square metre per day [lb/d/1 000 sq ft]) and nitrification (grams $NH_4$-N oxidized per square metre per day [lb/d/1 000 sq ft]) increase with the increase of influent concentrations. The increases are, however, limited and the maximum overall removal rates have been found to be, respectively, 12 g $SBOD_5/m^2 \cdot d$ (2.5 lb $SBOD_5$/d/1 000 sq ft) and 1.5 g $NH_4$-N/$m^2 \cdot d$ (0.3 lb $NH_4$-N/d/1 000 sq ft) (Reh *et al.*, 1977). The corresponding loading to the first stage should not exceed 20 g $SBOD_5/m^2 \cdot d$ (4 lb $SBOD_5$/d/1 000 sq ft) (Envirex, 1983).

It may be noted that $NH_4$-N instead of TKN is used in the design of nitrification RBCs for the treatment of domestic wastewater. The curves in Figure 6.8 were developed in such a way that they included the effects of the presence of organic nitrogen and the hydrolysis of organic nitrogen to $NH_4$-N, which typically occurs in the treatment of domestic wastewater having a TKN–to–$NH_4$-N ratio lower than 2.0. It has been suggested that for the treatment of wastewater with a TKN–to–$NH_4$-N ratio higher than 2.0, a higher $NH_4$-N (for example, 0.5 × TKN) is to be used in the RBC design by using Figure 6.8 (Envirex, 1983).

The straight line on Figure 6.9 shows the percentage of media required for the first pass of the RBC tank, and Figure 6.10 shows the temperature correc-

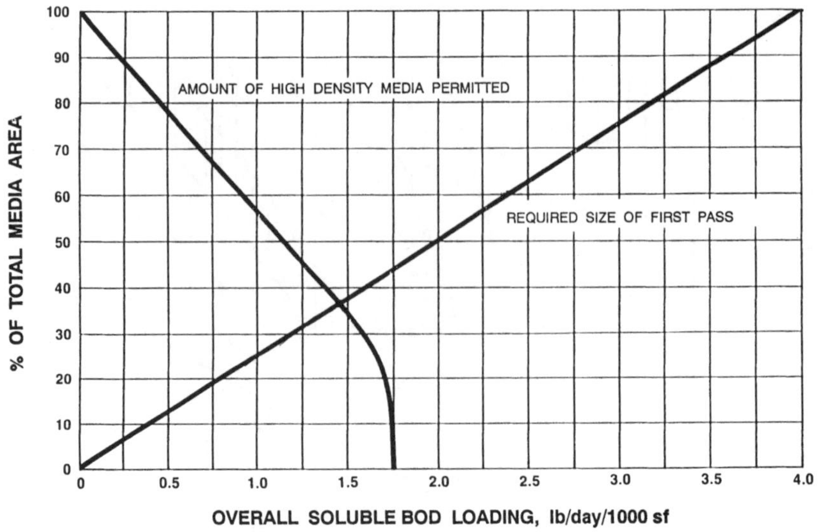


**Figure 6.9** **Rotating biological contactor design requirements (lb/d/ 1 000 sq ft × 4.882 = g/m²·d).**

tion factors. These correction factors are applied to the loadings shown on Figures 6.7 and 6.8 by multiplying those loadings (in cubic metres per square metre per day [gallons per day per square foot]) by the correction factor. The "standard" unit shaft is 8.2 m (27 ft) long consisting of a series of 3.7-m (12-ft) diameter corrugated polyethylene disks, giving a total available surface area of 4 300 m² (100 000 sq ft). Design curves specific for the media type

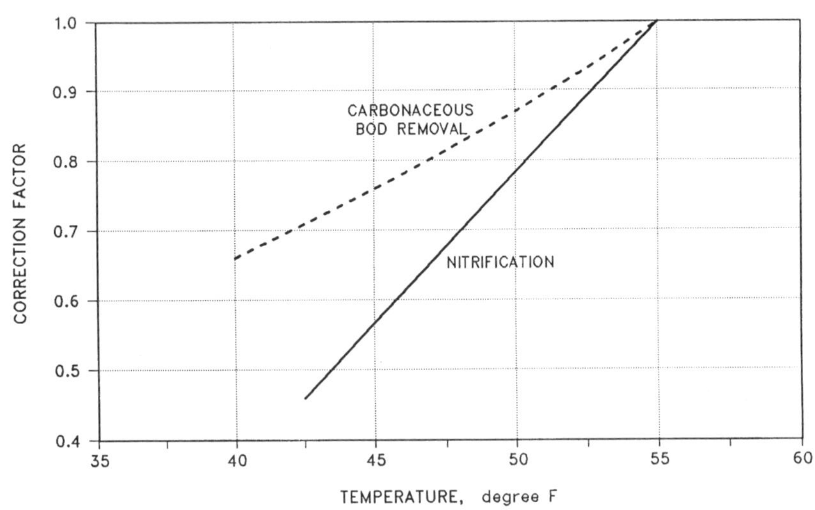


**Figure 6.10** **Rotating biological contactor design temperature correction factors ([°F − 32]0.555 = °C).**

may be available from the respective manufacturers (Brenner *et al.*, 1984). "High-density" RBC shafts may have surface areas of 14 000 $m^2$ (150 000 sq ft), but units with areas as high as 16 700 $m^2$ (180 000 sq ft) are also available. Because of possible oxygen-transfer limitations, excessive weight of biofilm developed, and potential bridging, the high-density units should preferably not be used in the front passes where the main process function is for the removal of BOD; the overall $SBOD_5$ loading should not exceed 8.3 $g/m^2 \cdot d$ (1.7 lb/d/1 000 sq ft) (Envirex, 1983). Designers should consider the use of diffused air, under the carbonaceous RBCs, to control biofilm growth. In nitrification passes, the high-density RBC units may offer some space-saving advantages, but care should be taken to ensure enough oxygen transfer for the required oxidation of ammonia to nitrate. Figure 6.9 also shows the percent of high-density media allowed as a function of overall SBOD loading.

**Two-Stage Rotating Biological Contactor.** In the two-stage nitrification RBC system, the RBCs are used mainly for converting ammonia to nitrate. The RBC influent is the settled secondary effluent generated from a biological treatment process such as the activated-sludge, trickling filter, or another RBC process. It has been mentioned previously that for successful development of nitrifiers, $SBOD_5$ of the nitrification stage influent should be higher than 15 mg/L, and $BOD_5$ should not be higher than the concentration of $NH_4$-N. Higher SBOD would result in the decrease of the available RBC surface area for nitrification. Figures 6.8 and 6.10 are typical graphs used in the design of the nitrification stage of a two-stage RBC system. High-density shafts may be used but preferably should be supplemented with air so that enough oxygen is present to support continuous nitrification.

**BIOLOGICAL FLUIDIZED-BED REACTORS.** Also attached-growth upflow systems, BFBs differs from packed-bed reactors in that a finer media, typically 0.5 mm sand, is used and the media is purposely fluidized. After the smaller sized media is in suspension, the specific surface area is available for the microorganisms to develop; thus, the population density of the organisms is greatly increased. These increases may allow the treatment, both BOD removal and nitrification, to be completed in a very short period of time. In comparison with the conventional suspended-growth systems, the size of the BFB biological reactor can be much smaller while still keeping the desired SRT for wastewater treatment. The use of BFBs in nitrification was first studied in the early 1970s (Jeris *et al.*, 1977). Most of the full-scale plants are used for the treatment of industrial wastewaters (Sutton and Mishra, 1990).

Figure 6.11 is a schematic diagram of a BFB plant. The reactor can be either a single-stage nitrification tank for the removal of both BOD and TKN or the second-stage tank for nitrification only in a two-stage system. A final clarifier typically is not required, but a sand trap may be used to remove and recycle any of the lost sand to the reactor. Solids are wasted by pumping a portion of the media to a sand-cleaning device such as a swirl concentrator

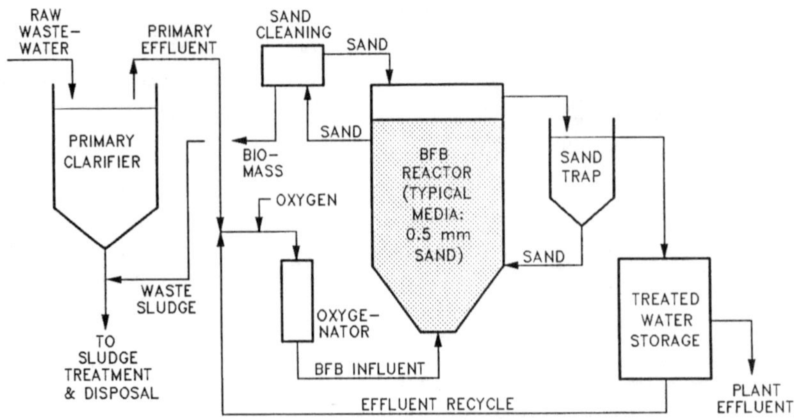


**Figure 6.11 Biological fluidized-bed treatment process.**

where the sand is cleaned and recycled while the washings are concentrated and wasted. To fluidize the media and to carry oxygen to the reactor, a portion of the effluent is recycled and, before entering the reactor, saturated with oxygen in an oxygenator. Whenever needed, an anoxic BFB reactor can be placed ahead of the aerobic BFB reactor to achieve denitrification. Nitrate in the recycle stream is removed as nitrogen gas through denitrification.

Performance data obtained by Jeris *et al.* (1977) from BFB pilot studies conducted in a municipal WWTP are presented in Table 6.1.

Similar to other biological processes, the performance decreases at low water temperatures, and depending on the influent characteristics, addition of alkalinity may be needed to support the development of nitrifiers.

The rates of $NH_4$-N removal reported from other studies include 3.8 $kg/m^3 \cdot d$ (240 lb/d/1 000 cu ft) at 24°C and 0.12 to 0.32 $kg/m^3 \cdot d$ (7.2 to 20.2 lb/d/1 000 cu ft) at 26 to 31°C (Cooper and Williams, 1990). For single-stage nitrification, loading of $BOD_5$ has been reported to be approximately 3 kg $BOD_5/m^3 \cdot d$ (185 lb $BOD_5$/d/1 000 cu ft) (Cooper and Wheeldon, 1981). This is still a developing technology, primarily applied to industrial systems, with no known full-scale installations designed primarily to nitrify (Cooper and Atkinson [Eds.], 1981; Jeris *et al.,* 1977; and Sutton and Mishra, 1990). Because there are still only a few full-scale installations, it is highly recommended that pilot studies be conducted for each specific application.

**BIOLOGICAL AERATED FILTERS.** The BAF combines the unit processes as found in most other biological reactors—aeration, biofilm kinetics, and filtration—in one compact reactor. The filter grain plays a dual role: it retains solid and biomass in the reactor and serves as media for attachment of the biofilm. Because the bacteria are fixed on the media, and aerated, high concentrations of active biomass can be reached. The relatively fine filter grain ranges in size from 2 to 10 mm and eliminates the need for downstream filtration and clarification. Figure 6.12 shows a simplified flow scheme of the BAF, as well as the unit processes it can replace.

The first BAF applied on a large scale was a downflow unit installed north of Paris in 1982 following long years of development work and demonstrations. To date, more than 200 plants have been built, mostly in Japan, France, Great Britain, and Scandinavia (see Table 6.2). The flow rate of a BAF ranges from small industrial facilities to large municipal installations close to 4.3 m/s (100 mgd). The small size of the bioreactor allows it to be covered, thereby allowing bioreactors to be built in sensitive urban sites or environments because odor is captured and removed (Rogalla *et al.*, 1992).

In North America, BAFs were first introduced in the early 1980s. In addition to a demonstration unit cofinanced by U.S. EPA (Condren, 1990), several small installation were built in southern locations. The technology did not find acceptance at the time in the U.S. because experience for correct design and operation was still lacking, and the need for compact reactors was not obvious in rural sites. In the late 1980s, more than 10 large plants using BAFs were built in Quebec, where harsh weather conditions favor covered installations. The largest BAF in Canada was started in Sherbrooke in 1991, with a peak flow of more than 0.9 $m^3/s$ (20 mgd). Several demonstration filters currently are under design for nitrogen removal in New York, New York.

Different technologies can be chosen depending on treatment objective and operating parameters. Flow into the biofilter can be upflow or downflow, whereas air can be introduced at the base of the filter or in the middle of the bed. The media is most often expanded shist or clay, but more recently lightweight plastic media has been introduced. In most designs, media is retained by fine nozzles adapted from drinking water filtration. Table 6.2 lists a few variations of BAFs, yet only three types have been used up to now on large scale: Biocarbone, Biofor, and Biostyr.

Biocarbone is a downflow filter, derived from the classic sand filter, where air is introduced approximately 0.3 m (1 ft) above the nozzled floor, as shown in Figure 6.13. An unaerated polishing section is therefore provided before the effluent is discharged, leading to low effluent solids. Backwash is countercurrent, allowing the area most heavily loaded with solids to be flushed most effectively, avoiding unnecessary solids accumulation. Because nozzles are protected by the filter media and only in contact with already treated effluent, no particular screening is necessary. One disadvantage of this design can be the countercurrent between air and water: While providing a good oxygen transfer, the fluids may block each other at water velocities above 120 $m^3/m^2{\cdot}d$ (3 000 gpd/sq ft), leading to more frequent backwashing, or bumping.

The Biofor reactor is an upflow reactor. Air is blown in at the base of the media, as shown in Figure 6.14. The influent, loaded with solids, has to cross the fine nozzles at the bottom of the reactor. This makes fine screening a necessity, even for the many plants that are designed downstream of physico-

**Table 6.1 Biological fluidized-bed pilot study results.**

| Site | Type | Year | Flow, mgd[a] | Influent[b] | Objective[c] |
|---|---|---|---|---|---|
| Soissons (France) | Biocarbone | 1982 | 2.2 | Primary | Nitrification |
| Le Touquet (France) | Biocarbone | 1984 | 2 | Primary lamella | Secondary |
| St. Palais (France) | Biofor | 1984 | 2 | Physicochemical | Secondary |
| Luneville (France) | Biocarbone | 1985 | 3 | Primary lamella | Nitrification |
| Ashikaga (Japan) | Biocarbone | 1985 | 3 | Paper mill | Secondary |
| Oneonta (Alabama) | Biocarbone | 1986 | 2 | Secondary | Nitrification |
| Perpignan (France) | Biocarbone | 1987 | 11 | Primary | Secondary |
| Ishikawa (Japan) | Biocarbone | 1987 | 6 | Paper mill | Secondary |
| Antibes (France) | Biocarbone | 1989 | 11 | Primary lamella | Secondary |
| Toulouse (France) | Biofor | 1989 | 8.5 | A-stage | Secondary |
| Boisbriand (Quebec, Can.) | Biocarbone | 1989 | 4 | Physicochemical | Secondary |
| Neidenfelds (Germany) | Biofor | 1989 | 7.5 | Paper mill | Secondary |
| Cloppenburg (Germany) | Biofor | 1990 | 4 | Activated sludge | 2-stage nitrification |
| North Bierley (Great Britain) | Biocarbone | 1990 | 3 | Secondary | Nitrification |
| Walsum (Germany) | Biofor | 1990 | 5 | Paper mill | Secondary |
| Tokyo-Nakagawa (Japan) | Biocarbone | 1990 | 9 | Paper mill | Tertiary |
| Corbeil (France) | Biofor | 1991 | 4 | A-stage | Nitrification |
| Sherbrooke (Quebec, Can.) | Biocarbone | 1991 | 23 | Primary lamella | Secondary |
| Porz-Wahn (Germany) | Biofor | 1991 | 5 | Activated sludge | TP + nitrification |
| Chateaugay (Quebec, Can.) | Biofor | 1991 | 25 | Physicochemical | Secondary |
| Ibiza (Spain) | Biocarbone | 1992 | 2.5 | Physicochemical | Secondary |
| Bielefeld | Biofor | 1992 | 12 | Activated sludge | TP + nitrification |
| Cergy (France) | Biostyr | 1992 | 10 | A-stage | Nitrification |
| Porcari (Italy) | Biofor | 1992 | 11 | Paper mill | Tertiary |
| Perrigueux (France) | Biofor | 1992 | 6.5 | Physicochemical | 2-stage nitrification |
| Auteuil (Quebec, Can.) | Biocarbone | 1992 | 6.5 | Primary | Secondary |
| Gutersloh (Germany) | Biofor | 1992 | 12 | Activated sludge | TP + nitrification |

| | | | | | |
|---|---|---|---|---|---|
| Grenoble (France) | Biofor | 1992 | 39 | Physicochemical | Secondary |
| Lagny (France) | Biocarbone | 1993 | 13 | Primary lamella | Nitrogen removal |
| Koln-Stammheim (Germany) | Biofor | 1993 | 82 | Activated sludge | TP + nitrification |
| Paris-Acheres (France) | Biofor/Biostyr | 1993 | 8 | Secondary | Nitrification |
| Nyborg (Denmark) | Biostyr | 1993 | 3.5 | Physicochemical | TP + TN |
| Santa Maria (Spain) | Biofor | 1993 | 2 | Primary lamella | 2-stage nitrification |
| Evreux (France) | Biostyr | 1993 | 5 | Secondary | Nitrification |
| St. Eustache (Quebec, Can.) | Biocarbone | 1993 | 8 | Physicochemical | Secondary |
| Hedon (Great Britain) | Biocarbone | 1994 | 3 | Primary lamella | Secondary |
| Frederikshavn (Denmark) | Biostyr | 1994 | 2.5 | Physicochemical | TP + TN |
| Grenoble (France) | Biofor | 1994 | 11 | Lamella primary | Secondary |
| Tokyo-Ariake (Japan) | Biocarbone | 1995 | 8 | Secondary | Nitrification |
| Melun (France) | Biostyr | 1995 | 6 | Physicochemical | TP + TN |
| Poole (Great Britain) | Biofor | 1995 | 7 | Primary | 2-stage nitrification |
| Rome (Italy) | Biostyr | 1995 | 50 | Primary | Nitrification |
| Puerto de la Cruz (Spain) | Biocarbone | 1995 | 4 | Physicochemical | Secondary |
| Rambouillet (France) | Biostyr | 1996 | 2.5 | Physicochemical | TP + TN |
| Rostock (Germany) | Biofor | 1996 | 17 | Lamella primary | Total nitrogen |
| Lyon St. Fons (France) | Biostyr | 1996 | 22 | Secondary | Nitrification |
| Oslo (Norway) | Biofor | 1996 | 94 | Physicochemical | TP + TN |
| Alanya (Turkey) | Biostyr | 1997 | 7.5 | A-stage | Nitrification |
| Herford (Germany) | Biostyr | 1997 | 9 | Physicochemical | TP + TN |
| Newtown Creek (New York) | Biofor/Biostyr | 1997 | 2 | A-stage | TN |
| Colombes (France) | Biofor/Biostyr | 1998 | 65 | Physicochemical | TP + TN |
| Davyhulme (Great Britain) | Biostyr | 1998 | 100 | Secondary | Nitrification |

[a] mgd × $(4.383 \times 10^{-2})$ = $m^3/s$.
[b] Preliminary treatment before activated sludge: physicochemical, chemically enhanced primary treatment, most often with lamella settling; A-stage, highly-loaded activated-sludge plant without primary treatment.
[c] Treatment objectives: TP = total phosphorus requirement, mostly below 1 mg/L; TN = total nitrogen requirement, mostly below 10 mg/L.

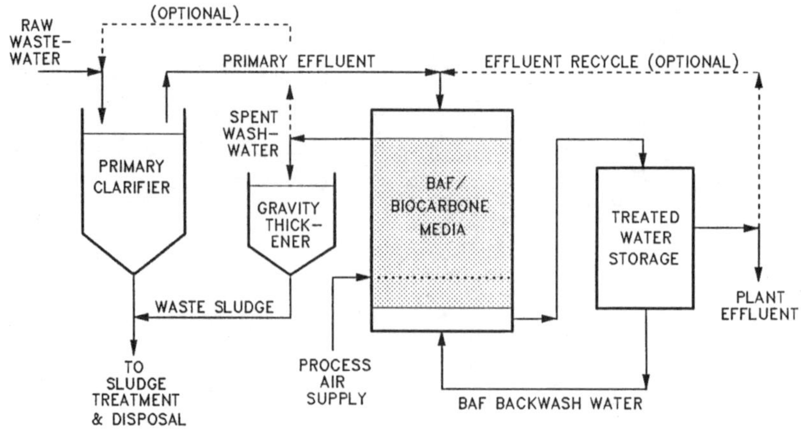


**Figure 6.12 Biological aerated filter/biocarbone wastewater treatment process.**

chemical treatment or after a first biological stage. Process air and water are required for expanding the media, which may lead to a loss of solids with the effluent at high velocities but prevents any head loss increase and unforeseen backwash. Backwash is performed concurrently by increasing air and water velocities.

The Biostyr filter uses lightweight floating polystyrene beads, which are easily available in large quantities and can be manufactured to meet specific

**Table 6.2 Biological aerated filter plants.**

| Type | Company | Media | Water | Backwash |
|---|---|---|---|---|
| Biocarbone | OTV (France) | Shale | Down | Countercurrent |
| Biofor | Degremont (France) | Clay | Up | Cocurrent |
| Biostyr | OTV (France) | Polystyrene | Up | Countercurrent |
| Biodrof | Degremont (France) | Clay | Down[a] | Countercurrent |
| SAFE | PWT (Great Britain) | Shale | Down | Countercurrent |
| Colox | Tetra (U.S.) | Clay | Up | Cocurrent |
| Biopur | Sulzer (China) | Structured plastic | Up | Cocurrent |
| BAF | Stereau (France) | Clay | Down | Countercurrent |
| Biofilter | CISE (France) | Pouzzolane | Up | Cocurrent |
| Biobead | Brightwater (Great Britain) | Polypropylene | Up | Mechanical |
| Biofilter | Wabag (Germany) | Structured plastic | Up | Cocurrent |

[a] Air is sucked downflow through the dry filter; all other air is blown up through media.

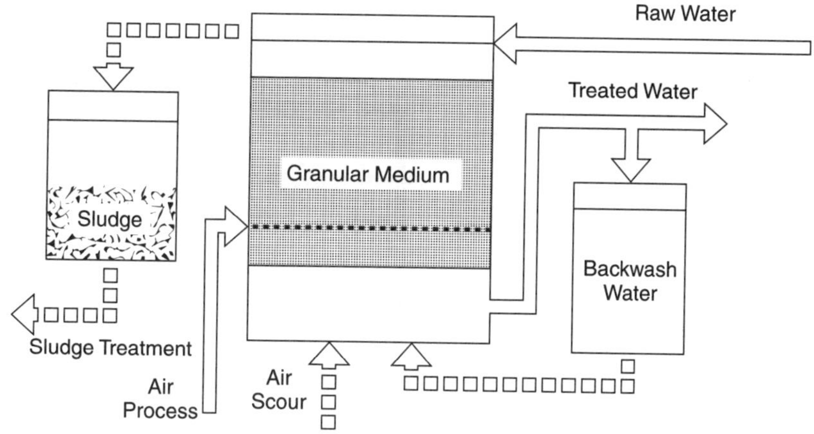


**Figure 6.13 Biocarbone downflow.**

criteria of grain size and density. The media is submerged by a nozzled ceiling, like a filter turned upside down. As shown in Figure 6.15, water passes through the filter in an upflow mode. Nozzles are again only in contact with already filtered water, and backwash is a simple countercurrent gravity flush. Treated water can be stored on top of the filter; no separate backwash pump or storage tank is necessary. The air can be introduced at the base of the media or inside, allowing an anoxic zone to be created in the lower part for denitrification. Both air and water are concurrent, but in the sense of media compression, allowing high filtration velocities while maintaining low suspended solids.

The critical operation of a BAF is periodic backwashing. Because it is a biological reactor, excess biomass accumulates, and because it is a filter, retained solids are kept within the bed. The head loss thus increases over time, up to the preset value, or up to the time set by the operator. Typical maxi-

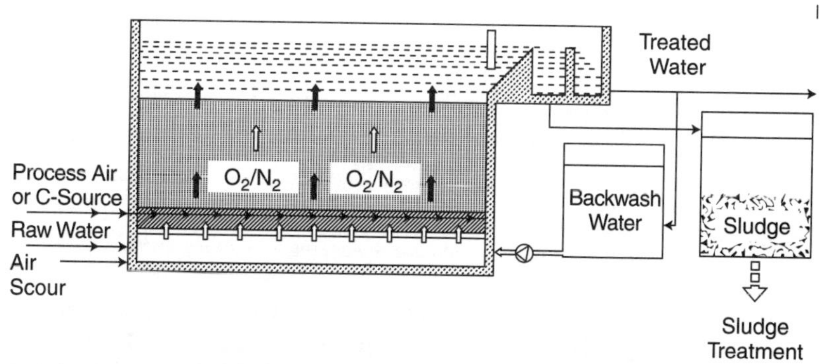


**Figure 6.14 Biofor upflow.**

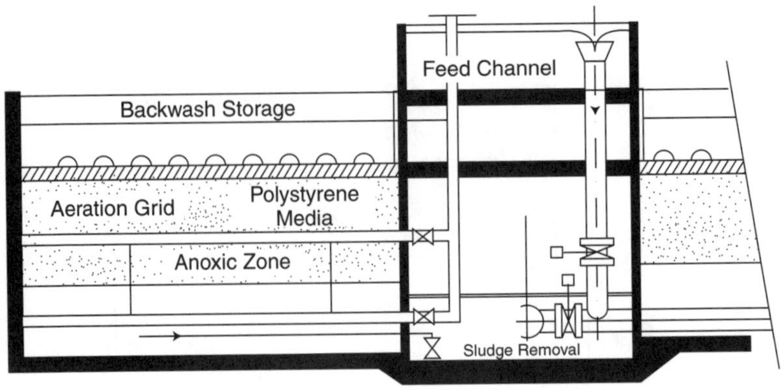


**Figure 6.15 Biostyr filter.**

mum head loss is 2 m (7 ft), and the time between backwashes varies between 24 and 48 hours under normal design conditions. Figure 6.16 shows increasing head loss in a downflow and an upflow system: whereas downflow head loss tends to be lower, its increase is more dramatic, while the upflow systems have more linear behavior.

Backwash is applied by vigorous water flushing and air scouring, as in sand filters, typically with flow rates up to 1 170 $m^3/m^2 \cdot d$ (20 gpm/sq ft) for both fluids. The sludge water generated from removing excess biomass and accumulated solids can be either thickened separately or returned to the primary settler, adding up to 20% of flow to the plant.

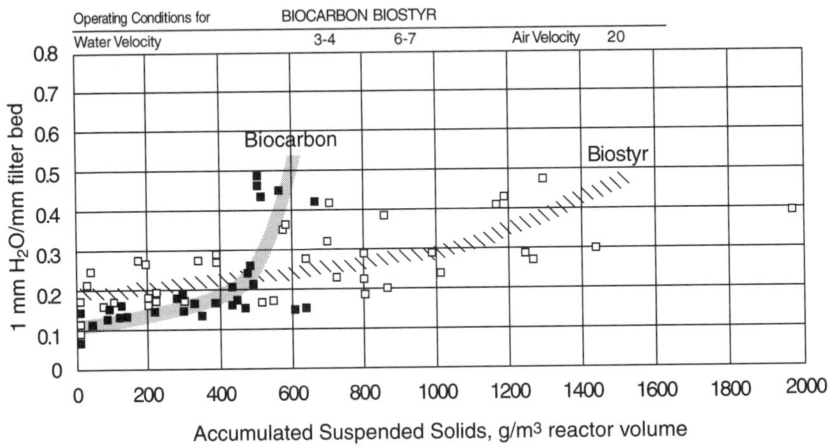


**Figure 6.16 Relationship between accumulated solids expressed as specific solids capture and the head loss gradient expressed as ratio of head loss to filter bed height for the Biocarbone and Biostyr filter.**

These BAFs can be designed for carbon removal, nitrification or denitrification, and even for all three reactions to be carried out in the same reactor (Rogalla and Bourbigot, 1990). The most often used design factor for BAFs is volumetric loading, expressed as kilograms per cubic metre per day (pounds per day per 1 000 cubic feet), with values up to ten times higher than classic biotowers. The loading will be set depending on the required effluent residual and will determine the necessary media volume. The other design value will be hydraulic loading and solids storage capacity, which will determine the cycle time between backwashes. The surface of the filter, and therefore media height, will be set according to these criteria.

For carbon removal, all BAFs have similar behavior and design procedures. Figure 6.17 shows effluent residal chemical oxygen demand (COD) dependent on influent organic loading, obtained on 12 full-scale plants with both Biorcarbon downflow and Biofor upflow reactors. For a municipal wastewater with a COD:BOD ratio between 2 and 2.5 and a filterable COD fraction of half the total, an effluent BOD below 30 mg/L is expected for loadings of less than 10 kg COD/m$^3$·d (Canler and Perret, 1994). For suspended solids, filtration in the sense of media compression favors lower residuals, as shown in Figure 6.18, where the upflow filters work at a velocity often twice as high as that of downflow systems.

Design for nitrification depends on four factors: carbon loading, nitrogen loading, desired effluent residual, and temperature. Figure 6.19 illustrates the influence of carbon loading on the efficiency of ammonia oxidation. Nitrifi-

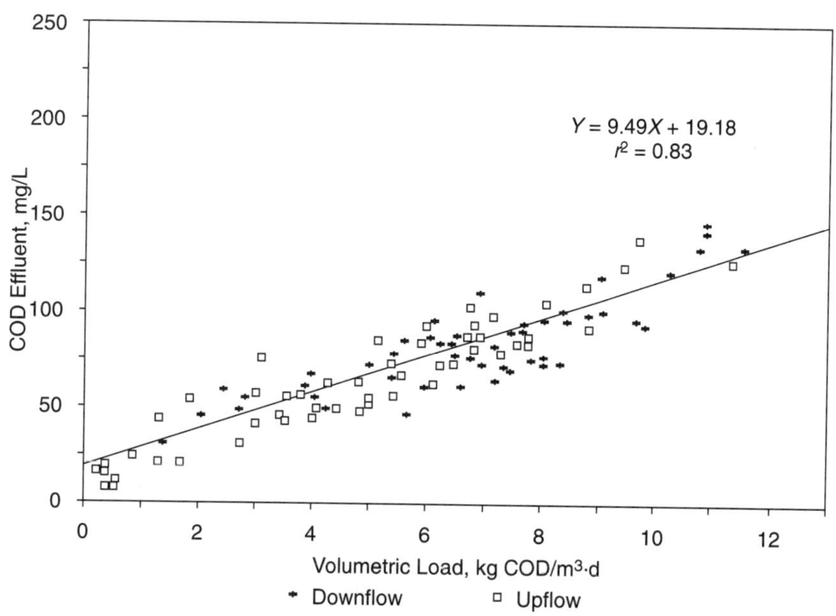


**Figure 6.17** **Changes in effluent COD concentration as a function of the applied load on the filters.**

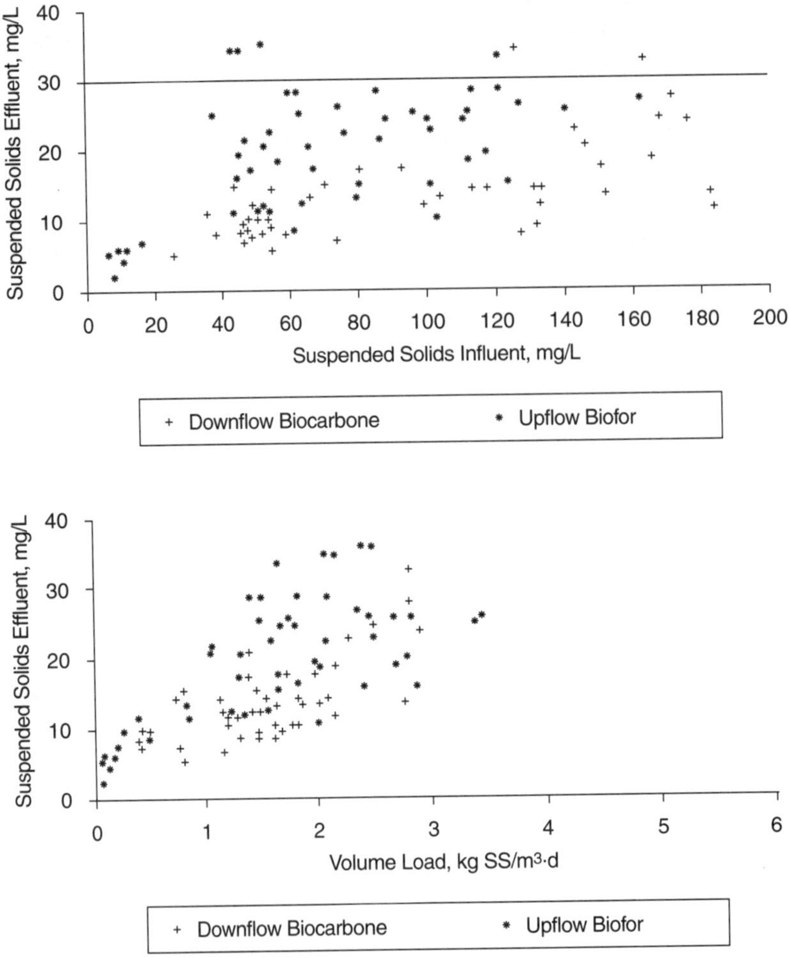


**Figure 6.18 Suspended solids (SS) balance on different BAFs.**

cation is lower if the COD load is above 2 kg COD/m$^3$·d, although significant simultaneous nitrification can be achieved (Rogalla *et al.*, 1990). Primary effluent with high carbon concentrations can be nitrified in two stages, leading to lower effluent ammonia residuals but increased media volume, surface, and filtration velocity.

Plastic media can enhance nitrification, as shown with parallel tests in Figure 6.20 (Tschui *et al.*, 1994). The polystyrene beads allow relatively high nitrification performance, which was confirmed on large prototype reactors with filter surfaces greater than 90 m$^2$ (1 000 sq ft) (Vedry *et al.*, 1994). Nitrification capacity is influenced by air and water velocities (Tschui *et al.*, 1994), and lower oxygen, alkalinity, and ammonia concentrations all reduce the diffusion into the biofilm and limit the performance on full scale.

Denitrification can be achieved on granular filters, regardless of the media and technology used, as long as unaerated zones or reactors are provided.

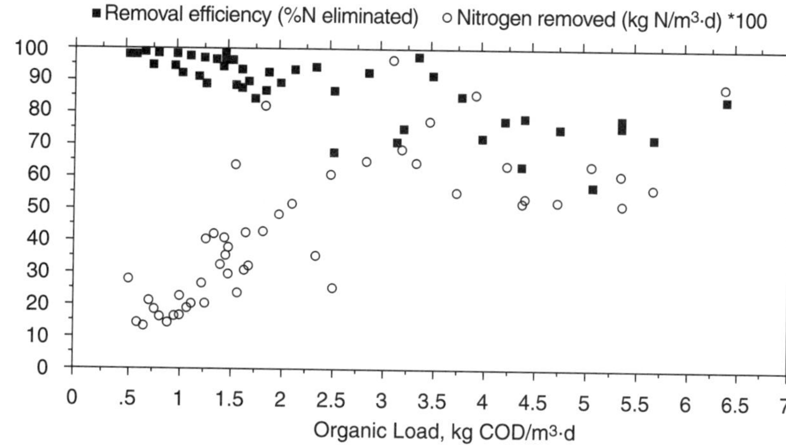


Figure 6.19 **Combined carbon and nitrogen treatment with BAF.**

Downflow filters with high denitrification loadings will have to be bumped occasionally to let the accumulated nitrogen gas escape. An anoxic zone can be placed as predenitrification comparable to the MLE configuration of activated sludge, or as postdenitrification downstream of aerated reactors if an external carbon source is fed.

As nitrate removal is a heterotrophic process, dominated by carbon removal, no significant difference in media or technology is observed, as shown in parallel tests in Figure 6.21 (Jepsen and Jansen, 1993). As illustrated in Figure 6.22, the main influence on denitrification kinetics is the

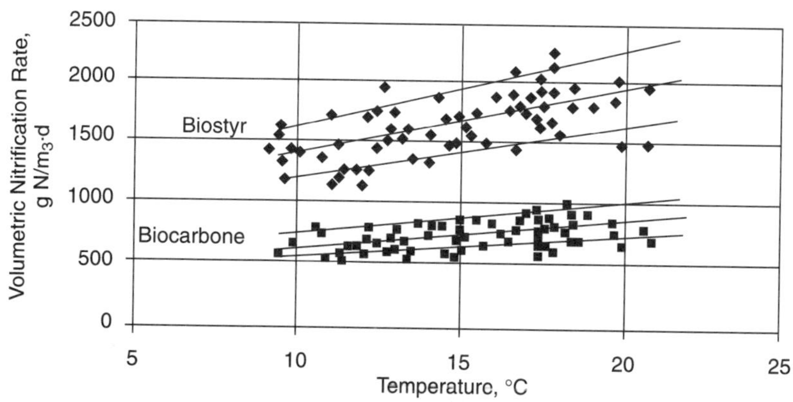


Figure 6.20 **Long-term temperature dependency of the volumetric nitrification rate. (Reprinted from *Water Science and Technology,* Vol. 29, Tschui, M., *et al.,* Tertiary Nitrification in Aerated Pilot Filters, p. 80, Copyright 1994, with kind permission from Elsevier Science Ltd, The Boulevard, Langford Lane, Kidlington OX5 1GB, UK.)**

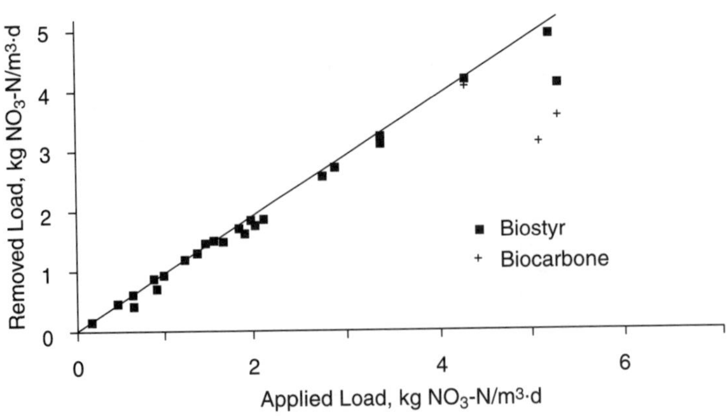


**Figure 6.21** **Applied and removed load, Biostyr and Biocarbone pilot plant. (Reprinted from *Water Science and Technology*, Vol. 27, Jepsen, S.E., and Jansen, J.L.C., Biological Filters for Post-Denitrification, p. 369, Copyright 1993, with kind permission from Elsevier Science Ltd, The Boulevard, Langford Lane, Kidlington OX5 1GB, UK.)**

quality and quantity of the carbon source driving the process. If predenitrification relies solely on primary effluent, the kinetics can be lowered to less than 1 kg N/m$^3$·d (Jepsen *et al.*, 1992), unless the primary effluent is enriched with fermentate or an outside carbon source. With upflow floating media, the anoxic zone can be integrated into the reactor by placing the air

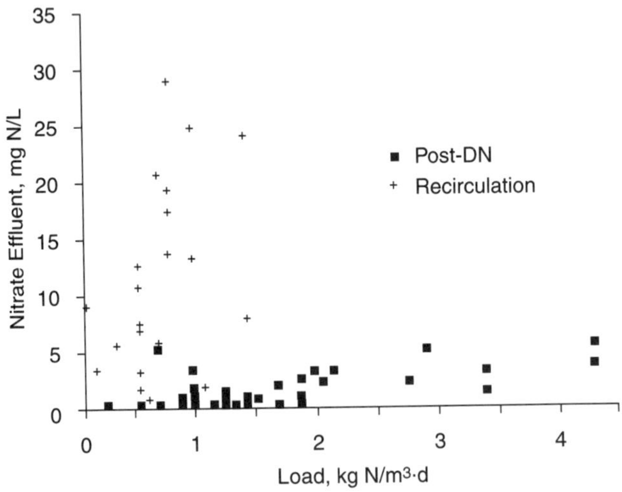


**Figure 6.22** **Effluent quality at different loading rates for Biostyr pilot plant when operated as postdenitrification reactor (C/N ratio ~ 6) and recirculation system (C/N ratio ~ 3.5).**

grid into the media and recirculating the nitrates from the top of the reactor, as shown in Figure 6.15 (Rogalla *et al.*, 1992).

The applications of BAF are growing fast, as the potential of the technology is confirmed on a large scale. The flexibility of the reactors make them attractive tools for plant extensions and upgrading, including the biological treatment of combined sewer overflows. One particular original example of this flexibility is the 250 000-$m^3$/d (65-mgd) Colombes plant near Paris, France, scheduled for start-up in 1998, located in an enclosed building in an urban neighborhood and consisting of sand and grit removal, fine screening, a high-rate physicochemical lamella settler for phosphorus removal, and the three stages of biofiltration, carbon removal, and nitrification–denitrification.

The plant is designed to handle a large flow from a combined sewer, multiplying its capacity more than fourfold from 2.8 $m^3$/s dry-weather flow to 12 $m^3$/s wet-weather flow, while maintaining full biological treatment. While the effluent is diluted, and the removal requirements are therefore reduced, a higher hydraulic flow can be accepted by the biofilters, since no washout of biomass is occurring. To increase capacity, the biofiltration stages are then switched from in-line to parallel treatment, while the heterotrophic biomass in the denitrification reactors will be aerated.

# DESIGN EXAMPLES

Design a nitrification facility for the treatment of a raw municipal wastewater with the following characteristics:

- Flow: 37 900 $m^3$/d.
- $BOD_5$: 200 mg/L.
- $SBOD_5$: 75 mg/L.
- TKN: 30 mg/L.
- $NH_4$-N: 20 mg/L.
- Temperature (minimum): 60°F (15.6°C) summer, 50°F (10.0°C) winter.

The permit effluent levels for $NH_4$-N are 1.5 mg/L for summer and 4.0 mg/L for winter. It is assumed that plant recycle streams would increase the flow by 10% and the strength of water by 15%. The primary treatment would provide a 35% removal of $BOD_5$ and organic nitrogen and no removals for $SBOD_5$ and $NH_4$-N.

**SINGLE-STAGE ACTIVATED-SLUDGE PROCESS.** This design aeration tank influent is as follows:

- Flow: 37 900 × 1. 1 = 41 600 $m^3$/d.
- $BOD_5$: 37 900 × 200 × 1.15 × (1 − 0.35)/1 000 = 5 666 kg/d, or 136 mg/L

- Organic nitrogen: 37 900 × (30 − 20) × 1.15 × (1 − 0.35)/1 000 = 283 kg/d, or 6.8 mg/L.
- $NH_4$-N: 37 900 × 20 × 1.15/1 000 = 872 kg/d, or 20.9 mg/L
- TKN: 872 = 283 = 1 155 kg/d, or 27.7 mg/L.
- Solids yield (0.65 kg/kg $BOD_5$) (see Wilson, 1992): 5 666 × 0.65 = 3 683 kg/d.
- Organic nitrogen in solids: 3 683 × 0.04 = 147 kg/d.
- TKN to be nitrified: 1 155 − 147 = 1 008 kg/d.
- Design SRT (at 10°C) (assumed): 10 days.
- Clarifier surface overflow rate: 20.4 $m^3/m^2 \cdot d$ (500 gpd/sq ft).
- Clarifier surface area: 41 600/20.4 = 2 039 $m^2$.
- Volume of clarifier: 2 039 × 3.7 = 7 544 $m^3$ (SWD = 12 ft).
- Clarifier safety factor (see Wilson, 1992): 2.0.
- Sludge settling characteristics:
  - "$V_{max}$": 23 ft/hr (7.0 m/h).
  - "$K$" value: 600.
  - Initial settling velocity (ISV): 2 × 500/(7.48 × 24) = 5.57 ft/hr (1.69 m/hr).
- Clarifier influent MLSS: $10^6$ × 1*n*(23/5.57)/600 = 2 363 mg/L.
- Volume of aeration tank: 3 683 × 10*d* × 1 000/2 363 = 15 600 $m^3$.
- Anoxic selector hydraulic detention time: 30 minutes.
- Anoxic selector volume: 41 600 × 30/1440 = 867 $m^3$.
- Total tankage (anoxic selector + aeration tank + clarifier): 867 + 15 600 + 7 544 = 24 011 $m^3$.
- Design RSS: 6 000 mg/L.
- Required return sludge flow: 2 363 × 41 600/(6 000 − 2 362) = 27 000 $m^3$/d.
- $Q_r/Q$: 27 000/41 600 = 65%.
- Required oxygen
  - For BOD removal: 1.2 × 5 666 = 6 799 kg/d.
  - For nitrification: 4.6 × 1 008 = 4 637 kg/d.
  - For aeration tank DO: 4 × (41 600 + 27 000)/1 000 = 274 kg/d (assuming 4.0 mg/L average).
  - Total: 11 710 kg/d.
- Percent oxygen transfer (AOTE for fine-pore diffusers): 10.0
- Required oxygen supply: 11 710 × 100/10 = 117 100 kg/d.
- Required air supply: 117 100 × 100/23.15 (20°C, 1 atm) = 505 832 kg/d = 505 832/(1.2 × 1 440) = 293 $m^3$/min.

It may be noted that the volume of aeration tank as determined above is for plug-flow operation mode where both the return sludge and tank influent are fed to the anoxic selector located at the beginning of the reactor. It has been shown that the aeration tankage can be significantly reduced if the influent is fed step-wise in a step-feed plant. Detailed discussion and design of the step-feed plants can be found elsewhere (Carrio *et al.*, 1992, and Wilson, 1992).

It may also be noted that although a return sludge flow of 65% plant flow as calculated is required, it is preferred to provide pumps with capacity as high as 100% plant flow. Moreover, to meet the peak demand, the capacity of air of the supply system should be at least two times that calculated above.

**TWO-STAGE ACTIVATED-SLUDGE PROCESS.** The design aeration tank influent is the same as for the single-stage activated-sludge process.

First-stage design:

- Solids yield (0. 9 kg/kg $BOD_5$): 5 666 × 0.9 = 5 099 kg/d.
- Organic nitrogen in solids: 5 099 × 0.04 = 204 kg/d.
- Design SRT: 1.5 days.
- Clarifier surface overflow rate: 32.6 $m^3/m^2 \cdot d$ (800 gpd/sq ft).
- Clarifier surface area: 41 600/32.6 = 1 276 $m^2$.
- Volume of clarifier (SWD = 12 ft [3.5 m]): 1 276 × 3.7 = 4 721 $m^3$.
- Clarifier safety factor: 1.4.
- Sludge settling characteristics:
  - — "$V_{max}$" value: 23 ft/hr (7.0 m/h).
  - — "$K$" value: 600.
  - — ISV: 1.4 × 800/(7.48 × 24) = 6.24 ft/hr (1.9m/h).
- Clarifier influent MLSS: $10^6 \times 1n(23/6.24)/600$ = 2 174 mg/L.
- Volume of aeration tank: 5 099 × 1.5$d$ × 1 000/2 174 = 3 518 $m^3$.
- Design RSS: 5 500 mg/L.
- Required return sludge flow: 2174 × 41 600/(5 500 − 2 174) = 27 200 $m^3$/d
- $Q_r/Q$: 27 200/41 600 = 65%.
- Required oxygen:
  - — For BOD removal: 1.0 × 5 666 = 6 799 k/d.
  - — For aeration tank DO: 2 × (41 600 + 27 200)/1 000 = 138 kg/d (assuming 2.0 mg/L average).
  - — Total: 6 937 kg/d.
- Percent oxygen transfer (AOTE for fine-pore diffusers): 8.0.
- Required oxygen supply: 6 937 × 100/8 = 86 713 kg/d.
- Required air supply: 86 713 × 100/23.15 (20°C, 1 atm) = 374 570 kg/d = 374 570/(1.2 × 1 440) = 217 $m^3$/min.

Second-stage design:

- Influent $BOD_5$ (typical first-stage effluent and second-stage yield values): 15 mg/L or 15 × 41 600/1 000 = 624 kg/d.
- Influent suspended solids (typical first-stage effluent and second-stage yield values): 25 mg/L or 25 × 41 600/1 000=1 040 kg/d.

- Solids yield (0.8 kg/kg suspended solids applied): 1 040 × 0.8 = 832 kg/d.
- Organic nitrogen in solids (for both steps): 0.04 (5 099 + 832) = 237 kg/d.
- TKN to be nitrified: 1 155 − 237 = 918 kg/d.
- Design SRT (same as for single stage): 10.0 day.
- Clarifier surface overflow rate: 20.4 $m^3/m^2 \cdot d$ (500 gpd/sq ft).
- Clarifier surface area: 41 600/20.4 = 2 039 $m^2$.
- Volume of clarifier: 2 039 × 3.7 = 7 457 $m^3$ (SWD = 12 ft).
- Clarifier safety factor: 2.0.
- Sludge settlings characteristics:
  - "$V_{max}$" value: 23 ft/hr (7.0m/hr).
  - "$K$" value: 600.
- ISV: 2 × 500/(7.48 × 24) = 5.57 ft/hr (1.7m/h).
- Clarifier influent MLSS: $10^6 \times 1n$ (23/5.57)/600 = 2 363 mg/L.
- Volume of aeration tank: 832 × 10$d$ × 1 000/2 363 = 3 521 $m^3$.
- Design RSS: 6 000 mg/L.
- Required return sludge flow: 2 363 × 41 600/(6 000 − 2 363) = 27 000$m^3$/d.
- $Q_r/Q$: 27 000/41 600 = 65%.
- Required oxygen:
  - For BOD removal: 1.2 × 624 = 749 kg/d.
  - For nitrification: 4.6 × 918 = 4 223 kg/d.
  - For aeration tank DO: (4 − 2) × 41 600 + 27 000/7 000 = 137 kg/d (assuming 4.0 mg/L average).
  - Total (second stage): 5 109 kg/d.
  - Percent oxygen transfer (AOTE for fine-pore diffusers): 10.0.
- Required oxygen supply: 5 109 × 100/10 = 51 090 kg/d.
- Required air supply: 51 090 × 100/23.15 (20°C, 1 atm) = 220 691 kg/d = 220 691/(1.2 × 1 440) = 128 $m^3$/min.

Overall design:

- Total tankage (aeration tanks + clarifiers): 3 518+ 4 721+3 521+ 7 457 = 19 217 $m^3$.
- Total oxygen required: 374 570 + 220 691 = 595 261 kg/d.
- Total air supply: 217 + 128 = 345 $m^3$/min.

It may be noted that, for this particular example, the total tankage for a two-stage system is much lower, approximately 20% less than that for a single-stage system. This difference would be greater for plants with higher influent $BOD_5$ and less for plants with lower influent $BOD_5$ as shown in Figure 6.5. The two-stage systems, however, required two sets of return sludge pumps, and the required air supply is approximately 5% greater. Note that oxygen requirements are approximately 10% lower for the two-stage system but that the poorer transfer efficiency (8%) in the first stage significantly affected energy requirements. To optimize energy use of a pure-oxygen or surface aeration system would be preferable.

# REFERENCES


Brenner, R.C., *et al.* (1984) Design Information on Rotating Biological Contactors. EPA-600/2-84-106, PB84-199561, U.S. EPA, Washington, D.C.

Buhr, H.O., *et al.* (1984) Making Full Use of Step Feed Capability. *J. Water Pollut. Control Fed.*, **56**, 325.

Canler, J.P., and Perret, J.M. (1994) BAF—Assessment of the Process on 12 Sewage Treatment Plants. *Water Sci. Technol.*, **29**, 10.

Carrio, L.A., *et al.* (1992) Advanced BNR—BNR for Large Plants. Paper presented at 65th Annu. Conf. Water Environ. Fed., New Orleans, La.

Condren, A.J. (1990) Technology Assessment of the Biological Aerated Filter. EPA/600/2-90-015, U.S. EPA, Washington, D.C.

Cooper, P.F., and Atkinson, B. (Eds.) (1981) *Biological Fluidised Bed Treatment of Water and Wastewater.* Ellis Harwood Ltd., New York, N.Y.

Cooper, P.F., and Williams, S.C. (1990) High-Rate Nitrification in a Biological Fluidized Bed. *Water Sci. Technol.*, **22,** 431.

Cooper, P.F., and Wheeldon, D.H.V. (1981) Treatment of Sewage in a Two-Fluidised Bed System. In *Biological Fluidised Bed Treatment of Water and Wastewater.* P.F. Cooper and B. Atkinson (Eds.), Ellis Harwood Ltd., New York, N.Y.

Eichinger, J., (1992) Denitrification on Deep Bed Filters at Munich. Paper presented at Eur. Conf. Nutrient Control, Wakefield, U.K.

Envirex (1983) *Design Manual for Rotating Biological Contactors.*

Great Lakes–Upper Mississippi River Board of State Public Health and Environmental Managers (1990) Recommended Standards for Wastewater Facilities (Ten State Standards).

Jepsen, S.E., and Jansen, J.L.C. (1993) Biological Filters for Post-Denitrification. *Water Sci. Technol.*, **27,** 5/6, 369.

Jepsen, S.E., *et al.* (1992) Denitrification in Submerged Filters. In *Chemical Wastewater Treatment 2.* R. Klute and H. Hahn (Eds.), Springer, Berlin, Ger.

Jeris, J.S., *et al.* (1977) Biological Fluidized-Bed Treatment for BOD and Nitrogen Removal. *J. Water Pollut. Control Fed.*, **49**, 816.

Lessel, T.H., (1993) Upgrading and Nitrification by Submerged Bio-Film Reactors—Experiences from a Large Scale Plant. Paper presented at 2nd Int. Spec. Conf. Biofilm Reactors, IAWQ, Paris, Fr.

Metcalf and Eddy, Inc. (1991) *Wastewater Engineering, Treatment, Disposal, and Reuse.* McGraw-Hill, Inc., New York, N.Y.

Mulbarger, M.C. (1991) Fundamental Secondary Treatment Insights. Paper presented at 64th Annu. Conf. Water Pollut. Control Fed., Toronto, Ont., Can.

Reh, C.W., *et al.* (1977) An Approach to Design of RBCs for Treatment of Municipal Wastewater. Paper presented at Am. Soc. Civil Eng. Natl. Environ. Eng. Conf.

Rogalla, R., and Bourbigot, M. (1990) New Developments in Complete Nitrogen Removal with Biological Aerated Filters. *Water Sci. Technol.*, **22**, 273.

Rogalla, F., *et al.* (1990) Reduced Hydraulic Detention Time for Complete Nutrient Removal with Innovative Biological Reactors. Paper presented at Int. Conf. North Sea Pollut. Tech. Strategies Improvement IAWPRC/EW-PCA/NVA.

Rogalla, F., *et al.* (1992) Upscaling a Compact Nitrogen Removal Process. *Water Sci. Technol.*, **26,** 1067.

Sen, D., *et al.* (1993) Full Scale Evaluation of Nitrification and Denitrification on Fixed Film Media (Ringlace) for Design of Single Sludge Nitrogen Removal System. Paper presented at 66th Annu. Conf. Water Environ. Fed., Anaheim, Calif.

Sen, D., *et al.* (1994) Performance of Fixed Film Media Integrated into Activated Sludge Reactors to Enhance Nitrogen Removal. Paper presented at Water Quality Int. '94, IAWQ 17th Biennial Int. Conf., Budapest, Hung.

Sutton, P.M., and Mishra, P.N. (1990) Fluidized Bed Biological Wastewater Treatment: Effects of Scale-Up on System Performance. *Water Sci. Technol.*, **22**, 419.

Tschui, M., *et al.* (1994) Tertiary Nitrification in Aaerated Pilot Filters. *Water Sci. Technol.*, **29,** 10/11, 80.

U.S. Environmental Protection Agency (1992) *Process Design Manual for Nitrogen Control.* U.S. EPA, Technol. Transfer, Washington, D.C.

Vedry, B., *et al.* (1994) Operating Two Biofilter Prototypes at Paris-Acheres Wastewater Treatment Plant. *Water Sci. Technol.*, **29,** 10/11, 80.

Water Environment Federation (1992) *Design of Municipal Wasteater Treatment Plants.* Manual of Practice No. 8, Alexandria, Va.

Wilson, T.E. (1992) Activated Sludge Treatment of Municipal Wastewater, USA Approach. In *Activated Sludge Process Design and Control: Theory and Practices*. IAWPRC, London, G.B.

# Chapter 7 Nitrogen Removal and Design of Combined Nitrification and Denitrification Processes

Nitrogen removal processes can be subdivided into the following two categories: separate-stage denitrification processes and single-sludge nitrification–denitrification processes. Separate-stage denitrification processes treat the effluent from a nitrifying activated-sludge process. Methanol typically is added as the carbon source for denitrification. Recently, single-sludge nitrification–denitrification processes have been used to a much greater extent for municipal wastewater. Fewer tanks and clarifiers are required when the processes are combined, and the need for methanol is eliminated.

Separate-stage processes have some advantages. The biggest advantage is that low effluent total nitrogen (TN) concentrations are attainable, mainly because the denitrification stage can be optimized separately. Also, separate-stage denitrification is needed for some industrial wastewater that is high in nitrogen but deficient in carbon.

The available alternatives within separate-stage and single-sludge processes will be discussed in detail in this chapter. Design criteria, design equations, example designs, and case studies will be presented for each alternative.

# SEPARATE-STAGE DENITRIFICATION

The sole purpose of a separate-stage denitrification process is to remove nitrate from wastewater, typically treating effluent from a nitrifying activated-sludge process. When carbon oxidation and nitrification are performed in one stage, the addition of separate-stage denitrification creates what is known as a two-stage nitrification–denitrification process. Each stage is operated independently. The division of carbon oxidation, nitrification, and denitrification into three separate stages is referred to as a three-stage process.

The design of nitrification processes is covered in Chapter 6. This section is devoted entirely to the design of separate-stage denitrification processes.

**CHARACTERISTICS.** There are three general types of separate-stage denitrification processes: suspended-growth, packed-bed attached-growth, and fluidized-bed attached-growth systems. A flow schematic diagram of each of the three types of processes is shown in Figure 7.1. Each will be discussed in detail in the following sections, but it should be noted that there are some characteristics that all separate-stage denitrification processes have in common. Denitrification is inhibited at colder temperatures and extremes of

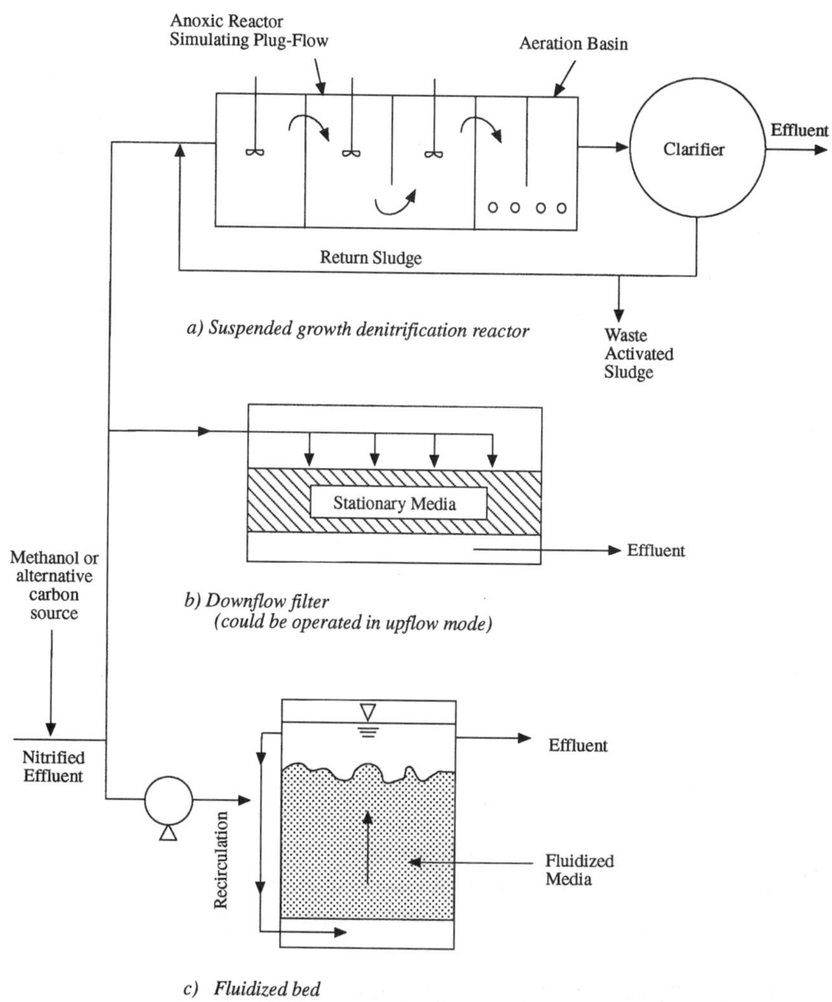


**Figure 7.1** **Alternative separate-stage denitrification processes.**

pH and by numerous chemical compounds. The effect of temperature can be accounted for as described in the design methodology that follows. The effect of pH is not clear because of conflicting results reported in the literature, but pH effects typically can be ignored if near-neutral pH (6.5 to 7.5) is maintained (Randall *et al.*, 1992). Alkalinity is produced by denitrification reactions, which typically serve to elevate the pH. This may be important if the nitrified effluent has a pH below 6.5.

**SUSPENDED GROWTH.** Figure 7.1a shows a schematic diagram of the separate-stage, suspended-growth denitrification process. The process is a variation of the activated-sludge process, which consists of a suspended-growth reactor followed by a clarifier. The primary operational difference (besides methanol addition) in suspended-growth denitrification compared to standard activated sludge is that either the majority of the reactor or the entire reactor is mixed rather than aerated. Either complete-mix or plug-flow reactors can be used for the anoxic reactor. The schematic in Figure 7.1a shows the anoxic reactor divided into sections, which simulates plug flow. An aeration basin with a short retention time (15 to 60 minutes) typically is provided directly before clarification for the following reasons: to strip any remaining nitrogen gas (also strips carbon dioxide); to add DO and help prevent denitrification (and possible rising sludge because of nitrogen gas bubbles adhering to the sludge floc) in the clarifier; and to oxidize any excess methanol that remains. Some settled sludge is wasted to maintain the desired solids retention time (SRT) or mixed liquor suspended solids (MLSS) concentration, and the rest is recycled to the anoxic portion of the reactor.

Mixing is required in the anoxic reactor. The proper intensity of mixing is needed to maintain the solids in suspension without mixing in too much DO. Mixing intensity required will vary from system to system, but it is reasonable to assume that 4 to 10 $W/m^3$ (20 to 50 hp/mil. gal) is required (Randall *et al.*, 1992). A better criterion is complete tank contents turnover every 20 minutes.

**Design Criteria.** The volume of a suspended-growth denitrification reactor can be determined based on denitrification kinetics or on full-scale or pilot experience. Hydraulic retention time (HRT), SRT, denitrification rate, and methanol requirements typically are the most important design considerations.

The design of suspended-growth denitrification reactors typically is based on either SRT considerations or a denitrification rate approach. There are three approaches, all of which are equally acceptable design methodologies, that will be mentioned here. The first is an SRT approach using kinetics coefficients determined on a nitrate basis. Design using this approach typically is accomplished using specific growth rate data for denitrifying bacteria and assumes non-carbon-limiting conditions. The second approach is the denitrification rate approach, which has historically been the most widely used approach. It is a relatively simple approach, but the specific denitrification rate (SDNR) must be known or assumed, and reported rates using methanol vary widely (0.10 to 1.2 g $NO_3^-$-N/g total suspended solids [TSS]·d) because of temperature and different organic substrate concentrations during the denitrification reactions (Parker *et al.*, 1975). These design approaches are discussed in detail in Metcalf and Eddy (1991) and in the *Process Design Manual for Nitrogen Removal* (Parker *et al.*, 1975).

The third approach is another SRT approach that applies basic kinetic information to provide perhaps the most rational approach to design (Randall *et al.*, 1992). The difference between this approach and the previously mentioned SRT approach is that the kinetic coefficients are on a chemical oxygen demand (COD) basis rather than a nitrate basis. This approach therefore accounts for the fact that COD, rather than nitrate, is typically the limiting factor for denitrification (McClintock *et al.*, 1988, and Stensel *et al.*, 1973). The half-rate constant for nitrate is low, approximately 0.1 mg/L; therefore, nitrate does not become limiting until it approaches very low concentrations.

Little information exists on kinetic coefficients on a COD basis in denitrification processes. McClintock *et al.* (1988), in a study treating a synthetic wastewater, reported that yield coefficients and decay rates were much lower when nitrate was used as electron acceptor compared to when oxygen was used. Stensel *et al.* (1973) have reported basic denitrification kinetic coefficients on a COD basis using methanol, and they are presented in Table 7.1. These coefficients will be used in the design example.

**Design Equations.** The design equations used for the SRT design approach using kinetic coefficients on a COD basis are illustrated here. The kinetics equations used are based on organic substrate removal in complete-mix systems and are also used in the kinetic approach to design for carbon oxidation activated-sludge processes (Lawrence and McCarty, 1970).

**Table 7.1 Denitrification kinetic coefficients using methanol (after Stensel *et al.*, 1973).**

| Parameter | Units | 20°C | 10°C |
|---|---|---|---|
| Synthesis yield coefficient, $Y$ | g VSS[a]/g COD | 0.18 | 0.17 |
| Endogenous decay coefficient, $b$ | $d^{-1}$ | 0.04 | 0.05 |
| Maximum specific substrate removal rate, $\hat{q}$ | g COD/g VSS·d | 10.3 | 3.1 |
| Half-velocity coefficient, $K_s$ | g COD/m$^3$ | 9.1 | 12.6 |

[a] Volatile suspended solids.

The specific substrate removal can be found from the maximum substrate removal rate according to the following:

$$q = \hat{q}\left(\frac{S_e}{K_s + S_e}\right) \tag{7.1}$$

Where

$q$ = specific substrate removal rate, $d^{-1}$;
$\hat{q}$ = maximum substrate removal rate, $d^{-1}$;
$S_e$ = effluent COD concentration, mg/L; and
$K_s$ = half-rate constant, mg/L;

Denitrification is adversely affected by colder temperatures. Specific substrate removal rates (and denitrification rates) can be corrected for temperature using the following equation:

$$q_T = q_{20}\Theta^{T-20} \tag{7.2}$$

Where

$\theta$ = temperature correction coefficient. Values for $\theta$ from 1.08 to 1.16 are commonly used (Metcalf and Eddy, Inc., 1991).
$T$ = temperature, °C

The minimum SRT required can be calculated from the temperature-corrected $q$ using the equation

$$\frac{1}{\text{SRT}} = Yq - b \tag{7.3}$$

Where

$Y$ = synthesis yield coefficient, g VSS/g COD used; and
$b$ = endogenous decay coefficient, $d^{-1}$.

The net biomass yield is a function of the synthesis yield coefficient, the endogenous decay coefficient, and SRT:

$$Y_N = \frac{Y}{1 + b(\text{SRT})} \tag{7.4}$$

After SRT is selected, $Y_N$ can be determined. Waste sludge can be determined from $Y_N$.

The concentration of COD from methanol maintained in the anoxic reactor must be selected. Select a low concentration of 3 to 5 mg COD/L. This amount will easily be oxidized in the postaeration part of the reactor. The denitrification rate could be increased by higher methanol concentrations if desired, resulting in a smaller anoxic basin. There is a tradeoff, however, between costs of higher methanol doses (and increased sludge production along with it) and capital costs for a larger reactor.

The value for the product of the denitrification biomass concentration and HRT can be determined from a mass balance on solids (Lawrence and McCarty, 1970):

$$(X)\text{HRT} = \frac{Y(S_o - S_e)\text{SRT}}{1 + b(\text{SRT})} \tag{7.5}$$

Where

HRT = hydraulic retention time, days;
$X$ = denitrification biomass concentration, mg/L; and
$S_o$ = influent COD concentration, mg/L.

The SDNR for the resultant design can then be calculated as follows:

$$\text{SDNR} = \frac{\text{N}_\text{o} - \text{N}}{(X)\text{HRT}} \tag{7.6}$$

Where

$\text{N}_\text{o}$ = influent nitrate-nitrogen ($NO_3$-N) concentration, mg/L; and
N = effluent $NO_3$-N concentration, mg/L.

**Design Example.** For illustration purposes, a typical nitrified effluent will be used for the influent to the denitrification process. Table 7.2 presents the nitrified effluent (influent to denitrification process) and the design final effluent characteristics. The design proceeds as follows:

1. Select the design kinetic coefficients and operational criteria. The coefficients used here were reported by Stensel *et al.* (1973) as shown in Table 7.1. These coefficients are not intended to be used universally. Pilot data should be obtained when possible.

   — Yield coefficient, $Y$ (g VSS/g COD) = 0.18;
   — Decay coefficient, $b$ ($\text{days}^{-1}$) = 0.04;
   — Half-rate constant, $K_s$ (g COD/$m^3$) = 9.1;
   — Maximum substrate removal rate, $q$ (g COD/g VSS·d) = 10.3 at 20°C;

**Table 7.2 Nitrified effluent and design final effluent characteristics.**

| Characteristic | Units | Influent | Nitrified effluent | Design final effluent |
|---|---|---|---|---|
| $NO_3^-$-N (plus $NO_2^-$-N) | mg/L | — | 17.5 | — |
| Total Kjeldahl nitrogen | mg/L | 30 | 2 | — |
| Total nitrogen | mg/L | 30 | 19.5 | 5 |
| Carbonaceous biochemical oxygen demant | mg/L | 200 | 5 | 10 |
| TSS | mg/L | 200 | 10 | 10 |
| Average flow | $m^3$/d | 38 000 | 38 000 | — |
| Minimum temperature | °C | 15 | 15 | 15 |

— Design MLSS (mg/L) = 3 500;
— VSS:TSS ratio = 0.75; and
— Temperature correction coefficient = 1.12.

2. Determine the maximum nitrate concentration allowed in the final effluent in order to stay below design effluent TN concentration. Assume soluble, nonbiodegradable organic nitrogen concentration in effluent is 1.0 mg/L. Assume total Kjeldahl nitrogen (TKN):VSS ratio in solids is 0.12 g N/g VSS. Then,

— Effluent particulate N = (TSS × VSS:TSS × 0.12) = (10 × 0.75 × 0.12) = 0.9 mg/L.
— Effluent TKN = (effluent particulate N + nonbiodegradable organic N + $NH_3$-N) = (0.9 mg/L + 1.0 mg/L + 0 mg/L) = 1.9 mg/L.
— Allowable nitrate (as nitrogen) in effluent = (design effluent TN − TKN) = (5 − 1.9) = 3.1 mg/L.

Therefore, base the design on an effluent $NO_3$-N of 3 mg/L.

3. Determine the methanol concentration that will be allowed in the anoxic part of the reactor. As a conservative estimate, a COD concentration of 5 mg/L (a biochemical oxygen demand [$BOD_5$] of approximately 3 mg/L) from added methanol will be assumed in the anoxic portion of the reactor. This is equivalent to 3.3 mg/L of methanol, based on the stoichiometric COD:methanol ratio of 1.5.
4. Calculate the organic substrate removal rate. Assume the methanol is the only rate-limiting factor:

$$q = \hat{q}\left(\frac{S_e}{K_s + S_e}\right) = 10.3\left(\frac{5}{9.1 + 5}\right) \qquad (7.7)$$

$$q = 3.65 \text{ g COD/g VSS·d at } 20°\text{C}$$

5. Calculate the substrate removal rate at the lowest expected reactor temperature:

$$q_T = q_{20}\Theta^{T-20}$$
$$q_{12} = 3.65(1.12)^{15-20} = 2.07 \text{ g COD/g VSS·d} \tag{7.8}$$

6. Calculate the minimum SRT required to sustain denitrification at this temperature:

$$\frac{1}{\text{SRT}_{\text{MIN}}} = Yq - b = (0.18)(2.07) - 0.04 = 0.333 \text{ days}$$
$$\text{SRT}_{\text{MIN}} = 3.0 \text{ days} \tag{7.9}$$

7. Determine a design SRT. This is done by multiplying the minimum SRT by an appropriate safety and peaking factor. A safety factor of 1.2 to 2.0 is adequate for most cases (Parker *et al.*, 1975). Peaking factor is based on peak loading to average loading to the process, typically 1.1 to 1.2. Select a safety factor of 1.66 and a peaking factor of 1.2:

$$\text{SRT}_{\text{Design}} = 3.0 \text{ days} \times (1.66) \times (1.2) = 6.0 \text{ days} \tag{7.10}$$

8. Calculate the methanol dose (McCarty *et al.*, 1969, and U.S. EPA, 1993). Assume the DO concentration entering the reactor is 2.0 mg/L:

$$\begin{aligned}
\text{Average } NO_3^-\text{-N removed} &= (\text{Influent } NO_3^-\text{-N}) - (\text{Effluent } NO_3^-\text{-N}) \\
&= 17.5 - 3.0 = 14.5 \text{ mg/L} \\
\text{Methanol dose} &= 2.47\ (NO_3^-\text{-N removed}) + 0.87 \text{ DO} \\
&\quad + \text{Design methanol in anoxic reactor} \\
\text{Methanol dose} &= 2.47\ (14.5) + 0.87\ (2) + 3.3 \\
&= 40.9 \text{ g methanol/m}^3
\end{aligned} \tag{7.11}$$

This is equivalent to $1.5 \times 40.9 = 61.4$ g COD/m$^3$.

9. Calculate the required anoxic HRT using a mass balance on the volatile suspended solids:

$$\begin{aligned}
\text{HRT} = \frac{V}{Q} &= \frac{Y(S_o - S_e)\text{SRT}}{(1 + b \cdot \text{SRT})\text{MLVSS}} + \frac{(\text{Inf. VSS})(\text{SRT})}{\text{MLVSS}} \\
&= \frac{0.18(61.4 - 5)(6.0)}{[1 + (0.04)(6.0)]2\,625} + \frac{(10)(0.75)(6.0)}{2\,625} \\
&= 0.018\,7 + 0.017\,1 = 0.035\,8 \text{ days} \\
&= 0.9 \text{ hours}
\end{aligned} \tag{7.12}$$

10. Select an aerobic/postaeration HRT. Use an HRT of 0.7 hours. An HRT of 0.25 to 1.0 hour is sufficient to perform the necessary functions described at the beginning of the Suspended Growth section of this chapter.
11. Calculate the volume and size of the reactors:

    — Total reactor HRT = 0.9 + 0.7 = 1.6 hours.
    — Total volume = 38 000 $m^3/d \cdot 1.6\ h / 24\ h/d$ = 2 533 $m^3$.

$$\text{Each reactor} = \frac{2\ 533\ \text{m}^3}{2} = 1\ 267\ \text{m}^3$$

$$\text{Width (each)} = \left[\frac{1\ 267\ \text{m}^3}{(3.6)(4{:}1)}\right]^{0.5} = 9.4\ \text{m} \qquad (7.13)$$

$$\text{Length (each)} = 4 \times 9.4 = 37.6\ \text{m}$$

    Choose a typical depth of 3.6 m and use two reactors operated in parallel, each being 56% anoxic and 44% aerobic and each with a length:width ratio of 4:1.
    Backmixing from the aerobic to the anoxic part of the reactor should be avoided. Baffle walls should be provided and designed to minimize backmixing.
12. Calculate the waste sludge production:

$$Y_N = \frac{Y}{1 + b\,(\text{SRT})} = \frac{0.18}{[1 + 0.04\,(6.0)]} = 0.145\,\frac{\text{g VSS}}{\text{g COD}}$$

$$\frac{0.145}{0.75} = 0.19\,\frac{\text{g TSS}}{\text{g COD}}$$

$$0.19\,\frac{\text{g TSS}}{\text{g COD}} \times 61.4\,\frac{\text{g COD}}{\text{m}^3} \times 38\ 000\,\frac{\text{m}^3}{\text{d}} \times 1\,\frac{\text{kg}}{1\ 000\ \text{g}} = 443\,\frac{\text{kg TSS}}{\text{day}} \qquad (7.14)$$

13. Design the clarifiers. Clarifier design should be based on currently accepted clarifier design procedures used for standard activated-sludge systems.

**Case History—River Oaks Advanced Wastewater Treatment Plant, Hillsborough County, Florida.** The River Oaks Advanced Wastewater Treatment Plant (AWTP), located in Hillsborough County, Florida, discharges into the ecologically sensitive Tampa Bay. The effluent limits are therefore strict. The maximum allowable annual average concentrations are 5, 5, 3, and 1 mg/L for BOD, TSS, TN, and TP, respectively. The plant was originally conservatively designed for limits of 2, 2, 1.2, and 0.4 mg/L (Tetreault and Parker, 1989).

The plant was recently upgraded from a 11 355-$m^3$/d (3-mgd) capacity to 37 850 $m^3$/d (10 mgd). As part of the upgrade, separate-stage suspended-growth denitrification with methanol was included. Typical preliminary treatment and flow equalization are followed by primary treatment, activated-sludge treatment for BOD removal and nitrification, suspended-growth denitrification with methanol addition, filtration, postaeration, and disinfection (Tetreault and Parker, 1989).

The denitrification system includes two tanks operated in parallel. Each tank is divided into sixteen 246-$m^3$ (65 000-gal) cells. This large number of cells was provided for maximum flexibility. The first 10 cells are anoxic, the next 4 can be operated anoxically or aerobically, and the last 2 cells are operated aerobically. The plant was designed for 45 420 $m^3$/d (12 mgd), although it is currently permitted for only 37 850 $m^3$/d (10 mgd). The current flow to the plant is approximately 27 252 $m^3$/d (7.2 mgd), and backwash flow from the filters is an additional 4 542 $m^3$/d (1.2 mgd). Methanol addition is controlled by a flow-proportioned feeding system (Henderson, S., River Oaks Advanced Wastewater Treatment Plant, Hillsborough County, Florida, personal communication, 1995).

Monthly plant data are shown in Table 7.3. It can be seen that the $NO_3$-N is reduced from levels of 10 to 16 mg/L down to an average 0.4 mg/L. The TN concentration of the denitrification process effluent before filtration averages 2 mg/L, while the final plant effluent TN following filtration is approximately 1.4 mg/L. It should be noted that the plant is well below its design flow and nitrogen loading, and is 63 to 79% of the design loading (Henderson, S., River Oaks Advanced Wastewater Treatment Plant, Hillsborough County, Florida, personal communication, 1995).

**ATTACHED GROWTH.** The attached-growth denitrification process may include packed-bed, fluidized-bed, moving-bed, and immobilized cell reactors (Reardon, 1993). The moving-bed and immobilized cell reactors are emerging processes with little operating experience and data available for evaluation. Therefore, this section will concentrate on the packed-bed technology, which is the most widely used process and has a long history of successful operating results.

The packed-bed process may further be classified into two types: gas filled and liquid filled (Soap and Detergent Association, 1989). In the gas-filled reactor, the enclosed vessel is used with the structured or random dumped type media as the growing site for denitrifying microorganisms. The open space is maintained with a nitrogen-enriched atmosphere that is essential for a denitrification reaction to take place. The nitrate-bearing wastewater is distributed evenly throughout the media in much the same fashion as the trickling filter process. Owing to a constant sloughing of biomass from the media during operation, a clarifier will be required to remove excess TSS from the denitrified effluent.

Table 7.3 River Oaks AWTP operating data (September 1990–July 1991).[a]

| | Plant influent | | | | | Process influent | | | | | Process effluent | | | | | Final plant effluent (following filtration) | | | | | |
|---|---|---|---|---|---|---|---|---|---|---|---|---|---|---|---|---|---|---|---|---|---|
| Month | Flow | BOD | TSS | TKN | $NH_4$ | BOD | TSS | TKN | $NH_4$ | $NO_X$ | BOD | TSS | TKN | $NH_4$ | $NO_X$ | BOD | TSS | TKN | $NH_4$ | $NO_X$ | TN |
| September 1990 | 6.96 | 167 | 183 | 31.6 | 23.7 | 2.0 | 0.5 | 1.27 | 0.11 | 13.20 | 2.0 | 7.0 | 0.93 | 0.10 | 0.56 | 1.00 | 1.00 | 0.64 | 0.10 | 0.47 | 1.12 |
| October 1990 | 7.21 | 174 | 156 | 34.0 | 23.4 | 2.0 | 0.7 | 1.59 | 0.12 | 11.37 | 2.0 | 6.0 | 0.42 | 0.10 | 0.07 | 1.00 | 0.00 | 0.73 | 0.10 | 0.10 | 0.80 |
| November 1990 | 6.66 | 212 | 173 | 34.9 | 24.5 | 3.0 | 0.8 | 1.47 | 0.13 | 16.30 | 3.0 | 9.0 | 1.04 | 0.10 | 0.33 | 1.00 | 1.00 | 0.74 | 0.10 | 0.30 | 1.01 |
| December 1990 | 7.03 | 237 | 200 | 36.5 | 28.0 | 3.0 | 0.8 | 1.56 | 0.14 | 15.90 | 4.0 | 11.0 | 2.40 | 0.10 | 0.56 | 1.00 | 1.00 | 3.08 | 0.10 | 0.58 | 1.33 |
| January 1991 | 7.54 | 215 | 185 | 33.4 | 27.3 | 3.0 | 0.1 | 6.00 | 2.39 | 13.32 | 2.0 | 6.0 | 4.62 | 2.35 | 0.71 | 1.00 | 1.00 | 3.08 | 2.28 | 0.72 | 3.80 |
| February 1991 | 7.62 | 194 | 133 | 35.6 | 19.6 | 5.6 | 10.7 | 2.65 | 0.55 | 13.36 | 4.2 | 8.0 | 1.23 | 0.15 | 0.76 | 1.00 | 0.60 | 0.88 | 0.17 | 0.83 | 1.71 |
| March 1991 | 7.23 | 213 | 167 | 38.7 | 26.2 | 3.6 | 7.4 | 2.09 | 0.26 | 14.79 | 4.3 | 8.1 | 1.36 | 0.11 | 0.48 | 0.83 | 0.55 | 0.72 | 0.10 | 0.55 | 1.21 |
| April 1991 | 6.61 | 200 | 142 | 38.0 | 26.0 | 3.7 | 4.9 | 1.64 | 0.17 | 14.54 | 3.6 | 5.2 | 1.41 | 0.10 | 0.14 | 0.87 | 0.13 | 0.75 | 0.10 | 0.13 | 0.89 |
| May 1991 | 7.23 | 173 | 146 | 37.7 | 24.7 | 2.4 | 3.7 | 1.79 | 0.18 | 12.74 | 6.0 | 5.7 | 1.28 | 0.10 | 0.19 | 0.80 | 0.32 | 0.89 | 0.10 | 0.25 | 1.14 |
| June 1991 | 7.04 | 177 | 130 | 34.5 | 23.4 | 3.2 | 3.6 | 1.53 | 0.13 | 12.39 | 4.8 | 8.2 | 1.40 | 0.10 | 0.23 | 1.09 | 0.50 | 0.87 | 0.10 | 0.34 | 1.21 |
| July 1991 | 7.77 | 167 | 127 | 31.5 | 21.5 | 2.9 | 7.1 | 1.31 | 0.14 | 10.21 | 3.9 | 5.8 | 1.04 | 0.10 | 0.24 | 1.65 | 0.40 | 0.79 | 0.10 | 0.20 | 0.98 |
| August 1991 | | | | | | | | | | | | | | | | | | | | | |
| September 1991 | | | | | | | | | | | | | | | | | | | | | |
| October 1991 | | | | | | | | | | | | | | | | | | | | | |
| November 1991 | | | | | | | | | | | | | | | | | | | | | |
| December 1991 | | | | | | | | | | | | | | | | | | | | | |
| Average | 7.17 | 194 | 158 | 35.1 | 24.4 | 3.1 | 3.7 | 2.08 | 0.39 | 13.47 | 3.6 | 7.3 | 1.56 | 0.31 | 0.39 | 1.02 | 0.59 | 0.98 | 0.30 | 0.41 | 1.38 |

[a] All values are expressed in mg/L, except flow, which is in mgd (mgd × [4.383 × $10^{-2}$] = $m^3/s$).

The liquid-filled reactor may further be divided into two categories: high-porosity and low-porosity media. The high-porosity-type reactor typically is operated in an upflow or cross-flow mode. Because a constant discharge of excess biomass is unavoidable, a subsequent clarification step such as clarifier and/or filter is required to produce a clear effluent with low TSS. In addition, a periodic backwash of the reactor will be required to prevent the reactor from clogging. The low-porosity type reactor typically is operated in a downflow mode. The use of fine media affords this type of bioreactor to achieve a dual function of denitrification and filtration. Therefore, the treated effluent will be low in TSS and oxidized nitrogen. However, the fine media will retain not only the removed suspended solids and biomass, but also the generated nitrogen gas. A periodic nitrogen gas release operation (bumping) and filter backwashing will have to be performed to relieve the buildup of head loss caused by trapped gas and solids.

The most popular and cost-effective packed-bed denitrification process is the downflow fine-media deep-bed filter for both denitrification and filtration. Therefore, this denitrification process, as shown in Figure 7.1b, will be used to illustrate the attached-growth denitrification.

**Design Criteria.** Because the denitrification filter is not just a filter for removing TSS but also a biological reactor for denitrification, some special considerations must be made in the design (Chen and Slack, 1991).

First, a nozzleless filter underdrain system should be used for the denitrification filter application. At the Howard F. Curren AWTP in Tampa, Florida, a nozzle-type filter bottom caused biofouling, clogging, and breakage that eventually led to short-circuiting of backwash air and water and loss of filter media (Pickard *et al.*, 1985). After conversion to a nozzleless underdrain, these undesirable problems were eliminated. Second, a coarse (effective size 1.8 to 2.3 mm), uniform (uniformity coefficient at 1.35), and spherical (sphericity of 0.8 to 0.9) media should be used for this application. Again, the Tampa plant experience proved that replacement of angular media with the rounded media greatly reduced the frequency of gas bumping and backwashing. Third, air and water backwash is required to keep the denitrification filter clean and free of biosolids accumulation. However, the degree of filter cleaning should be carefully controlled to avoid any long lag time of denitrification reaction when the filter is put back into operation after a vigorous backwash. Otherwise, a reseed of biomass with backwash water may be required. Fourth, the influent to the filter vessel should be introduced with the least amount of turbulence to minimize the increase of DO in the filter influent, which will affect the consumption of the carbon source used for denitrification. A specially designed weir block for the influent trough has been used successfully for this purpose.

Figure 7.2 is the cross section of a deep-bed filter and depicts the essential elements of a downflow denitrification filter designed to address the aforementioned considerations. Three parts of methanol are added per part $NO_3$-N

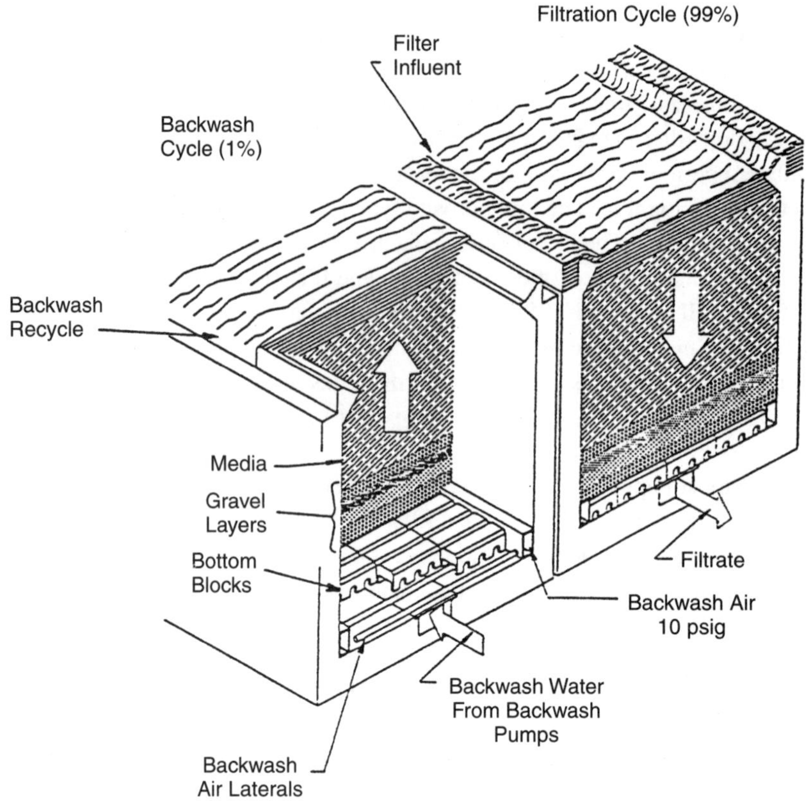


**Figure 7.2** **Cross section of a downflow denitrification filter (psi × 6 895 = Pa).**

to provide sufficient carbon for complete denitrification (Savage and Chen, 1975). If complete denitrification is not a requirement, the quantity of methanol as depicted in Figure 7.3 can be used (Redd *et al.*, 1992).

The method of distributing nitrified flow to the filter cells is important to the denitrification filter system. The HRT in every filter cell should be equal at any given time. Therefore, an even distribution of filter influent should be made to all filter vessels. The incoming flow may vary, but the flow is split evenly to every individual filter in operation. It is not desirable to use a declining rate filter system for denitrification because of the uneven filtration rate through individual filters.

In the following section, the sizing of a denitrification filter system will be discussed.

**Design Equations.** Several quantitative parameters that are important to the sizing and operation of a denitrification filter system will be discussed in this section. There are two different but related methods available to engineers for sizing a denitrification filter system, namely, the volumetric nitrogen loading and hydraulic contact time methods.

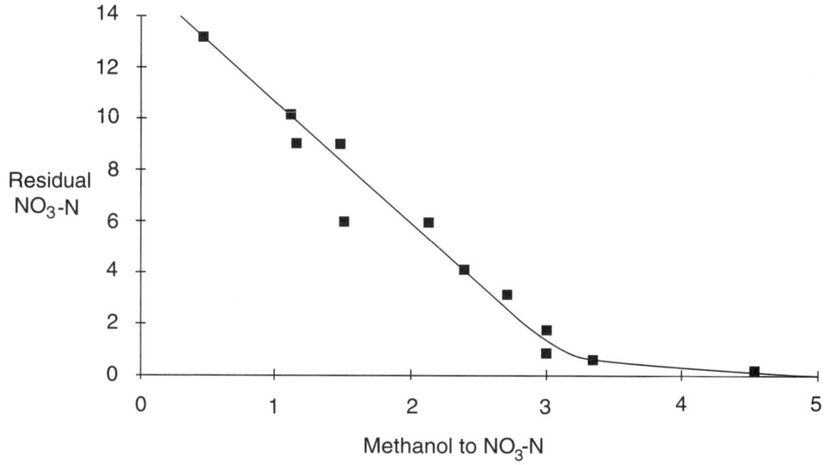


**Figure 7.3 Nitrate removal efficiency at various methanol dosages.**

The volumetric approach is probably the most typically used method of the two in that the denitrification reactor size is calculated based on an appropriate $NO_3$-N loading per unit reactor volume. It is a "rule of thumb" approach when denitrification kinetic information is lacking or a conservative design is tolerable. The range for the unit volumetric loading rate is between 240 and 3 200 kg $NO_x$-N/1 000 $m^3$/d (15 to 200 lb $NO_x$-N/1 000 cu ft/d). The selection of a loading value for reactor sizing will be dictated by the type of reactor and media used, characteristics of the nitrified water, and reaction temperature.

The hydraulic contact time method applies the principle that for a biological system to remove $NO_3$-N from water the removal efficiency is directly proportional to the HRT in the reactor until a maximum denitrification is approached when the rate of removal is leveling off. Typical denitrification design curves for a deep-bed Denite™ filter using coarse sand as the filter media are shown in Figure 7.4 (Savage, 1983). It is evident that these design curves are only applicable to a specific type of reactor or media and within a certain range of $NO_3$-N levels. It is advisable that for different types of reactors or possibly various ranges of nitrate levels, separate design curves may have to be derived from a pilot study for their specific denitrification applications.

Sometimes the size of a denitrification filter system will not be determined by the reaction kinetics but by the hydraulic limitation according to the site-specific regulations. For example, in California, the Title 22 regulations require that the maximum filtration rate for a water reclamation filter system should not be more than 294 $m^3/m^2 \cdot d$ (5 gpm/sq ft) with one of the filter cells out of service (State of California, 1978). The kinetic rate may allow a filtration rate to be more than 294 $m^3/m^2 \cdot d$ (5 gpm/sq ft); however, the Title 22 regulations will dictate the size of the filter system.

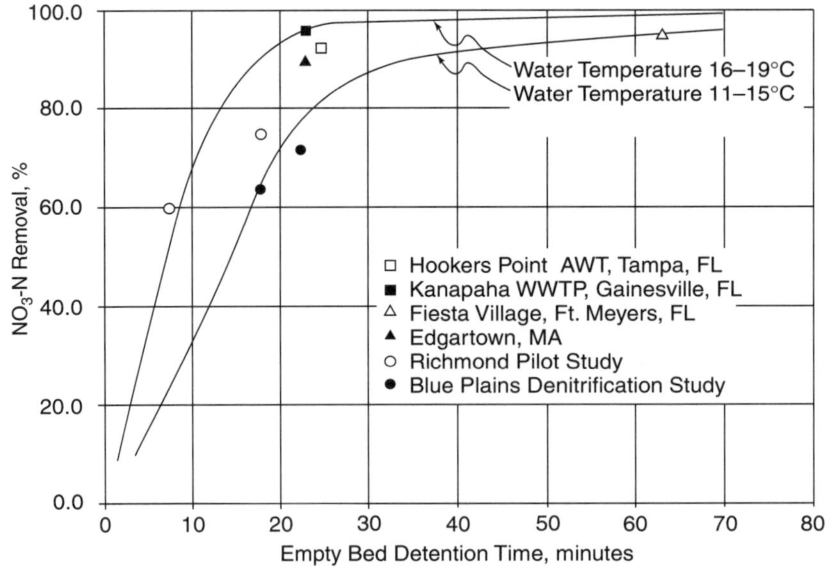


**Figure 7.4 Typical design curves using hydraulic contact time for a downflow deep-bed filter (Savage, 1983).**

Other factors that may affect the sizing of a filter system are the limitation of the filter backwashing and gas bumping frequency. Although backwashing and bumping operations will not greatly affect denitrification efficiency of a deep-bed filter, in practice, the frequency of filter backwashing should be limited to no more than once per day and gas bumping to no more than once per hour (Savage and Chen, 1972). For a deep-bed denitrification filter, after the solids loading exceeds 5 to 10 kg/m$^2$/cycle (1 to 2 lb/sq ft/cycle) the filter will require a backwashing. When the $NO_3$-N removal exceeds 0.25 to 0.5 kg/m$^2$/cycle (0.05 to 0.1 lb/sq ft/cycle), the filter will need a bumping. When treating a nitrified stream with high TSS and/or nitrate concentrations, filter sizing should be reviewed with these limitations in mind.

**Design Example.** A design example will be used to illustrate the application of the design equations for a complete denitrification process. Table 7.4 shows the influent characteristics and effluent requirements for the design examples.

The design proceeds as follows:

1. Select the appropriate design parameters:

   — Solids loading/backwash (BW) cycle = 7.3 kg/m$^2$ (1.5 lb/sq ft).
   — $NO_3$-N loading/bumping cycle = 0.3 kg/m$^2$ (0.06 lb/sq ft).

**Table 7.4 Influent characteristics and effluent requirements for denitrification by deep-bed filter, design example.**

| Parameter | Nitrified influent | Effluent requirements |
|---|---|---|
| Minimum water temperature | 15°C | 15°C |
| Average flow, $m^3/d$ | 38 000 | 38 000 |
| mgd | 10.0 | 10.0 |
| Peak flow, $m^3/d$ | 57 000 | 57 000 |
| mgd | 15.0 | 15.0 |
| TSS, mg/L | 10 | 5 |
| CBOD,[a] mg/L | 5 | 10 |
| COD, mg/L | 35 | 50 |
| $NO_x$-N, mg/L | 17.5 | 3 |
| $NH_3$-N, mg/L | 1 | 1 |
| TKN, mg/L | 2 | 2 |
| TN, mg/L | 19.5 | 5 |

[a] Carbonaceous BOD.

- Biomass synthesis = 0.53 kg/kg $NO_3$-N removed (McCarty *et al.*, 1969).
- Methanol: $NO_3$-N = 3:1 for complete denitrification.
- Initial backwash water rate = 14.6 m/h (6 gpm/sq ft) (for 15 minutes).
- Final backwash water rate = 19.5 m/h (8 gpm/sq ft) (for 5 minutes).
- Backwash air rate = 1.52 m/min (5 cfm/sq ft) (for 15 minutes).

2. Complete denitrification:

   a. Calculate filter area required based on hydraulic contact time:

      - $NO_3$-N removal = (17.5 mg/L – 3 mg/L)/17.5 mg/L = 0.83 (83%).
      - Empty bed detention time for 83% $NO_3$-N removed = 24 minutes (from Figure 7.4).
      - Use 1.83 m sand media.
      - Filtration rate = (1.83 m × 60)/24 = 4.58 m/h (1.9 gpm/sq ft).
      - Denitrification area required = (38 000/24) × (1/4.58) = 346 $m^2$ (3 725 sq ft).
      - Use five filter cells each approximately 364/5 = 69.2 $m^2$ (745 sq ft).

— Use 3 m × 23 m (10 ft × 74.5 ft).

— Check maximum filtration rate with one filter cell out of service:

$$\frac{1.5(38\ 000)}{24} \times \frac{1}{4 \times (3 \times 23)} = 8.6 \text{ m/h (3.5 gpm/sq ft)}$$
$$< 12.2 \text{ m/h (5 gpm/sq ft)}$$

— Check backwash and bumping frequencies:

TSS removed = (10 − 5) × 38 000/1 000 = 190 kg/d (418 lb/d)
$NO_3$-N removed = (17.5 − 3) × 38 000/1 000 = 551 kg/d (1 212 lb/d)
Biomass yield = 0.53 × 551 = 292 kg/d (642 lb/d)
Total solids loading = 190 + 292 = 482 kg/d (1 060 lb/d)
Average specific loading = 482/346 = 1.39 kg/m$^2$·d (0.28 lb/sq ft/d)
Period between backwashings = 7.3/1.39 = 5.3 days > 1 day
Period between bumpings = (0.3 × 346/551 = 0.188 day = 4.5 hours > 1 hour

— Use five filter cells, each 3 m × 23 m (10 ft × 74.5 ft).

b. Calculate filter backwashing requirements:

— Backwashing a deep-bed filter requires 15 minutes of combined air and water backwashing with a water rate at 14.6 m/h (6 gpm/sq ft) and air rate at 1.52 m/min (5 cfm/sq ft), followed by 5 minutes of water rinse at 19.5 m/h (8 gpm/sq ft).

— Backwash air blower at 1.52 m/min × 69.2 m$^2$ = 105.2 m$^3$/min (3 720 cfm).

— Total backwash (BW) water used per filter backwash is

$$69.2 \times \left(\frac{14.6}{60} \times 15 + \frac{19.5}{60} \times 5\right) = 365 \text{ m}^3\text{/BW (96 500 gal/BW)} \tag{7.15}$$

— Backwash water as percent of filter throughput is

$$\left(365 \text{ m}^3\text{/BW} \times \frac{1 \text{ BW}}{5.3 \text{ day}}\right) \div \frac{38\ 000}{5} \text{ m}^3\text{/d} = 0.009 \text{ or } 0.9\% \tag{7.16}$$

— The cost of backwash water treatment = \$0.75/1 000 gal × 10 000 000 gal × 0.009 × 365 = \$24 600/year.

- Assuming a capitalization rate of 5% and a useful life of 20 years, the present worth factor will be 12.46.
- The present worth of the backwash water treatment will be \$24 600 × 12.46 = \$307 000, or \$341 000 for each percentage increase in backwash water rate.

c. Calculate mudwell and clearwell sizes. Normally, the clearwell will be sized to hold sufficient water for a minimum of one backwashing and the mudwell for 1.5 backwashings:

Clearwell size = 365 $m^3$ (96 500 gal)

Mudwell size = 365 × 1.5 = 548 $m^3$ (145 000 gal)

d. Calculate methanol feed rate and storage:

Feed rate—use methanol to $NO_3$-N ratio at 3:1
3 × 551 kg/d = 1 653 kg/d (3 636 lb/d)
1 L/0.79 kg × 1 653 = 2 092 L/d (553 gpd)
Storage—use 30 days storage capacity
30 × 2 092 = 62 760 L (16 600 gal)
Methanol cost − \$0.70/gal × 553 gal/d = \$387/day = 387 × 365 = \$141 300/year

**Case Histories.** *CASE I: HOWARD F. CURREN ADVANCED WASTEWATER TREATMENT PLANT, TAMPA, FLORIDA.* The Tampa-St. Petersburg area in Florida has been among the fastest growing regions in the U.S. Concerned with eutrophication problems of Hillsborough Bay associated with this dynamic growth, the city initiated an upgrade and expansion program for the Hooker's Point WWTP (now Howard F. Curren AWTP) in the early 1970s, including nitrogen removal (Chen and Slack, 1991). Completed with a two-stage, high-purity oxygen aerobic treatment and Denite™ filters, the AWTP can process an average wastewater flow of 227 100 $m^3$/d (60 mgd) with more than 90% nitrogen removal. It is probably the largest municipal treatment plant in the world with such stature. Another plant expansion was completed in 1990 to further increase the treatment capacity to 363 360 $m^3$/d (96 mgd).

As shown in Figure 7.5, the pretreated wastewater flow is first processed by the first-stage oxygen activated sludge with an HRT of 1 hour for carbonaceous BOD removal, followed by a second-stage oxygen activated-sludge process with an HRT of 2 hours for nitrification. Flexibility is designed so that the two-stage operation may be converted to a single-stage system where carbonaceous oxidation and nitrification will occur in the same reactor.

After aerobic treatment the nitrified element fortified with methanol as the carbon source is filtered through 32 Denite™ filters for denitrification and

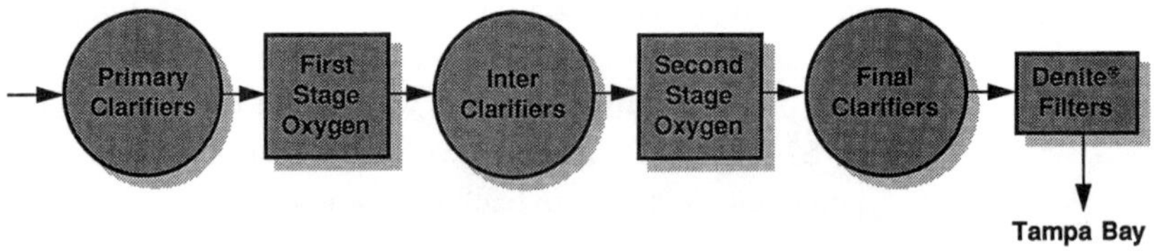


**Figure 7.5** **Howard F. Curren AWTP, Tampa, Florida (363 000 m³/d [96 mgd]) (1978).**

TSS removal. Each Denite™ filter has a size of 3 m wide × 32 m long (10 ft wide × 105 ft long). The denitrification reaction happens very fast, with an empty bed detention time of approximately 20 minutes.

This AWTP was designed to remove 90% of the $BOD_5$, suspended solids, TN, and total phosphate. However, the requirement of phosphate removal was waived a year after start-up because it demonstrated that nitrogen nutrient was the limiting growth factor in the bay ecological system. Since the start-up of the AWTP in January 1978, the performance of the WWTP has been exceeding the original goal. As indicated in Table 7.5, the filtered effluent has consistently met the permit requirements of 5 mg/L $BOD_5$, 5 mg/L TSS, and 3 mg/L TN limit. The important fact is that the bay has rebounded from eutrophication, with great biodiversity detected as early as 1989, in just 10 years (*Tampa Tribune,* 1992).

**Table 7.5** **Howard F. Curren AWTP effluent analysis (1980–1993).**

| | Flow, mgd[a] | BOD | TSS | TN | $NO_X$ | MEOH: $NO_2$-N |
|---|---|---|---|---|---|---|
| 1980 | 45.01 | 4.5 | 3.0 | 2.7 | 0.8 | 3.1 |
| 1981 | 47.50 | 2.0 | 2.2 | 2.5 | 1.05 | 3.1 |
| 1982 | 52.49 | 4.8 | 2.7 | 2.8 | 1.23 | 2.9 |
| 1983 | 55.61 | 4.6 | 3.9 | 2.9 | 0.9 | 3.0 |
| 1984 | 49.87 | 2.6 | 2.5 | 2.4 | 0.86 | 3.1 |
| 1985 | 45.71 | 2.2 | 2.6 | 2.6 | 0.83 | 3.2 |
| 1986 | 51.16 | 2.2 | 2.4 | 2.6 | 1.10 | 3.0 |
| 1987 | 58.18 | 3.5 | 2.7 | 3.0 | 1.14 | 2.60 |
| 1988 | 56.82 | 4.2 | 3.3 | 3.5 | 1.66 | 2.60 |
| 1989 | 56.92 | 2.6 | 1.5 | 2.9 | 1.12 | 2.67 |
| 1990 | 52.86 | 2.6 | 1.9 | 2.6 | 0.83 | 3.2 |
| 1991 | 54.36 | 2.1 | 1.6 | 2.3 | 0.74 | 3.19 |
| 1992 | 56.27 | 2.5 | 1.9 | 2.6 | 0.83 | 2.74 |
| 1993 | 52.46 | 2.6 | 1.8 | 2.5 | 0.84 | 2.74 |
| Average | 52.5 | 3.1 | 2.4 | 2.7 | 1.00 | 2.94 |

[a] mgd × $(4.383 \times 10^{-2})$ = m³/s.

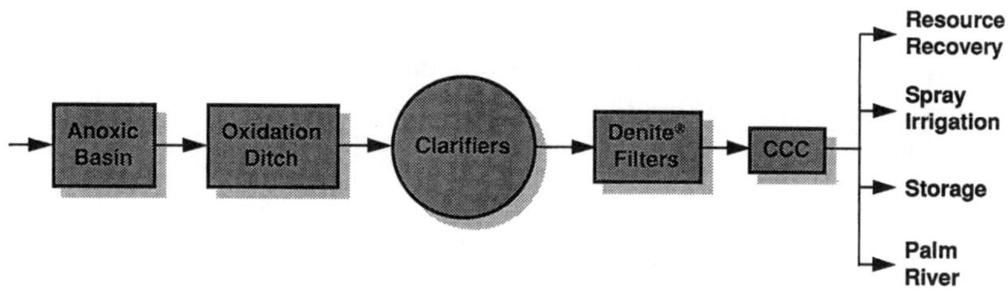


**Figure 7.6 Falkenburg Road plant, Brandon, Florida (23 000 m$^3$/d [6 mgd]) (1988).**

*CASE II: FALKENBURG ROAD ADVANCED WASTEWATER TREATMENT PLANT, HILLSBOROUGH COUNTY, FLORIDA.* The Falkenburg Road AWTP (Chen and Slack, 1991) is a regional WWTP in Hillsborough County, Florida, that uses Denite™ filters for suspended solids removal and denitrification. The original treatment plant, with a capacity of 11 355 m$^3$/d (3 mgd) was operational in April 1988. In May 1990, the plant was expanded to a treatment capacity of 22 710 m$^3$/d (6 mgd).

As shown in Figure 7.6, the treatment system consists of two oxidation ditches including anoxic treatment (HRT approximately 4.8 hours) and aerobic treatment (HRT approximately 19.2 hours), two clarifiers each at 30.5 m diameter (100 ft diameter), five Denite™ filters each at 3 m wide × 15.2 m long (10 ft wide × 50 ft long), and chlorination–dechlorination. The treatment facility can process an average wastewater flow of 22 710 m$^3$/d (6 mgd). Extensive reuse of the treated wastewater has been practiced for cooling makeup water at the neighboring county resource recovery power plant, for spray irrigation of landscape, and for other uses. The excess treated effluent is discharged to the Palm River.

The performance of the treatment plant has been excellent as indicated in Table 7.6. The 4-year results show an effluent consistently meeting the AWTP standards of 5:5:3:1 for $BOD_5$:TSS:TN:TP.

**Table 7.6 Falkenburg road AWTP yearly results.**[a]

| Year | BOD | TSS[b] | TN | $NO_3^-$ | TP |
|---|---|---|---|---|---|
| 1990 | 3.14 | 0.90 | 1.94 | 1.02 | 0.79 |
| 1991 | 1.10 | 0.51 | 1.70 | 0.80 | 0.47 |
| 1992 | 1.10 | 0.21 | 1.15 | 0.38 | 0.23 |
| 1993 | 1.36 | 0.43 | 1.17 | 0.33 | 0.29 |
| Average | 1.68 | 0.51 | 1.49 | 0.63 | 0.45 |

[a] Denite used for polishing.
[b] Inlet TSS ≤ 15 ppm.

In addition, the Denite™ filters produce a clear effluent low in turbidity, typically below 0.5 nephelometric turbidity units. The backwash water usage has been less than 2% of the filter throughput on average, which demonstrates the production efficiency of the deep-bed filter system.

# SINGLE-SLUDGE NITRIFICATION–DENITRIFICATION

**GENERAL CHARACTERISTICS.** This section discusses single-sludge nitrification–denitrification suspended-growth systems. Single-sludge systems are those in which nutrient removal is achieved in a single basin and clarifier. The basin may be divided into several zones to achieve anaerobic or anoxic conditions depending on the treatment goal. Several configurations have been used including predenitrification, postdenitrification, and combined pre- and postdenitrification systems. The more commonly used single-sludge processes are presented in *Design of Municipal Wastewater Treatment Plants* (WEF, 1992). Various design procedures have been developed for designing nitrogen removal systems (Bidstrup and Grady, 1988; Randall *et al.*, 1992; Soap and Detergent Association, 1989; TREEO, 1988; U.S. EPA, 1993; and WEF, 1992). The objective of these processes is the removal of nitrogen from the wastewater biologically. As discussed in Chapter 6, a nitrifying population of microorganisms must first be established for the oxidation of ammonium-nitrogen ($NH_4^+$-N) to nitrate-nitrogen ($NO_3^-$-N). A pH range of 6.5 to 8.0 should be maintained in the aeration basin to prevent nitrifier inhibition (Parker *et al.*, 1975). As presented in Chapter 6, the minimum required SRT must be calculated, and a total processing peaking factor of 1.5 to 2.0 should be used to ensure that nitrification can be achieved throughout the year. As with any type of biological process, a significant amount of nitrogen is removed from the wastewater through incorporation into the biomass. Based on the empirical formula $C_5H_7O_2N$ for biomass, nitrogen makes up approximately 12% by weight (Hoover and Porges, 1952). Therefore, the maximum amount of nitrogen that can be removed by wasting of the activated sludge ranges from approximately 7.4 mg/L for high-rate activated-sludge systems to approximately 3.7 mg/L for extended aeration facilities based on the TN influent loading.

To meet stringent effluent requirements for TN removal, biological denitrification systems using one or more anoxic zones are required. A single-sludge system employing one anoxic zone can achieve an effluent TN concentration of 4 to 11 mg/L as nitrogen. Two or more anoxic zones are required to reduce the TN concentration in the effluent from 1 to 3 mg/L as nitrogen (Morales *et al.*, 1991; Soap and Detergent Association, 1989; and

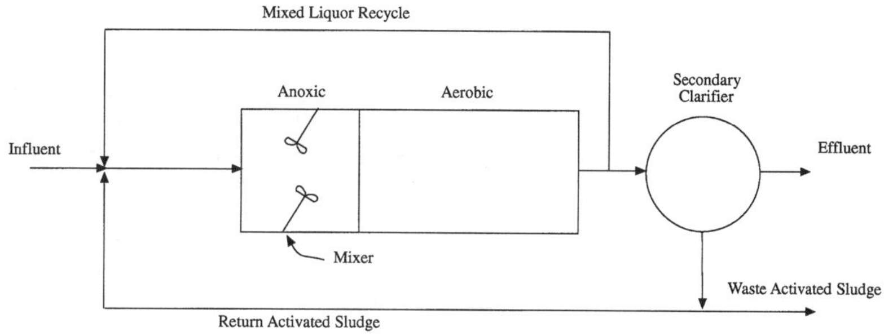


**Figure 7.7 Predenitrification/nitrification system.**

U.S. EPA, 1993). Oxidation ditches preceded by anaerobic/anoxic selectors have consistently achieved effluent TN concentrations of less than 3 mg/L as nitrogen (Mines and Woods, 1994). Typically, supplemental carbon is only needed when postdenitrification systems are used.

**PREDENITRIFICATION SYSTEM.** This section presents design criteria, design equations, a design example, and a case history for predenitrification systems. Figure 7.7 is a schematic of a typical predenitrification system. For the predenitrification (anoxic/oxic) mode of operation, an anoxic zone precedes the aerobic or oxic zone. The carbonaceous and nitrogenous demand of the wastewater is primarily met in the aerobic zone. After the nitrates have been formed, mixed liquor must be recycled back to the anoxic zone to allow the denitrifiers to accomplish denitrification. Predenitrification systems are the most economical single-sludge systems to operate because a supplemental carbon source typically is not required. An exogenous carbon source, such as methanol, may have to be added to the wastewater entering the predenitrification zone, should the raw wastewater have a low $BOD_5$. Well-known single-sludge processes such as sequencing batch reactors and Bio-Denitro™ can be used for removing nitrogen from wastewater. Designing these processes is similar to designing predenitrification, single-sludge systems. Design criteria for these systems are presented elsewhere (U.S. EPA, 1993, and WEF, 1992).

**Design Criteria.** Design criteria used in sizing predenitrification systems are presented in Table 7.7. The size of the anoxic zone depends on the amount of nitrogen to be removed from the wastewater and the denitrification rate of the denitrifiers. For most applications, the anoxic zone represents approximately 20 to 40% of the total volume (anoxic plus aerobic). Equations for sizing the anoxic zone will be presented in the next section of this chapter. It is imperative that the biomass in the anoxic zone be kept in suspension without entraining DO, which inhibits the process. The mixed liquor recycle (MLR) pump intakes should be located in such a manner as to reduce the return of DO to the anoxic zone.

Table 7.7 **Design criteria for nitrification/denitrification systems.**

| Parameter | Pre-denitrification | Post-denitrification | Combined pre- and post-denitrification |
|---|---|---|---|
| System SRT, days | 6–30 | 6–30 | 8–40 |
| HRT, hours | | | |
| First anoxic zone | 2–5 | — | 2–5 |
| First aerobic zone | 4–12 | 4–12 | 4–12 |
| Second anoxic zone | — | 2–5 | 2–5 |
| Reaeration zone | — | — | 0.5–1.0 |
| MLSS, mg/L | 1 500–4 000 | 1 500–4 000 | 2 000–5 000 |
| Mixed liquor recycle | (2–4) *Q* | — | (4–6) *Q* |
| RAS[a] flow | 100% (*Q* design) | 100% (*Q* design) | 100% (*Q* design) |
| DO, mg/L | | | |
| Anoxic zones | 0 | 0 | 0 |
| Aerobic zones | 1–2 | 1–2 | 1–2 |
| Mixing Requirements | | | |
| Anoxic zones, $kw/10^3\ m^3$[b] | 4–10 | 4–10 | 8–13 |
| Aerobic zones, $kw/10^3\ m^3$ | 20–40 | 20–40 | 20–40 |
| Aerobic zones, $m^3/min \cdot 10^3\ m^3$ | 10–30 | 10–30 | 10–40 |

[a] Return activated sludge.
[b] Energy input is an important parameter; however, the manufacturer should be consulted for determining the number and placement of mixers. Propeller and turbine mixers have been used successfully.

**Design Equations.** There are two procedures that can be used for designing prenitrification–denitrification systems. Each method uses a similar approach for sizing the aerobic treatment zone, whereas the sizing of the anoxic zone uses different approaches. The first method is that presented in the U.S. Environmental Protection Agency's (U.S EPA's) *Design Manual for Phosphorus Removal* (Bowker and Stensel, 1987) and is similar to the procedure presented in *Design of Municipal Wastewater Treatment Plants* (WEF, 1992), whereas the second approach bases the sizing of the anoxic zone on matching the respiration requirements (oxygen equivalents) of the denitrifiers with the oxygen equivalents supplied by the nitrates produced during nitrification that are recycled back to the anoxic zone. The details of the second approach are presented elsewhere (Mines *et al.*, 1992). The procedure used in sizing the aerobic or oxic portion of the biological process is based on the model developed by Lawrence and McCarty (1970). Examples are presented in *Wastewater Engineering: Treatment, Disposal and Reuse* (Metcalf & Eddy, Inc., 1991). These examples fail to account for inert solids that enter the biological system, which have a tremendous effect on the sizing of the aerobic zone. In the design examples presented in this chapter, 60 mg/L of inert solids is assumed to enter the biological system. Biokinetic constants used in the examples were taken from Lawrence and McCarty (1970), Metcalf &

Eddy, Inc. (1991), and WEF (1992). Temperature correction coefficients ($\Theta$) were taken from WEF (1992). The procedure for the conventional approach is presented below.

*AEROBIC ZONE CALCULATIONS.* The aerobic zone calculations are as follows:

1. Select minimum weekly temperature, minimum pH, and minimum DO level to be maintained.
2. Calculate the maximum growth rate of *Nitrosomonas* using the following equation developed from the data of Knowles *et al.* (1965) corrected for pH and DO as presented in U.S. EPA's *Process Design Manual for Nitrogen Control* (Parker *et al.*, 1975):

$$(\mu_{max})_{NS} = 0.47e^{0.098(T-15)}\left[\frac{DO}{K_{DO} + DO}\right][1 - 0.833(7.2 - pH)] \quad (7.17)$$

Where

$T$ = mixed liquor temperature, °C;
$(\mu_{max})_{NS}$ = maximum growth rate of *Nitrosomonas*, day$^{-}$1;
pH = pH of mixed liquor, standard pH units;
DO = DO level in aeration basin or oxic zone, mg/L;
and
$K_{DO}$ = half-saturation constant for DO, mg/L ($K_{DO}$ = 1.0 mg/L [Bidstrup and Grady, 1988]).

3. Calculate the minimum SRT required for a given temperature, DO, pH, and influent ammonia concentration using the following equation:

$$\frac{1}{(SRT)_{MIN}} = \frac{(\mu_{max})_{NS}(NH_4^+\text{-N})_0}{K_{NS} + (NH_4^+\text{-N})_0} - b_{NS} \quad (7.18)$$

Where

$(SRT)_{MIN}$ = minimum solids retention time required for nitrification based on the *Nitrosomonas* growth rate, days$^{-1}$;
$(NH_4^+\text{-N})_o$ = influent ammonium nitrogen concentration, mg/L;
$b_{NS}$ = endogenous decay coefficient for *Nitrosomonas*, days$^{-1}$;
$K_{NS}$ = half-saturation constant for ammonium-nitrogen, mg/L; and
$K_{NS}$ = $10^{0.015\,1T-1.148}$ (Knowles *et al.*, 1965).

4. Calculate the design SRT based on a safety factor and peaking factor as follows:

$$(\mathrm{SRT})_{\mathrm{DESIGN}} = (\mathrm{SRT})_{\mathrm{MIN}}(\mathrm{SF})(\mathrm{PF}) \tag{7.19}$$

Where

$(\mathrm{SRT})_{\mathrm{DESIGN}}$ = design SRT for aerobic or oxic reactor, days;
SF = safety factor based on uncertainty of performance, typically 1.2 to 2.0; and
PF = peaking factor based on the peak nitrogen loading to average loading to the treatment unit, typically 1.1 to 1.2.

5. Estimate the overall SRT of the biological system using the following equation:

$$(\mathrm{SRT})_{\mathrm{OVERALL}} = (\mathrm{SRT})_{\mathrm{DESIGN}}(\mathrm{MF}) \tag{7.20}$$

Where
$(\mathrm{SRT})_{\mathrm{OVERALL}}$ = The overall SRT of the biological system, days; and
MF = Multiplication factor to account for the anoxic volume which will increase the SRT of the system. For most BNR systems, this parameter will vary from approximately 1.4 to 2.0. The multiplication factor is calculated as follows:

$$\mathrm{MF} = \frac{1}{1 - Z_1} \tag{7.21}$$

where $Z_1$ is the preanoxic zone fraction of total reactor volume.

6. For a given set of heterotrophic biokinetic constants, calculate the effluent soluble $BOD_5$ for the $(\mathrm{SRT})_{\mathrm{OVERALL}}$ as follows:

$$S_e = \frac{K_s[1 + b(\mathrm{SRT})_{\mathrm{OVERALL}}]}{[(\mathrm{SRT})_{\mathrm{OVERALL}}(Yk - b) - 1\,]} \tag{7.22}$$

Where

$S_e$ = effluent soluble $BOD_5$, mg/L;
$b$ = endogenous decay coefficient for organic removal, $\text{days}^{-1}$;

$Y$ = heterotrophic yield coefficient, g VSS/g $BOD_5$;
$k$ = maximum substrate ($BOD_5$) utilization rate, $days^{-1}$; and
$K_S$ = half-rate constant, mg/L.

7. For a given set of autotrophic biokinetic constants, calculate the effluent ammonium-nitrogen concentration using the following equation:

$$(NH_4^+\text{-N})_e = \frac{K_N[1 + b_{NS}(SRT)_{DESIGN}]}{(SRT)_{DESIGN}[Y_{NS}\, k_{NS} - b_{NS}] - 1} \quad (7.23)$$

Where

$(NH_4{}^+\text{-N})_e$ = effluent ammonium nitrogen concentration, mg/L;
$Y_{NS}$ = *Nitrosonomas* yield coefficient, VSS/g $NH_4{}^+$-N

8. Determine the amount of nitrogen to be oxidized as follows:

$$NO = TKN_o - (NH_4^+)_e - N_{SYN} \quad (7.24)$$

Where
NO = influent nitrogen converted to oxidized nitrogen, mg/L;
$TKN_o$ = influent total Kjeldahl nitrogen, mg/L;
$(NH_4{}^+\text{-N})_e$ = effluent ammonium nitrogen concentration, mg/L; and
$N_{SYN}$ = influent nitrogen used in synthesis of biomass, mg/L.

The nitrogen used in synthesis neglecting the small amount of nitrogen synthesized by the nitrifiers is calculated using the following equation:

$$N_{SYN} = \frac{Y(S_o - S_e)F_N}{[1 + b(SRT)_{OVERALL}]} + (X_e)F_N \quad (7.25)$$

Where

$X_e$ = effluent VSS concentration, mg/L;
$F_N$ = fraction of nitrogen in volatile suspended solids, fraction; and
$F_N$ = 5 to 12%.

9. The volume of the aerobic zone (neglecting synthesis by the nitrifiers) necessary to achieve nitrification for a given temperature, pH, and DO is calculated as follows:

$$V_{OXIC} = \frac{Q(\text{SRT})_{DESIGN}}{X}\left[\frac{Y\,(S_o - S_e)}{1 + b(\text{SRT})_{DESIGN}} + X_L\right] \quad (7.26)$$

Where

$V_{OXIC}$ = volume of aerobic or oxic zone, $m^3$;
$S_o$ = influent $BOD_5$ to biological treatment system, mg/L;
$S_e$ = effluent $BOD_5$ from biological treatment system, mg/L;
$X$ = MLSS concentration, mg/L;
$Q$ = volumetric flow rate into the biological treatment process, $m^3/d$;
FSS = fixed suspended solids, mg/L; and
$X_L$ = Inert solids in the influent (FSS + nondegradable VSS), mg/L.

*ANOXIC ZONE CALCULATIONS.* Anoxic zone calculations are as follows:

1. Estimate the nitrate nitrogen concentration in the effluent and MLR returned to the anoxic zone using the following equation:

$$\text{N} = \frac{\text{NO}(Q)}{(\text{MLR} + \text{RAS} + Q)} \quad (7.27)$$

Where

N = nitrate-nitrogen concentration in the effluent and MLR, mg/L;
MLR = mixed liquor recycle flow, $m^3/d$;
RAS = return activated sludge flow, $m^3/d$; and
$Q$ = influent wastewater flow rate, $m^3/d$.

Alternatively, the RAS and MLR flows can be determined for a specified effluent nitrate concentration.

2a. Estimate the mass of nitrates to be removed in the anoxic zone. First, calculate the nitrate equivalence of DO in the MLR as follows:

$$(NO_3^- \text{-N})_{EQ} = (\text{DO})_{MLR}\left(0.35\,\frac{\text{g } NO_3^- \text{-N}}{\text{g } O_2}\right)(\text{MLR})\left(\frac{1 \text{ kg}}{1\,000 \text{ g}}\right) \quad (7.28)$$

where $(NO_3^- \text{-N})_{EQ}$ = nitrogen equivalence of DO in the MLR, kg/d.

2b. Next, calculate the mass of nitrates to be removed in the anoxic zone to achieve a specific nitrate limit in the effluent using the following equation:

$$\text{NOR} = [(\text{RAS} + \text{MLR})\text{N}]\left(\frac{1\ \text{kg}}{1\ 000\ \text{g}}\right) \tag{7.29}$$

where NOR = mass of nitrates to be removed in the anoxic zone, kg/d.

2c. Calculate the total mass of nitrates to be removed in the anoxic zone as follows:

$$\text{TNOR} = \text{NOR} + (\text{NO}_3^-\text{-N})_{\text{EQ}} \tag{7.30}$$

where TNOR = mass of nitrates to be removed in the anoxic zone, kg/d.

3. Estimate the anoxic volume required to remove the mass of nitrates and nitrate equivalence of DO calculated in step 2c, using the following equation:

$$V_{\text{ANOXIC}} = \frac{\text{TNOR}(1\ 000\ \text{g/kg}) - 0.03(S_o)(Q)(1.06)^{T-20}}{0.029(X)(1.06)^{T-20}} \tag{7.31}$$

where $V_{\text{ANOXIC}}$ = anoxic volume, m$^3$.

4. The quantity of sludge that must be wasted from this system can be calculated using the following equation, which neglects the small amount of biomass synthesized by the nitrifiers:

$$P_x = \left[\frac{Y(S_o - S_e)}{1 + b\ (\text{SRT})_{\text{OVERALL}}} + X_L\right](Q)\left(\frac{1\ \text{kg}}{1\ 000\ \text{g}}\right) \tag{7.32}$$

where $P_X$ = waste sludge production, kg/d.

5. Calculate the overall SRT of the biological system using the following equation:

$$(\text{SRT})_{\text{OVERALL}} = \frac{X(V_{\text{OXIC}} + V_{\text{ANOXIC}})}{P_x} \tag{7.33}$$

where $(\text{SRT})_{\text{OVERALL}}$ = overall SRT of the biological system, days.

If the calculated $(\text{SRT})_{\text{OVERALL}}$ is not within 5% of the assumed $(\text{SRT})_{\text{OVERALL}}$ in Step 5 of the aerobic zone calculations, the entire design procedure must be repeated.

*OXYGEN REQUIREMENTS.* Oxygen is required to meet both the carbonaceous and nitrogenous oxygen demands. The total kilograms of oxygen required can be estimated using the following equations:

$$O_2 = \text{CBOD} + \text{NOD} - \text{DOC} \tag{7.34}$$

$$\text{CBOD} = \{Q[(1 - 1.42Y)(S_o - S_e)] + 1.42\,(b)\,(X)V_{\text{OXIC}}\} \times \left(\frac{1\ \text{kg}}{1\ 000\ \text{g}}\right) \tag{7.35}$$

$$\text{NOD} = Q(4.57)(\text{NO})\left(\frac{1\ \text{kg}}{1\ 000\ \text{g}}\right) \tag{7.36}$$

$$\text{DOC} = Q\left(2.86\,\frac{\text{g O}_2}{\text{g NO}_3^-\text{-N}}\right)[\text{NO} - (\text{NO}_3^-\text{-N})_e]\left(\frac{1\ \text{kg}}{1\ 000\ \text{g}}\right) \tag{7.37}$$

Where

$O_2$ = total quantity of oxygen that must be supplied to the oxic zone, kg/d;
CBOD = carbonaceous oxygen demand, kg/d;
NOD = nitrogenous oxygen demand, kg/d; and
DOC = denitrification oxygen credit, kg/d.

Best design practice is to ignore the denitrification oxygen credit.

*ALKALINITY REQUIREMENTS.* Sufficient alkalinity must be maintained so that the pH does not drop during nitrification, thereby inhibiting the process. Typically, 50 to 100 mg/L of alkalinity as calcium carbonate ($CaCO_3$) is maintained in the effluent from biological nutrient removal (BNR) systems. The effluent alkalinity can be calculated as follows from the anoxic/oxic process:

$$\text{ALK}_e = \text{ALK}_o - 7.14\,(\text{NO}) + 3.57\,[\text{NO} - (\text{NO}_3^- - \text{N})_e] \tag{7.38}$$

Where

$\text{ALK}_e$ = effluent alkalinity, mg/L as $CaCO_3$; and
$\text{ALK}_o$ = influent alkalinity, mg/L as $CaCO_3$.

**Predenitrification Design Example.** A predenitrification/nitrification WWTP is to be sized based on the information presented in Table 7.8. Effluent TSS is assumed to be 20 mg/L and effluent nitrate-nitrogen concentration to range from 4 to 10 mg/L.

Table 7.8 **Wastewater characterization for design examples.**[a]

| **Influent** | | **Mixed liquor** | |
|---|---|---|---|
| $Q$ | = 38 000 $m^3/d$ | DO | = 2 mg/L |
| pH | = 7.1 | MLSS | = 3 500 mg/L |
| Minimum temperature | = 15°C | MLVSS | = 0.75 MLSS |
| | | $K_{DO}$ | = 1.0 mg/L |
| TKN | = 30 mg/L | MLSS | = 2 625 mg/L |
| Alkalinity | = 200 mg/L | | |
| $BOD_5$ | = 200 mg/L | | |
| Inerts | = 60 mg/L | | |
| **Heterotrophic biokinetic constants at 20°C** | | | |
| $Y$ | = 0.6 g VSS/g $BOD_5$ | $b$ | = 0.06 $days^{-1}$ |
| $k$ | = 5 $days^{-1}$ | $K_s$ | = 60 mg/L $BOD_5$ |
| ***Nitrosomonas* biokinetic constants at 20°C** | | | |
| $Y_{NS}$ | = 0.15 g VSS/g $NH_4^+$-N | $b_{NS}$ | = 0.05 $days^{-1}$ |
| $k_{NS}$ | = 3 $days^{-1}$ | $K_{NS}$ | = $10^{0.051T-1.148}$ = 0.74 mg/L |
| **Heterotrophic biokinetic constants at 15°C** | | | |
| $Y$ | = 0.6 g VSS/g $BOD_5$ | $b$ | = 0.05 $days^{-1}$ |
| $k$ | = 3.2 $days^{-1}$ | $K_s$ | = 39 mg/L |
| ***Nitrosomonas* biokinetic constants at 15°C** | | | |
| $Y_{NS}$ | = 0.15 g VSS/g $NH_4^+$-N | $b_{NS}$ | = 0.04 $days^{-1}$ |
| $k_{NS}$ | = 1.95 $days^{-1}$ | $K_{NS}$ | = $10^{0.051T-1.140}$ = 0.41 mg/L |

[a] A temperature correction coefficient ($\theta$) of 1.04 was used for correcting $b$, whereas a $\theta$ of 1.09 was used for correcting $k$ and $K_s$.

Calculate the maximum growth rate for *Nitrosomonas* for the ambient conditions:

$$(\mu_{max})_{NS} = 0.47e^{0.098(T-15)}\left[\frac{DO}{K_{DO} + DO}\right][1 - 0.833\,(7.2 - pH)]$$

$$(\mu_{max})_{NS} = 0.47e^{0.098(T-15)}\left[\frac{2}{1 + 2}\right][1 - 0.833\,(7.2 - 7.1)]$$

$$(\mu_{max})_{NS} = 0.28\,days^{-1} \quad (7.3\text{[illegible]})$$

Correct the decay coefficient $b_{NS}$ for temperature:

$$b_{NS(T°C)} = b_{NS(20°C)}(1.04)^{(T-20°C)}$$

$$b_{NS(15°C)} = 0.05\ (1.04)^{(15-20)} = 0.04\ days^{-1} \quad (7.40)$$

Calculate the minimum SRT for nitrification:

$$\frac{1}{(\mathrm{SRT})_{\mathrm{MIN}}} = \frac{(\mu_{\mathrm{MAX}})_{\mathrm{NS}}\ (\mathrm{NH_4^+\text{-}N})_o}{K_{\mathrm{NS}} + (\mathrm{NH_4^+\text{-}N})_o} - b_{\mathrm{NS}}$$

$$K_{\mathrm{NS}} = 10^{0.051T-1.148} = 10^{0.051(15)-1.148} = 0.41 \text{ mg/L}$$

$$\frac{1}{(\mathrm{SRT})_{\mathrm{MIN}}} = \frac{(0.28)\ (30)}{0.41 + 30} - 0.04$$

$$(\mathrm{SRT})_{\mathrm{MIN}} = 4.2 \text{ days} \qquad (7.41)$$

Estimate the design SRT using a safety factor of 1.25 and a peaking factor of 1.2:

$$(\mathrm{SRT})_{\mathrm{DESIGN}} = (\mathrm{SRT})_{\mathrm{MIN}}\ (\mathrm{SF})\ (\mathrm{PF})$$

$$(\mathrm{SRT})_{\mathrm{DESIGN}} = (4.2 \text{ days})\ (1.25)\ (1.2) = 6.3 \text{ days} \qquad (7.42)$$

Estimate the overall SRT for the system assuming an anoxic volume of 34%. This will increase the $(\mathrm{SRT})_{\mathrm{DESIGN}}$ by approximately 52%:

$$(\mathrm{SRT})_{\mathrm{OVERALL}} = (\mathrm{SRT})_{\mathrm{DESIGN}}\ (\mathrm{MF})$$

$$\mathrm{MF} = \frac{1}{1 - 0.34} = 1.52 \qquad (7.43)$$

$$(\mathrm{SRT})_{\mathrm{OVERALL}} = (6.3 \text{ days})\ (1.52) = 9.6 \text{ days}$$

Correct the heterotrophic biokinetic constants for temperature except for the yield coefficient, which has been found to be independent of temperature (Benefield *et al.*, 1975):

$$b_{(T°\mathrm{C})} = b_{(20°\mathrm{C})}(1.04)^{(T°\mathrm{C}-20°\mathrm{C})} \qquad (7.44)$$

$$b_{(15°\mathrm{C})} = 0.06\ (1.04)^{(15-20)} = 0.05 \text{ days}^{-1}$$

$$k_{(T°\mathrm{C})} = k_{(20°\mathrm{C})}(1.09)^{(T-20)}$$

$$k_{(15°\mathrm{C})} = 5\ (1.09)^{(15-20)} = 3.2 \text{ days}^{-1} \qquad (7.45)$$

$$K_{s(T°\mathrm{C})} = K_{s(20°\mathrm{C})}(1.09)^{(T-20°\mathrm{C})}$$

$$K_{s(15°\mathrm{C})} = 60(1.09)^{(15-20)} = 39 \text{ mg/L} \qquad (7.46)$$

Calculate the effluent soluble $\mathrm{BOD_5}$ for the overall SRT:

$$S_e = \frac{K_s\ [1 + b\ (\mathrm{SRT})_{\mathrm{OVERALL}}]}{(\mathrm{SRT})_{\mathrm{OVERALL}}(Yk - b) - 1}$$

$$S_e = \frac{39\ [1 + 0.05(9.6)]}{9.6\ (0.6 \cdot 3.2 - 0.05) - 1} = 3.4 \text{ mg/L} \qquad (7.47)$$

Calculate the effluent $NH_4^+$-N concentration for the design SRT assuming all of the influent TKN can be oxidized to nitrate:

$$(NH_4^+\text{-N})_e = \frac{K_{NS}\,[1 + b_{NS}(SRT)_{DESIGN}]}{(SRT)_{DESIGN}[Y_{NS}k_{NS} - b_{NS}] - 1} \tag{7.48}$$

Recall the following:

$$K_{NS} = 10^{0.051(T)-1.148} = 10^{0.051(15)-1.148} = 0.41 \text{ mg/L}$$

$$K_{NS(15°C)} = 3.0(1.09)^{(15-20)} = 1.95 \text{ days}^{-1}$$

$$(NH_4^+\text{-N})_e = \frac{0.41[1 + 0.04(6.3)]}{6.3(0.15 \times 1.95 - 0.04) - 1} = 0.87 \text{ mg/L}$$

Calculate the synthesis TKN assuming the nitrogen ($F_N$) content of the biomass is approximately 12% by weight of the VSS:

$$\begin{aligned} N_{SYN} &= \frac{Y\,(S_o - S_e)F_N}{1 + b\,(SRT)_{OVERALL}} + X_e(F_N) \\ &= \frac{0.6(200 - 3.4)(0.12)}{1 + 0.05(9.6)} + (0.75)(20)(0.12) = 11.4 \end{aligned} \tag{7.49}$$

The amount of nitrogen to be oxidized can now be calculated:

$$\begin{aligned} NO &= TKN_o - (NS_4^+\text{-N})_e - N_{SYN} \\ NO &= 30 - 0.87 - 11.4 = 17.7 \text{ mg/L} \end{aligned} \tag{7.50}$$

The required oxic volume to achieve nitrification is calculated as follows:

$$V_{OXIC} = \frac{Q(SRT)_{DESIGN}}{X}\left[\frac{Y(S_o - S_e)}{1 + b(SRT)_{DESIGN}} + X_L\right] \tag{7.51}$$

$$\begin{aligned} Y &= \left(\frac{0.6 \text{ g VSS}}{\text{g BOD}_5}\right)\left(\frac{1 \text{ g TSS}}{0.75 \text{ g VSS}}\right) = \frac{0.8 \text{ g TSS}}{\text{g BOD}_5} \\ V_{OXIC} &= \frac{38\,000(6.3)}{3\,500}\left[\frac{0.8(200 - 3.4)}{1 + 0.05(6.3)} + 60\right] = 12\,285 \text{ m}^3 \end{aligned} \tag{7.52}$$

Assuming an MLR flow of $3Q$ ($3 \times 38\,000$ m$^3$/d) and a return sludge flow of $1Q$ (38 000 m$^3$/d), the $NO_3^-$-N concentration in the MLR can be calculated as follows:

$$\begin{aligned} N &= \frac{(NO)\,Q}{Q + RAS + MLR} \\ N &= \frac{(17.7)(38\,000)}{[38\,000 + 38\,000 + 3(38\,000)]} = 3.54 \text{ mg/L} \end{aligned} \tag{7.53}$$

Assuming that the DO in the RAS flow is negligible and the DO in the MLR is 0.5 mg/L, calculate the nitrate equivalence of the DO in the MLR as follows:

$$(NO_3^- \text{-N})_{EQ} = (DO)_{MLR} \left(0.35 \, \frac{\text{g } NO_3^- \text{-N}}{\text{g } O_2}\right) (MLR) \left(\frac{1 \text{ kg}}{1\ 000 \text{ g}}\right)$$

$$(NO_3^- \text{-N})_{EQ} = (0.5)\ (0.35)(3 \times 38\ 000) \left(\frac{1 \text{ kg}}{1\ 000 \text{ g}}\right) = 20 \text{ kg/d} \tag{7.54}$$

Calculate the mass of nitrates to be removed (NOR) in the anoxic zone as follows:

$$NOR = [(RAS + MLR)\,N]\left(\frac{1\,\text{kg}}{1\,000\,\text{g}}\right)$$

$$NOR = [38\,000 + (3 \times 38\,000)]\,(3.54)\left(\frac{1\,\text{kg}}{1\,000\,\text{g}}\right) = 538\,\text{kg/d} \tag{7.55}$$

The effluent nitrate concentration is the same as that in the MLR flow. Calculate the total mass of nitrates to be removed in the anoxic zone as follows:

$$TNOR = NOR + (NO_3^- \text{-N})_{EQ}$$

$$TNOR = 538 + 20 = 558 \text{ kg/d} \tag{7.56}$$

The volume of the anoxic zone is calculated as follows:

$$V_{ANOXIC} = \frac{TNOR(1\,000\,\text{g/kg}) - 0.03\,(S_o)(Q)(1.06)^{T-20}}{0.029(X)(1.06)^{T-20}}$$

$$V_{ANOXIC} = \frac{(558)(1\,000\,\text{g/kg}) - 0.03(200)(38\,000)(1.06)^{15-20}}{0.029(3\,500)(1.06)^{15-20}} = 5\ 111\,\text{m}^3 \tag{7.57}$$

Calculate the waste sludge production from the entire biological system:

$$P_x = \left[\frac{Y(S_o - S_e)}{1 + b(SRT)_{OVERALL}} + X_L\right](Q)\left[\frac{1 \text{ kg}}{1\ 000 \text{ g}}\right]$$

$$P_x = \left[\frac{0.8\ (200 - 3.4)}{1 + 0.05\ (9.6)} + 60\right](38\ 000)\left[\frac{1 \text{ kg}}{1\ 000 \text{ g}}\right] = 6\ 320 \text{ kg TSS/d} \tag{7.58}$$

Check the new overall system SRT to see if it is within 5% of the initial value assumed at the beginning of the procedure:

$$(\mathrm{SRT})_{\mathrm{OVERALL}} = \frac{X(V_{\mathrm{OXIC}} + V_{\mathrm{ANOXIC}})}{P_x}$$

$$(\mathrm{SRT})_{\mathrm{OVERALL}} = \frac{(3\ 500)(12\ 285 + 5\ 111)}{(6\ 320)(1\ 000\ \mathrm{g/kg})} = 9.6\ \mathrm{days} \tag{7.59}$$

Because the calculated $(\mathrm{SRT})_{\mathrm{overall}}$ is equal to the initial estimate of $(\mathrm{SRT})_{\mathrm{overall}}$, it is not necessary to repeat the procedure.

$$\mathrm{ALK_e} = \mathrm{ALK_o} - 7.14\ (\mathrm{NO}) + 3.57\ [\mathrm{NO} - (\mathrm{NO_3^-\text{-}N})_e] \tag{7.60}$$

Check the alkalinity balance:

$$\mathrm{ALK_e} = 200 - 7.14(17.7) + 3.57(17.7 - 3.54) = 124\ \mathrm{mg/L}$$

The oxygen requirements for the oxic portion of the process are calculated as follows:

$$O_2 = \mathrm{CBOD} + \mathrm{NOD} - \mathrm{DOC} \tag{7.61}$$

$$\mathrm{CBOD} = \{Q[(1 - 1.42Y)(S_o - S_e)] + 1.42(b)(X)\ V_{\mathrm{OXIC}}\}\left(\frac{1\ \mathrm{kg}}{1\ 000\ \mathrm{g}}\right)$$

$$\mathrm{CBOD} = \{38\ 000[(1 - 1.42 \times 0.6)(200 - 3.4)] + 1.42(0.05)(2\ 625)(12\ 285)\}\left[\frac{1}{1\ 000}\right] = 3\ 395\ \mathrm{kg/d} \tag{7.62}$$

$$\mathrm{NOD} = Q(4.57)(\mathrm{NO})\frac{1}{1\ 000}$$

$$\mathrm{NOD} = 38\ 000(4.57)(17.7)\left[\frac{1\ \mathrm{kg}}{1\ 000\ \mathrm{g}}\right] = 3\ 074\ \mathrm{kg/d} \tag{7.63}$$

$$\mathrm{DOC} = Q\left(2.86\ \frac{\mathrm{g}\ O_2}{\mathrm{g\ NO_3^-\text{-}N}}\right)[\mathrm{NO} - (\mathrm{NO_3^-\text{-}N})_e]\left(\frac{1\ \mathrm{kg}}{1\ 000\ \mathrm{g}}\right)$$

$$\mathrm{DOC} = 38\ 000\left[2.86\ \frac{\mathrm{g}\ O_2}{\mathrm{g\ NO_3^-\text{-}N}}\right][(17.7) - (3.54)]\left[\frac{1\ \mathrm{kg}}{1\ 000\ \mathrm{g}}\right] = 1\ 539\ \mathrm{kg/d}$$

$$O_2 = 3\ 395 + 3\ 074 - 1\ 539 = 4\ 930\ \mathrm{kg/d} \tag{7.64}$$

Ignoring the DOC, size the aeration system for providing 6 469 kg/d of oxygen.

**Metro Denver Wastewater Treatment Plant Case History.** The Metro Wastewater Reclamation District's WWTP in Denver, Colorado, consists of two separate plants on the same site: a 265 000-$m^3$/d (70-mgd) nitrification–denitrification activated-sludge treatment system (North Complex) and a 435 275-$m^3$/d (115-mgd) high-purity-oxygen secondary treatment system (South Complex). Both systems were recently modified, with the South Complex undergoing improvements to increase capacity and the North Complex converted from the conventional plug-flow activated-sludge process to nitrification–denitrification. This case study will focus on the evaluation of the North Complex treatment system.

Major modifications included the conversion of the first 32 m (105 ft) of the 64 m $\times$ 9.1 m $\times$ 4.6 m (210 ft $\times$ 30 ft $\times$ 15 ft) sidewater depth first pass (pass A) to an anoxic zone, retrofitting the remainder of pass A and all of the second and third passes (passes B and C) with fine-bubble ceramic disc aeration, installation of MLR pumping to return approximately half of the aeration basin effluent to the anoxic zone, and adding instrumentation to control the process. Figure 7.8 is a schematic of a typical basin. At the design flow rate of 5.83 mgd, the anoxic detention time is approximately 1.5 hours, and the aerobic detention time is 7.3 hours. The 12 aeration basins are divided into four quadrants of three basins each. Each quadrant is coupled with three circular final clarifiers, which measure 39.6 m (130 ft) in diameter by 3 m (10 ft) sidewater depth.

Data were collected on the basins from October 1990 through August 1991. The average influent wastewater characteristics for the specified time period to the North Complex treatment facilities were 141 mg/L $BOD_5$, 96 mg/L TSS, 25.2 mg/L $NH_4^+$-N, and a temperature of 18°C. Average effluent characteristics for this same time period were 6 mg/L $BOD_5$ , 15 mg/L TSS, 0.9 mg/L $NH_4^+$-N , and 10.4 mg/L $NO_3$-N, respectively. Therefore, approximately 55% of the incoming nitrogen is removed through nitrification–denitrification.

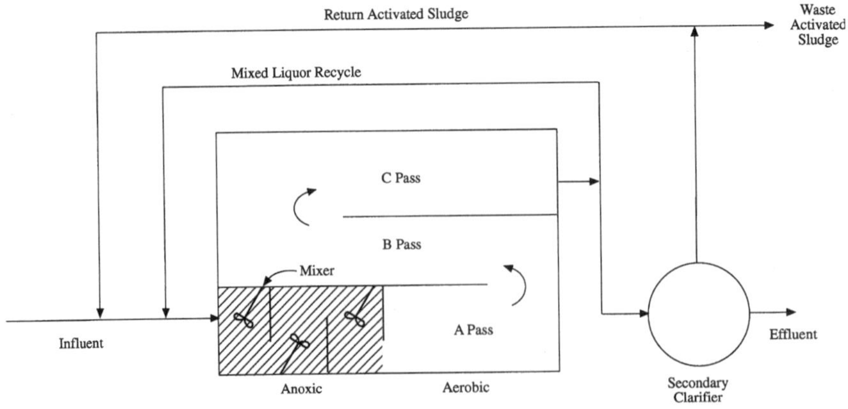


**Figure 7.8 Metro Denver North Complex Anoxic/Oxic process.**

**POSTDENITRIFICATION SYSTEMS.** This section presents design criteria, design equations, a design example, and a case history for a typical postdenitrification system. Figure 7.9 is a process flow schematic for a postdenitrification system. In the oxic/anoxic mode of operation, the carbonaceous and nitrogenous oxygen demands of the wastewater are exerted in the aeration basin by the heterotrophic and autotrophic bacteria, whereas biological denitrification occurs in the anoxic zone producing alkalinity and providing an oxygen credit. By the time the wastewater reaches the anoxic zone, the exogenous carbon is depleted and denitrification occurs by endogenous respiration. An external carbon source such as raw or primary effluent, methanol, or acetate can be added to stimulate denitrification. The external carbon dosage to the anoxic zone will vary depending on the wastewater characteristics entering the oxic portion of treatment, pH, temperature, and detention time. Typically, a dosage of approximately 3 kg of methanol is required per kilogram of nitrate-nitrogen to be reduced. The major disadvantage to this type of process results from a low COD:N ratio lowering the rate of denitrification. Another major disadvantage is extremely high operations and maintenance costs associated with methanol addition. As a result, the anoxic volume is larger than the volume required for a predenitrification system. High DO levels in the mixed liquor that are carried over from the oxic zone will also inhibit denitrification in the anoxic zone.

**Design Criteria.** Design criteria for sizing postdenitrification systems are presented in Table 7.7. The size of the anoxic zone is governed by the quantity of nitrogen to be removed from the wastewater, the denitrification rate of the denitrifiers, and the organic carbon concentration. Chemical feed facilities are required for supplemental carbon addition, usually methanol. The methanol may vary from 3 to 12 kg methanol/kg nitrate reduced; therefore, chemical feed facilities must be sized accordingly.

**Design Equations.** A similar approach using SRT as presented for the predenitrification system will be used for the postdenitrification system.

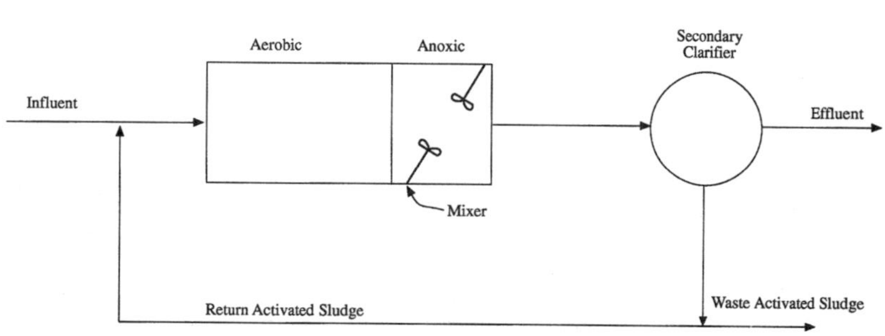


**Figure 7.9 Postdenitrification/nitrification system.**

*AEROBIC ZONE CALCULATIONS.* The equations previously presented under the aerobic zone calculations for predenitrification systems are used for calculating the aerobic volume required for nitrification at specified ambient conditions. They will not be repeated in this section; however, they will be illustrated again in the design example.

*ANOXIC ZONE CALCULATIONS.* Anoxic zone calculations are as follows:

1. Estimate the mass of nitrates to be removed in the anoxic zone. First, calculate the nitrate equivalence of DO in the mixed liquor from the aerobic zone:

$$(NO_3^- \text{-N})_{EQ} = (DO)_{ML} \left(0.35 \frac{\text{g } NO_3^- \text{-N}}{\text{g } O_2}\right)(Q + RAS)\left(\frac{1 \text{ kg}}{1\,000 \text{ g}}\right) \quad (7.65)$$

   Where

   $(NO_3^- \text{-N})_{EQ}$ = nitrogen equivalence of DO in the mixed liquor from the aerobic zone, kg/d; and

   $(DO)_{ML}$ = dissolved oxygen concentration in aerobic zone, mg/L.

2. Calculate the mass of nitrates to be removed in the anoxic zone using the following equation:

$$NOR = Q[NO - (NO_3^- \text{-N})_e]\left(\frac{1 \text{ kg}}{1\,000 \text{ g}}\right) \quad (7.66)$$

   where NOR = mass of nitrates to be removed in the anoxic zone, kg/d.

3. Calculate the total mass of nitrates to be removed in the anoxic zone using the following equation:

$$TNOR = [NOR + (NO_3^- \text{-N})_{EQ}]\left(\frac{1 \text{ kg}}{1\,000 \text{ g}}\right) \quad (7.67)$$

   where TNOR = total mass of nitrates to be removed in the anoxic zone, kg/d.

4. Estimate the specific denitrification rate corrected for ambient temperature in the postdenitrification, anoxic zone using the following equation (Burdick *et al.*, 1982):

$$SDNR_2 = 0.12\,(SRT)_{OVERALL}^{-0.706}\,(1.02)^{T-20} \quad (7.68)$$

   where $SDNR_2$ = specific denitrification rate in the postdenitrification, anoxic zone, days$^{-1}$.

5. Calculate the volume of the anoxic zone as follows:

$$V_{\text{ANOXIC}} = \frac{(\text{TNOR})(1\ 000\ \text{g/kg})}{(X)(\text{SDNR}_2)} \tag{7.69}$$

where $V_{\text{ANOXIC}}$ = volume of the anoxic zone, $m^3$.

6. The total volume of the biological system can now be calculated using the following equation:

$$V_{\text{TOTAL}} = V_{\text{OXIC}} + V_{\text{ANOXIC}} \tag{7.70}$$

where $V_{\text{TOTAL}}$ = total volume of the biological system, $m^3$.

7. The quantity of sludge that must be wasted from this system can be calculated using the following equation, which neglects the small amount of biomass synthesized by the nitrifiers:

$$P_X = \left[\frac{Y(S_o - S_e)}{1 + b(\text{SRT})_{\text{OVERALL}}} + X_L\right](Q)\left(\frac{1\ \text{kg}}{1\ 000\ \text{g}}\right) \tag{7.71}$$

Where

$P_X$ = waste sludge production, kg/d; and
$X_L$ = inert solids in the influent, mg/L.

8. Calculate the overall SRT of the biological system as follows:

$$(\text{SRT})_{\text{OVERALL}} = \frac{X(V_{\text{OXIC}} + V_{\text{ANOXIC}})}{P_X(1\ 000\ \text{g/kg})} \tag{7.72}$$

where $(\text{SRT})_{\text{OVERALL}}$ = overall SRT of the biological system, days. If the calculated SRT is not within 5% of the assumed $(\text{SRT})_{\text{OVERALL}}$ in step 5 of the aerobic zone calculations, the entire design procedure must be repeated.

*OXYGEN REQUIREMENTS.* The equations previously presented can also be used in computing the oxygen required for postdenitrification systems.

*ALKALINITY REQUIREMENTS.* The equation previously presented can be used for calculating the effluent alkalinity concentration for postdenitrification systems.

**Postdenitrification Design Example.** A postdenitrification/nitrification WWTP is to be sized. Influent wastewater characteristics and biokinetic constants used in this example design are presented in Table 7.8. The $(\text{SRT})_{\text{DESIGN}}$ in the previous example was 6.3 days. An effluent suspended solids concentration of 20 mg/L is assumed, and the effluent TN concentration is greater than 4 mg/L but less than 10 mg/L.

Estimate the overall SRT for the system assuming an anoxic volume of 30%. $Z_z$ is equal to the postanoxic zone fraction of the total reactor volume. This will increase the $(SRT)_{DESIGN}$ by approximately 43%.

$$(SRT)_{OVERALL} = (SRT)_{DESIGN}\,(MF)$$
$$MF = \frac{1}{1 - Z_Z} \tag{7.73}$$

$$MF = \frac{1}{1 - 0.30} = 1.43$$
$$(SRT)_{OVERALL} = (6.3\text{ days})(1.43) = 9.0\text{ days} \tag{7.74}$$

Calculate the effluent soluble $BOD_5$ for the overall SRT:

$$S_e = \frac{K_s[1 + b\,(SRT)_{OVERALL}]}{(SRT)_{OVERALL}\,(Yk - b) - 1}$$
$$S_e = \frac{39[1 + 0.05(9)]}{9(0.6 \times 3.2 - 0.05) - 1} = 3.6\text{ mg/L} \tag{7.75}$$

Calculate the effluent $NH_4^+$-N concentration for the design SRT assuming all of the influent TKN can be oxidized to nitrate:

$$(NH_4^+\text{-N})_e = \frac{K_{NS}\,[1 + b_{NS}\,(SRT)_{DESIGN}]}{(SRT)_{DESIGN}\,[Y_{NS}k_{NS} - b_{NS}] - 1}$$
$$(NH_4^+\text{-N})_e = \frac{0.41[1 + 0.04(6.3)]}{6.3(0.15 \times 1.95 - 0.04) - 1} = 0.87\text{ mg/L} \tag{7.76}$$

Calculate the synthesis TKN assuming the nitrogen ($F_N$) content of the biomass is approximately 12% by weight of the VSS:

$$N_{SYN} = \left[\frac{Y(S_o - S_e)F_N}{1 + b(SRT)_{OVERALL}}\right] + (X_e)(F_N)$$
$$N_{SYN} = \frac{0.6(200 - 3.6)}{1 + 0.05(9.0)}(0.12) + (0.75)(20)(0.12) = 11.6\,\text{mg/L} \tag{7.77}$$

The amount of nitrogen to be oxidized can be calculated as follows:

$$NO = TKN_o - (NH_4^+\text{-N})_e - N_{SYN}$$
$$NO = 30 - 0.87 - 11.6 = 17.5\text{ mg/L} \tag{7.78}$$

The required oxic volume to achieve nitrification is calculated as follows:

$$V_{\text{OXIC}} = \frac{Q(\text{SRT})_{\text{DESIGN}}}{X}\left[\frac{Y(S_o - S_e)}{1 + b(\text{SRT})_{\text{DESIGN}}} + X_L\right] \tag{7.79}$$

$$Y = \left(\frac{0.6 \text{ g VSS}}{\text{g BOD}_5}\right)\left(\frac{1 \text{ g TSS}}{0.75 \text{ g VSS}}\right) = \frac{0.8 \text{ g TSS}}{\text{g BOD}_5}$$

$$V_{\text{OXIC}} = \frac{38\ 000\ (6.3)}{3\ 500}\left[\frac{0.8\ (200 - 3.6)}{1 + 0.05\ (6.3)} + 60\right] = 12\ 277 \text{ m}^3 \tag{7.80}$$

Assuming a DO of 0.5 mg/L in the mixed liquor and a RAS flow of $1Q$, the nitrate equivalence of the DO in the mixed liquor from the aerobic zone is calculated as follows:

$$(\text{NO}_3^-\text{-N})_{\text{EQ}} = (\text{DO}_{\text{ML}})\left[0.35\frac{\text{g NO}_3^-\text{-N}}{\text{g O}_2}\right](Q + \text{RAS})\left[\frac{1 \text{ kg}}{1\,000 \text{ g}}\right]$$

$$(\text{NO}_3^-\text{-N})_{\text{EQ}} = \frac{(0.5)(0.35)(38\,000 + 38\,000)}{1\,000 \text{ g/kg}} = 13 \text{ kg/d} \tag{7.81}$$

Calculate the nitrate level in the effluent:

$$\text{N} = \frac{Q(\text{NO})}{[Q + \text{RAS}]}$$

$$\text{N} = \frac{38\ 000 \times (17.5)}{[38\ 000 + 38\ 000]} = 8.8 \text{ mg/L} \tag{7.82}$$

Alternately, the RAS flow could be calculated for a specified effluent nitrate concentration.

Using the following equation, calculate the mass of nitrates to be removed in the anoxic zone recalling that 8.8 mg/L of nitrates remains in the effluent:

$$\text{NOR} = \frac{Q[\text{NO} - (\text{NO}_3^-\text{-N})_e]}{1\ 000 \text{ g/kg}}$$

$$\text{NOR} = \frac{38\ 000\ (17.5 - 8.8)}{1\ 000 \text{ g/kg}} = 331 \text{ kg/d} \tag{7.83}$$

The total mass of nitrates to be removed is calculated below:

$$\text{TNOR} = \text{NOR} + (\text{NO}_3^-\text{-N})_{\text{eq}}$$

$$\text{TNOR} = 331 + 13 = 344 \text{ kg/d} \tag{7.84}$$

The SDNR in the postdenitrification anoxic zone corrected for ambient temperature conditions is calculated as follows:

$$\begin{aligned} \text{SDNR}_2 &= 0.12\ (\text{SRT})_{\text{OVERALL}}^{-0.706}(1.02)^{T-20} \\ \text{SDNR}_2 &= 0.12\ (9)^{-0.706}(1.02)^{15-20} = 0.023\ \text{days}^{-1} \end{aligned} \quad (7.85)$$

The anoxic volume can be calculated as follows:

$$\begin{aligned} V_{\text{ANOXIC}} &= \frac{(\text{TNOR})\ (1\ 000\ \text{g/kg})}{(X)\ (\text{SDNR}_2)} \\ V_{\text{ANOXIC}} &= \frac{(344)\ (1\ 000\ \text{g/kg})}{(3\ 500)\ (0.023)} = 4\ 273\ \text{m}^3 \end{aligned} \quad (7.86)$$

The total volume of the biological system is calculated below:

$$\begin{aligned} V_{\text{TOTAL}} &= V_{\text{OXIC}} + V_{\text{ANOXIC}} \\ V_{\text{TOTAL}} &= 12\ 277 + 4\ 273 = 16\ 550\ \text{m}^3 \end{aligned} \quad (7.87)$$

Calculate the waste sludge production as follows:

$$\begin{aligned} P_X &= \left[\frac{Y(S_o - S_e)}{1 + b(\text{SRT})_{\text{OVERALL}}} + X_L\right](Q)\left(\frac{1\ \text{kg}}{1\ 000\ \text{g}}\right) \\ P_X &= \left[\frac{0.8\ (200 - 3.6)}{1 + 0.05\ (9)} + 60\right](38\ 000)\left(\frac{1\ \text{kg}}{1\ 000\ \text{g}}\right) = 6\ 398\ \text{kg/d} \end{aligned} \quad (7.88)$$

Calculate the overall SRT as follows:

$$\begin{aligned} (\text{SRT})_{\text{OVERALL}} &= \frac{(X)\ (V_{\text{TOTAL}})}{(P_X)\ (1\ 000\ \text{g/kg})} \\ (\text{SRT})_{\text{OVERALL}} &= \frac{(3\ 500)\ (16\ 550)}{(6\ 398)\ (1\ 000\ \text{g/kg})} = 9.1\ \text{days} \end{aligned} \quad (7.89)$$

The calculated $(\text{SRT})_{\text{OVERALL}}$ is 9.1 days, which is approximately equal to the assumed $(\text{SRT})_{\text{OVERALL}}$ value; therefore, another iteration is not necessary. Check the alkalinity balance as follows:

$$\begin{aligned} \text{ALK}_e &= \text{ALK}_o - 7.14(\text{NO}) + 3.57[\text{NO} - (\text{NO}_3^- \text{-N})_e] \\ \text{ALK}_e &= 200 - 7.14(17.5) + 3.57(17.5 - 8.8) = 106\ \text{mg/L} \end{aligned} \quad (7.90)$$

The oxygen requirement for the oxic portion of the process is calculated as follows:

$$O_2 = \text{CBOD} + \text{NOD} - \text{DOC}$$

$$\text{CBOD} = \frac{38\,000[(1-1.42(0.6)](200-3.6)+1.42(0.05)(2\,625)(12\,277)}{1\,000\,\text{g/kg}}$$

$$= 3\,393\,\text{kg/d}$$

$$\text{NOD} = 38\,000(4.57)(17.5)\left(\frac{1\,\text{kg}}{1\,000\,\text{g}}\right) = 3\,039\,\text{kg/d}$$

$$\text{DOC} = 38\,000\left(2.86\frac{\text{g}\,O_2}{\text{g}\,NO_3^-\text{-N}}\right)[17.5 - 8.8]\left(\frac{1\,\text{kg}}{1\,000\,\text{g}}\right) = 946\,\text{kg/d}$$

$$O_2 = 3\,393 + 3\,039 - 946 = 5\,486\,\text{kg/d} \tag{7.91}$$

**Main Street Wastewater Treatment Plant Case History.** The Main Street WWTP in Pensacola, Florida, is a 75 700-m$^3$/d (20-mgd) AWTP using a modified pure-oxygen activated-sludge process. The major liquid train components include grit removal, primary clarification, biological treatment, secondary clarification, chlorination, dechlorination, and discharge to Pensacola Bay. Figure 7.10 is a schematic of the existing facility. The National Pollutant Discharge Elimination System permit sets the maximum annual average load discharged to Pensacola Bay at 606 kg/d (1 334 ppd) for both $BOD_5$ and TSS, 455 kg/d (1 000 ppd) for TN, and 76.4 kg/d (168 ppd) for TP, respectively. At 75 700 m$^3$/d, this equates to effluent concentrations of 8 mg/L for $BOD_5$ and TSS, 6 mg/L for TN, and 1 mg/L for TP. The permitted flow capacity of 75 700 m$^3$/d is based on the entire biological process being operated as pure-oxygen activated sludge. In the existing configuration, the back half of the aeration basin is operated as an anoxic zone, resulting in a process capacity of approximately 68 100 m$^3$/d (18 mgd) or 80% of capacity.

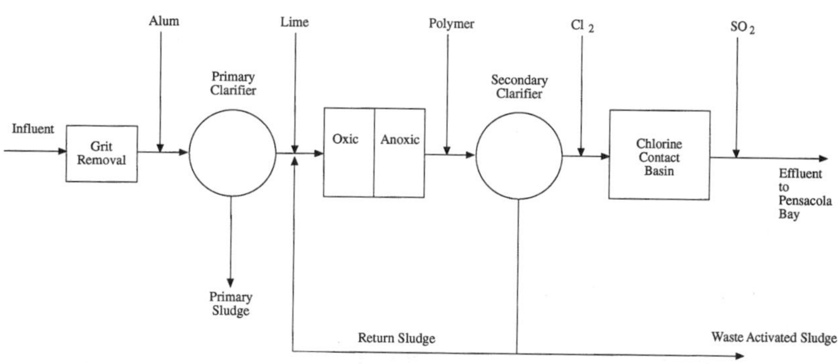


**Figure 7.10 Existing Main Street WWTP flow schematic.**

Phosphorus removal is accomplished in the primaries through alum addition at approximately 110 mg/L to the primary influent. Lime at 13 mg/L is added to the primary effluent to increase the alkalinity for biological treatment. The original pure-oxygen OASES process has been modified to accomplish biological denitrification in the second half of the basin. The biological process consists of five trains with four compartments per train. The first two compartments of each train are fed pure oxygen in which surface aerators with dual impellers are used for entraining oxygen to the mixed liquor and keeping the biomass in suspension. The surface impellers on the aerators in compartments three and four of each train have been removed to keep oxygen from being entrained into the mixed liquor; however, the lower impellers remain in place for mixing these compartments. Methanol addition is provided, and it is sometimes added during the winter months or when the influent $BOD_5$ is low to enhance denitrification. Typically, methanol has been added a dose of 3 kg methanol/kg nitrate-nitrogen reduced. This ratio has been as high as 12 kg methanol/kg nitrate-nitrogen reduced. Nonionic polymer at approximately 0.5 mg/L is added to the secondary influent for enhancing TSS removal in the secondaries and to eliminate the need for effluent filtration.

Data from the facility were evaluated from April 1991 through March 1992. The monthly average values for $BOD_5$, TSS, TN, and TP were 214 mg/L, 263 mg/L, 30.4 mg/L, and 7.0 mg/L, respectively. During this 12-month period, effluent $BOD_5$, TSS, TN, and TP averaged 3.0 mg/L, 3.0 mg/L, 6.7 mg/L, and 0.5 mg/L, respectively. The average wastewater flow rate for this period averaged 54 740 $m^3/d$ (14.5 mgd), resulting in a nominal detention time of 3.6 hours for both the oxic and anoxic zones in the modified pure-oxygen process. The SRT for the entire system averaged 17.9 days.

**COMBINED PRE- AND POSTDENITRIFICATION SYSTEMS.** This section presents design criteria, design equations, a design example, and a case history for a combined pre- and postdenitrification system. Figure 7.11 is a schematic of a typical combined pre-and postdenitrification system. This layout is typical of a modified Bardenpho™ system. The first-stage anoxic zone precedes the first-stage aerobic or oxic zone. The first-stage anoxic zone generally removes approximately 60 to 75% of the nitrate-nitrogen. The carbonaceous and nitrogenous oxygen demand of the wastewater is primarily met in the aerobic zone. The MLR is recycled from the end of the aerobic zone back to the head of the first-stage anoxic zone to allow biological denitrification to occur. The second-stage anoxic zone follows the first-stage aerobic zone. The second-stage anoxic zone is required to consistently meet effluent requirements of less than 3 mg/L TN. This zone is sized to remove the remaining nitrates resulting in an 80 to 90% removal of nitrates through this zone. A second-stage oxic or aerobic zone typically follows the second-stage anoxic zone with a detention time of approximately 30 to 60 minutes. The aeration is necessary to prevent denitrification in the secondary clarifiers and

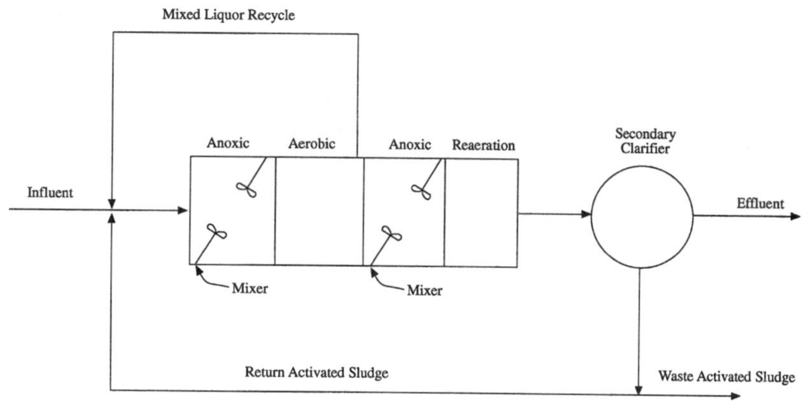


**Figure 7.11** **Combined pre- and postdenitrification/nitrification system.**

to convert any residual ammonia to nitrate. Methanol, acetate, raw wastewater, or primary effluent can be added to the second-stage anoxic zone to enhance biological denitrification, although it is seldom practiced.

**Design Criteria.** Design criteria used in sizing combined pre- and postdenitrification systems are presented in Table 7.7. For combined systems, the total anoxic volume typically represents 30 to 50% of the total volume of the biological process. Mixed liquor recycle typically is returned at a rate of approximately 4:1 and up to 6:1 of the incoming wastewater flow rate to the plant. Submerged vertical turbine mixers on platforms or submerged propeller mixers typically are used to keep the biomass suspended in the anoxic zone. Either diffused aeration or surface aerators are used in the first-stage aerobic zone. Typically, fine-bubble diffusers are used in the second-stage aerobic zone for reaerating the mixed liquor before clarification. As mentioned previously, the MLR pump intakes should be located such that minimal DO is recycled back to the first-stage anoxic zone.

**Design Equations.** The procedure used in designing a combined pre- and postdenitrification system is similar to that already presented for sizing predenitrification and postdenitrification systems. It is also similar to the approach presented in *Design of Municipal Wastewater Treatment Plants* (WEF, 1992).

*FIRST-STAGE AEROBIC ZONE CALCULATIONS.* First-stage aerobic zone calculations are as follows:

1. Calculate the $(SRT)_{DESIGN}$.
2. Estimate the overall SRT of the biological system using the following equation:

$$(SRT)_{OVERALL} = (SRT)_{DESIGN}\ (MF) \tag{7.92}$$

Where

$(SRT)_{OVERALL}$ = The overall SRT of the biological system, days; and

MF = multiplication factor to account for the anoxic volume, which will increase the SRT of the system. For most BNR systems, this parameter will vary from approximately 1.4 to 2.0.

$$\text{MF} = \frac{1}{[(1 - Z_1)(1 - Z_2)(1 - Z_3)]} \tag{7.93}$$

Where

$Z_1$ = preanoxic zone fraction of total reactor volume;
$Z_2$ = postanoxic zone fraction of total reactor volume; and
$Z_3$ = reaeration zone fraction of total reactor volume.

3. For a given set of heterotrophic biokinetic constants, calculate the effluent soluble $BOD_5$ for the $(SRT)_{OVERALL}$ as follows:

$$S_e = \frac{K_s[1 + b(SRT)_{OVERALL}]}{(SRT)_{OVERALL}(Yk - b) - 1} \tag{7.94}$$

4. For a given set of autotrophic biokinetic constants, calculate the effluent ammonium-nitrogen concentration using the following equation:

$$(NH_4^+\text{-N})_e = \frac{K_{NS}[1 + b_{NS}(SRT)_{DESIGN}]}{(SRT)_{DESIGN}[(Y_{NS})(k_{NS}) - b_{NS}] - 1} \tag{7.95}$$

where $(NH_4^+\text{-N})_e$ = effluent ammonium nitrogen concentration, mg/L.

5. Determine the amount of nitrogen to be oxidized as follows:

$$\text{NO} = \text{TKN}_0 - (NH_4^+\text{-N})_e - N_{SYN} \tag{7.96}$$

where NO = influent nitrogen converted to oxidized nitrogen, mg/L.

The nitrogen used in synthesis neglecting the small amount of nitrogen synthesized by the nitrifiers is calculated using the following equation:

$$N_{SYN} = \frac{Y(S_o - S_e)F_N}{[1 + b(SRT)_{OVERALL}]} + (X_e)F_N \tag{7.97}$$

Where

$X_e$ = effluent VSS concentration, mg/L;
$F_N$ = fraction of nitrogen in VSS; and
$F_N$ = 5 to 12%.

6. Calculate the volume of the aerobic or oxic zone necessary to achieve nitrification for a given temperature, pH, and DO as follows:

$$V_{OXIC_1} = \frac{Q(SRT)_{DESIGN}}{X}\left[\frac{(S_o - S_e)}{1 + b(SRT)_{DESIGN}} + X_L\right] \tag{7.98}$$

where $V_{OXIC_1}$= volume of aerobic or oxic zone, $m^3$.

*FIRST-STAGE ANOXIC ZONE CALCULATIONS.* First-stage anoxic zone calculations are as follows:

1. Estimate the nitrate-nitrogen concentration in the MLR returned to the first-stage anoxic zone using the following equation:

$$N = \frac{NO(Q)}{(MLR + RAS + Q)} \tag{7.99}$$

where N = nitrate-nitrogen concentration returned to anoxic zone, mg/L.

2. Estimate the nitrate equivalence of DO in the MLR as follows:

$$(NO_3^- \text{-N})_{EQ_1} = (DO)_{MLR}\left(0.35\,\frac{g\ NO_3^- \text{-N}}{g\ O_2}\right)(MLR)\left(\frac{1\ kg}{1\ 000\ g}\right) \tag{7.100}$$

where $(NO_3^-\text{-N})_{EQ_1}$ = nitrogen equivalence of DO in the MLR, kg/d.

3. Calculate the mass of nitrates to be removed in the first-stage anoxic zone using the following equation:

$$NOR_1 = [RAS(NO_3^- \text{-N})_e + MLR(N)]\left(\frac{1\ kg}{1\ 000\ g}\right) \tag{7.101}$$

where $NOR_1$ = mass of nitrates to be removed in the anoxic zone, kg/d.

4. The total mass of nitrates to be removed in the first-stage anoxic zone is calculated as follows:

$$TNOR_1 = NOR_1 + (NO_3^- \text{-N})_{EQ_1} \tag{7.102}$$

where $TNOR_1$ = total mass of nitrates to be removed in the first-stage anoxic zone, kg/d.

5a. Calculate the volume of the first-stage anoxic zone as outlined below. First, calculate the SDNR in the anoxic zone as a function of the volume as given in the following equation:

$$SDNR_1 = \frac{TNOR_1(1\ 000\ g/kg)}{V_{ANOXIC_1}(X)} \quad (7.103)$$

where $SDNR_1$ = specific denitrification rate in the first-stage anoxic zone, $days^{-1}$.

5b. Estimate the SDNR in the anoxic zone corrected for ambient temperature using the following equation (Burdick *et al.*, 1982):

$$SDNR_1 = [0.03(F{:}M) + 0.029](1.06)^{T-20} \quad (7.104)$$

recalling that:

$$F{:}M = \frac{Q(S_o)}{X(V_{ANOXIC_1})}$$

where F:M = food-to-microorganism ratio, $days^{-1}$.

5c. The volume of the first-stage anoxic zone can be calculated by equating the two equations presented in steps 5a and 5b as follows:

$$V_{ANOXIC_1} = \frac{TNOR_1(1\ 000\ g/kg) - 0.03\ (Q)(S_o)(1.06)^{T-20}}{0.029(X)(1.06)^{T-20}} \quad (7.105)$$

6a. Calculate the quantity of nitrates to be removed in the second anoxic zone as follows:

$$NOR_2 = Q\frac{[NO - (NO_3^- \text{-}N)_e]}{1\ 000\ g/kg} - NOR_1 \quad (7.106)$$

where $NOR_2$ = mass of nitrates to be removed in the second anoxic zone, kg/d.

6b. The nitrate equivalence of DO in the mixed liquor from the aerobic zone is calculated as follows:

$$(NO_3^- \text{-}N)_{EQ_2} = (Q + RAS)(DO_{MLR})\left(0.35\frac{g\,NO_3^- \text{-}N}{g\,O_2}\right)\left(\frac{1\,kg}{1\,000\,g}\right) \quad (7.107)$$

where $(NO_3^- \text{-}N)_{EQ_2}$ = nitrate equivalence of DO in the mixed liquor, kg/d.

6c. The total mass of nitrates to be removed in the second-stage anoxic zone can be calculated as follows:

$$TNOR_2 = NOR_2 + (NO_3^- \text{-}N)_{EQ_2} \quad (7.108)$$

where $TNOR_2$ = total mass of nitrates to be removed in the second-stage anoxic zone, kg/d.

7. The SDNR in the second-stage anoxic zone can be calculated using the following equation corrected for ambient temperature (Burdick *et al.*, 1982):

$$SDNR_2 = 0.12(SRT)_{OVERALL}{}^{-0.706}(1.02)^{T-20} \tag{7.109}$$

Where

$SDNR_2$ = specific denitrification rate in the second-stage anoxic zone, $days^{-1}$; and
$T$ = wastewater temperature, °C.

8. The volume of the second-stage anoxic zone is calculated as follows:

$$V_{ANOXIC_2} = \frac{TNOR_2(1\ 000\ g/kg)}{(X)(SDNR_2)} \tag{7.110}$$

where $V_{AOXIC2}$ = volume of the second-stage anoxic zone, $m^3$.

*CALCULATIONS FOR SECOND AEROBIC ZONE.* The second-stage aerobic zone is sized according to detention time. Typically, a detention time of 30 to 60 minutes is maintained at the design flow. The second-stage aerobic zone volume is calculated as follows:

$$V_{OXIC_2} = (\Theta)\left(\frac{1\ hr}{60\ min}\right)\left(\frac{1\ day}{24\ hr}\right)(Q) \tag{7.111}$$

Where

$V_{OXIC2}$ = volume of the second stage oxic zone, $m^3$, and
$\theta$ = detention time in the second stage oxic zone, minutes.

*CHECK THE OVERALL DESIGN.* The overall volume of the biological system must be calculated and the overall SRT compared with the initial assumption. Alkalinity and oxygen requirements should also be computed.

1. First, calculate the total volume of the biological system using the following equation:

$$V_{TOTAL} = V_{ANOXIC_1} + V_{OXIC_1} + V_{ANOXIC_2} + V_{OXIC_2} \tag{7.112}$$

where $V_{TOTAL}$ = total volume of the biological system, $m^3$.

2. Calculate the total quantity of sludge produced as follows:

$$P_X = \left[\frac{Y(S_o - S_e)}{1 + b(SRT)_{OVERALL}} + X_L\right](Q)\left(\frac{1\ kg}{1\ 000\ g}\right) \tag{7.113}$$

where $P_X$ = waste sludge production, kg/d.

3. Compute the overall SRT as follows:

$$(\text{SRT})_{\text{OVERALL}} = \frac{(V_{\text{TOTAL}})(X)}{P_X(1\ 000\ \text{g/kg})} \qquad (7.114)$$

where $(\text{SRT})_{\text{OVERALL}}$ = the overall solids retention time of the biological system, days. If the calculated $(\text{SRT})_{\text{OVERALL}}$ is not within 5% of the assumed $(\text{SRT})_{\text{OVERALL}}$ in step 2 of the aerobic calculations, the entire design procedure must be repeated.

4. Calculate the oxygen requirements.
5. The alkalinity requirements can be calculated. Typically, 50 to 100 mg/L of alkalinity as calcium carbonate is maintained in the effluent from BNR systems.

**Combined Pre- and Postdenitrification Design Example.** A combined pre- and postdenitrification WWTP must be sized. Influent wastewater characteristics and biokinetic constants used in the design example are presented in Table 7.8. The only exception is that the effluent TN concentration to be achieved is equal to or less than 3 mg/L and the effluent TSS concentration is equal to or less than 5 mg/L.

To meet the effluent requirements listed above, a filtration system must be used to eliminate particulate $BOD_5$ and nitrogen that would otherwise escape treatment in the effluent solids. Listed below are the major steps necessary to complete the process design with exception to the effluent filters. The design SRT from previous calculations is 6.3 days.

Estimate the overall SRT for the biological system assuming the preanoxic volume, postanoxic volume, and reaeration volume are 27%, 30%, and 6%, respectively, of the total reactor volume.

Calculate the effluent soluble $BOD_5$ for the overall SRT:

$$\text{MF} = \frac{1}{[(1 - Z_1)(1 - Z_2)(1 - Z_3)]}$$

$$\text{MF} = \frac{1}{[(1 - 0.27)(1 - 0.30)(1 - 0.06)]} = 2.08 \qquad (7.115)$$

$$(\text{SRT})_{\text{OVERALL}} = (\text{SRT})_{\text{DESIGN}}(\text{MF})$$

$$(\text{SRT})_{\text{OVERALL}} = (6.3\ \text{days})(2.08) = 13.1\ \text{days} \qquad (7.116)$$

$$S_e = \frac{K_s[1 + b(\text{SRT})_{\text{OVERALL}}]}{(\text{SRT})_{\text{OVERALL}}(Yk - b) - 1}$$

$$S_e = \frac{39[1 + 0.05(13.1)]}{13.1[(0.6)(3.2) - 0.05] - 1} = 2.7\ \text{mg/L} \qquad (7.117)$$

Calculate the effluent $NH_4^+$-N concentration for the design SRT assuming all of the influent TKN can be oxidized to nitrate:

$$(NH_4^+\text{-N})_e = \frac{K_{NS}[1 + b_{NS}(SRT)_{DESIGN}]}{(SRT)_{DESIGN}[Y_{NS}k_{NS} - b_{NS}] - 1}$$

$$(NH_4^+\text{-N})_e = \frac{0.41[1 + 0.04(6.3)]}{6.3[(0.15)(1.95) - 0.04] - 1} = 0.87 \text{ mg/L} \tag{7.118}$$

Calculate the synthesis TKN assuming the nitrogen ($F_N$) content of the biomass is approximately 12% by weight of the VSS:

$$N_{SYN} = \frac{Y(S_o - S_e)F_N}{1 + b(SRT)_{overall}} + (X_e)F_N$$

$$N_{SYN} = \frac{0.6(200 - 2.7)(0.12)}{1 + 0.05(13.1)} + (0.75)(5)(0.12) = 9.0 \text{ mg/L} \tag{7.119}$$

The amount of nitrogen to be oxidized can now be calculated:

$$NO = TKN_o - (NH_4^+\text{-N})_e - N_{SYN}$$

$$NO = 30 - 0.87 - 9.0 = 20.1 \text{ mg/L} \tag{7.120}$$

The required oxic volume to achieve nitrification is calculated as follows:

$$V_{OXIC_1} = \frac{Q(SRT)_{DESIGN}}{X}\left[\frac{Y(S_o - S_e)}{1 + b(SRT)_{DESIGN}} + X_L\right]$$

$$Y = \left(\frac{0.6 \text{ g VSS}}{\text{g BOD}_5}\right)\left(\frac{1 \text{ g TSS}}{0.75 \text{ g VSS}}\right) = \frac{0.8 \text{ g TSS}}{\text{g BOD}_5} \tag{7.121}$$

$$V_{OXIC_1} = \frac{38\ 000(6.3)}{3\ 500}\left[\frac{0.8(200 - 2.7)}{1 + 0.05(6.3)} + 60\right] = 12\ 314 \text{ m}^3$$

Assuming an MLR flow of $3Q$ ($3 \times 38\ 000$ m$^3$/d) and a return sludge flow of $1Q$ (38 000 m$^3$/d), the $NO_3^-$-N concentration in the MLR can be calculated as follows:

$$N = \frac{(NO)Q}{Q + RAS + MLR}$$

$$N = \frac{(20.1)(38\ 000)}{[38\ 000 + 38\ 000 + 3(38\ 000)]} = 4.0 \text{ mg/L} \tag{7.122}$$

Assuming that the DO in the RAS flow is negligible and the DO in the MLR is 0.5 mg/L, calculate the nitrate equivalence of the DO in the MLR as follows:

$$(NO_3^- \text{-N})_{EQ_1} = (DO)_{MLR}\left(0.35\,\frac{\text{g } NO_3^-\text{-N}}{\text{g } O_2}\right)(MLR)\left(\frac{1 \text{ kg}}{1\ 000 \text{ g}}\right)$$
$$(NO_3^- \text{-N})_{EQ_1} = (0.5)(0.35)(3 \times 38\ 000)\left(\frac{1 \text{ kg}}{1\ 000 \text{ g}}\right) = 20 \text{ kg/d} \tag{7.123}$$

The mass of nitrates to be removed (NOR) in the first-stage anoxic zone assuming an effluent nitrate concentration of 1.0 mg/L is calculated as follows:

$$NOR_1 = [RAS(NO_3^- \text{-N})_e + MLR(N)]\left(\frac{1 \text{ kg}}{1\ 000 \text{ g}}\right)$$
$$NOR_1 = [38\ 000(1.0) + (3 \times 38\ 000)(4.0)]\left(\frac{1 \text{ kg}}{1\ 000 \text{ g}}\right) = 494 \text{ kg/d} \tag{7.124}$$

Calculate the total mass of nitrates to be removed in the first-stage anoxic zone as follows:

$$TNOR_1 = NOR_1 + (NO_3^- \text{-N})_{EQ_1}$$
$$TNOR_1 = 494 + 20 = 514 \text{ kg/d} \tag{7.125}$$

The volume of the first-stage anoxic zone is calculated as follows:

$$V_{ANOXIC_1} = \frac{TNOR_1(1\,000 \text{ g/kg}) - 0.03(Q)(S_o)(1.06)^{T-20}}{0.029(X)(1.06)^{T-20}}$$
$$V_{ANOXIC_1} = \frac{(514)(1\,000 \text{ g/kg}) - 0.03(38\,000)(200)(1.06)^{(15-20)}}{0.029(3\,500)(1.06)^{(15-20)}} = 4\,531 \text{ m}^3 \tag{7.126}$$

Calculate the nitrate to be removed in the second anoxic zone as follows:

$$NOR_2 = \frac{Q[NO - (NO_3^-)_e]}{1\ 000 \text{ g/kg}} - NOR_1$$
$$NOR_2 = \frac{38\ 000(20.1 - 1.0)}{1\ 000 \text{ g/kg}} - 494 = 232 \text{ kg/d} \tag{7.127}$$

The nitrate equivalence of DO in the mixed liquor from the aerobic zone is calculated as follows:

$$(NO_3^--N)_{EQ_2} = (Q + RAS)(DO_{ML})\left(0.35\ \frac{g\ NO_3^--N}{g\ O_2}\right)\left(\frac{kg}{1\ 000\ g}\right)$$

$$(NO_3^--N)_{EQ_2} = \frac{[38\ 000 + 38\ 000](0.5)\left(0.35\ \frac{g\ NO_3^--N}{g\ DO}\right)}{1\ 000\ g/kg} = 13\ kg/d \qquad (7.128)$$

The total mass of nitrates to be removed in the second-stage anoxic zone can be calculated as follows:

$$TNOR_2 = NOR_2 + (NO_3^--N)_{EQ_2}$$
$$TNOR_2 = 232 + 13 = 245\ kg/d \qquad (7.129)$$

The SDNR in the second-stage anoxic zone can be calculated using the following equation corrected for ambient temperature (Burdick *et al.*, 1982):

$$SDNR_2 = 0.12(SRT)_{OVERALL}^{-0.706}[1.02]^{T-20}$$
$$SDNR_2 = 0.12(13.1)^{-0.706}(1.02)^{15-20} = 0.017\ 7\ days^{-1} \qquad (7.130)$$

The volume of the second-stage anoxic zone is calculated as follows:

$$V_{ANOXIC_2} = \frac{TNOR_2(1\ 000\ g/kg)}{(X)(SDNR_2)}$$
$$V_{ANOXIC_2} = \frac{245(1\ 000\ g/kg)}{(3\ 500)(0.017\ 7\ days^{-1})} = 3\ 955\ m^3 \qquad (7.131)$$

*CALCULATIONS FOR SECOND AEROBIC ZONE.* The second-stage aerobic zone volume is calculated as follows assuming a detention time of 45 minutes:

$$V_{OXIC_2} = (\Theta)\left(\frac{1\ hr}{60\ min}\right)\left(\frac{1\ day}{24\ hr}\right)(Q)$$
$$V_{OXIC_2} = (45\ min)\left(\frac{1\ hr}{60\ min}\right)\left(\frac{1\ day}{24\ hr}\right)(38\ 000) = 1\ 188\ m^3 \qquad (7.132)$$

*CHECK THE OVERALL DESIGN.* First, calculate the total volume of the biological system using the following equation:

$$V_{TOTAL} = V_{ANOXIC_1} + V_{OXIC_1} + V_{ANOXIC_2} + V_{OXIC_2}$$
$$V_{TOTAL} = 4\ 531 + 12\ 314 + 3\ 955 + 1\ 188 = 21\ 988\ m^3 \qquad (7.133)$$

Calculate the total quantity of sludge produced as follows:

$$P_X = \left[\frac{Y(S_o - S_e)}{1 + b(\mathrm{SRT})_{\mathrm{OVERALL}} + X_L}\right](Q)\left(\frac{1\ \mathrm{kg}}{1\ 000\ \mathrm{g}}\right)$$

$$P_X = \left[\frac{0.8(200 - 2.7)}{1 + 0.05(13.1)} + 60\right](38\ 000)\left(\frac{1\ \mathrm{kg}}{1\ 000\ \mathrm{g}}\right) = 5\ 904\ \mathrm{kg/d} \quad (7.134$$

Compute the overall SRT as follows:

$$(\mathrm{SRT})_{\mathrm{OVERALL}} = \frac{(V_{\mathrm{TOTAL}})(X)}{P_X(1\ 000\ \mathrm{g/kg})}$$

$$(\mathrm{SRT})_{\mathrm{OVERALL}} = \frac{(21\ 988\ \mathrm{m}^3)(3\ 500\ \mathrm{mg/L})}{5\ 904(1\ 000\ \mathrm{g/kg})} = 13.0\ \mathrm{days} \quad (7.135)$$

Because $(\mathrm{SRT})_{\mathrm{OVERALL}}$ calculated is approximately equal to $(\mathrm{SRT})_{\mathrm{OVERALL}}$ assumed, a second iteration is not required.

Calculate the oxygen requirements using the following equation:

$$O_2 = \mathrm{CBOD} + \mathrm{NOD} - \mathrm{DOC} \quad (7.136)$$

$$\mathrm{CBOD} = \frac{\{Q[(1 - 1.42Y)(S_o - S_e)] + 1.42(b)(X)V_{\mathrm{OXIC}}]\}}{1\ 000\ \mathrm{g/kg}}$$

$$\mathrm{CBOD} = \frac{38\ 000\{[1 - 1.42(0.6)](200 - 2.7)\} + 1.42(0.05)(2\ 625)12\ 314}{1\ 000\ \mathrm{g/kg}} = 3\ 405\ \mathrm{kg/d} \quad (7.137)$$

$$\mathrm{NOD} = Q(4.57)\ (\mathrm{NO})\left(\frac{1\ \mathrm{kg}}{1\ 000\ \mathrm{g}}\right)$$

$$\mathrm{NOD} = 38\ 000(4.57)\ (20.1)\left(\frac{1\ \mathrm{kg}}{1\ 000\ \mathrm{g}}\right) = 3\ 491\ \mathrm{kg/d} \quad (7.138)$$

$$\mathrm{DOC} = Q\left(2.86\ \frac{\mathrm{g\ O_2}}{\mathrm{g\ NO_3^-\text{-}N}}\right)[\mathrm{NO} - (\mathrm{NO_3^-\text{-}N})_e]\left(\frac{1\ \mathrm{kg}}{1\ 000\ \mathrm{g}}\right)$$

$$\mathrm{DOC} = 38\ 000\left(2.86\ \frac{\mathrm{g\ O_2}}{\mathrm{g\ NO_3^-\text{-}N}}\right)[20.1 - 1]\left(\frac{1\ \mathrm{kg}}{1\ 000\ \mathrm{g}}\right) = 2\ 076\ \mathrm{kg/d}$$

$$O_2 = 3\ 405 + 3\ 491 - 2\ 076 = 4\ 820\ \mathrm{kg/d} \quad (7.139)$$

The alkalinity requirements can be calculated using the following equation:

$$\mathrm{ALK_e} = \mathrm{ALK_o} - 7.14(\mathrm{NO}) + 3.57[\mathrm{NO} - (\mathrm{NO_3^-\text{-}N})_e]$$

$$\mathrm{ALK_e} = 200 - 7.14(20.1) + 3.57(20.1 - 1) = 125\ \mathrm{mg/L\ as\ CaCO_3} \quad (7.140)$$

**Case History.** The Washington Suburban Sanitary Commission operates the 31 000-$m^3$/d (8.19-mgd) Parkway WWTP in Laurel, Maryland. The liquid process train consists of bar screens, aerated grit chambers, Parshall flumes, primary clarification, a four-stage Bardenpho™ process, secondary clarification, chlorination, and postaeration. Alum is added to the raw influent and secondary influent for removing phosphorus. Caustic is added to the secondary influent for alkalinity and pH adjustment. Effluent is discharged to the Patuxent River. Effluent design criteria were established as follows: 20 mg/L $BOD_5$ and TSS, 1 mg/L TP, 7 mg/L TN (from April 1 to October 15), 5 mg/L DO, and 6.5 to 8.5 for pH. The Parkway facility also was obligated to achieve the lowest level of nitrogen removal technically possible with BNR and attempt to meet a TN level of 3.0 mg/L during April 1 through October 15 under SB262.

Data from the facility from August 1992 through December 1992 have been evaluated. Monthly removal aerages for $BOD_5$, TSS, TN, and TP were 95.2%, 98%, 84.8% and 83.4%, respectively. The average flow rate for this time period was 5.65 mgd, and the average MLSS concentration was 2 876 mg/L. The sludge age varied from 9 to 21 days with an average value of 14 days. Approximately 15 mg/L of alum and caustic was used during the evaluation period.

At the average design flow of 28 400 $m^3$/d (7.50 mgd), the nominal detention times in the first-stage anoxic zone, first-stage aerobic zone, second-stage anoxic zone, and second-stage aerobic zone were 2.4, 9.7, 1.3, and 0.5 hours, respectively. The RAS pumping capacity provides 0 to 100% of average plant influent flow, and the internal MLR pumps are capable of providing 160 to 400% of the average influent flow.

# References


Benefield, L.D., *et al.* (1975) Temperature Considerations in the Design and Control of Completely Mixed Activated Sludge Plants. *Proc. 2nd Annu. Natl. Conf. Environ. Eng. Res., Develop., Design, Am. Soc. Civ. Eng.,* Univ. Fla., Gainesville.

Bidstrup, S.M., and Grady, C.P.L. (1988) SSSP-Simulation of Single-Sludge Processes. *J. Water Pollut. Control Fed.,* **60**, 351.

Bowker, R.P.G., and Stensel, H.D. (1987) *Design Manual for Phosphorus Removal.* EPA-625/1-87-001, U.S. EPA, Washington, D.C.

Burdick, C.R., *et al.* (1982) Advanced Biological Treatment to Achieve Nutrient Removal. *J. Water Pollut. Control Fed.*, **54**, 1078.

Chen, J.J., and Slack, D.C. (1991) Deep-Bed Tertiary Filter Systems. Paper presented at Calif. Water Pollut. Control Fed. Annu. Conf., Pasadena, Calif.

Hoover, S.R., and Porges, N. (1952) Assimilation of Dairy Wastes by Activated Sludge. *Sew. Ind. Wastes,* **24**, 306.

Knowles, G., *et al.* (1965) Determination of Kinetic Constants for Bacteria in a Mixed Culture, with the Aid of an Electronic Computer. *J. Gen. Microbiol.,* **38**, 263.

Lawrence, A.W., and McCarty, P.L. (1970) Unified Basis for Biological Treatment Design and Operation. *J. Sanit. Eng. Div., Proc. Am. Soc. Civ. Eng.,* **96**, 757.

McCarty, P.L., *et al.* (1969) Biological Denitrification of Wastewaters by Addition of Organic Materials. *Proc. 24th Ind. Waste Conf., Purdue Univ.,* West Lafayette, Ind.

McClintock, S.A., *et al.* (1988) Nitrate Versus Oxygen Respiration in the Activated Sludge Process. *J. Water Pollut. Control Fed.,* **60**, 342.

Metcalf & Eddy, Inc. (1991) *Wastewater Engineering: Treatment, Disposal and Reuse*. McGraw Hill, Inc., New York, N.Y.

Mines, R.O., and Woods, C.W. (1994) Assessment of BNR Systems in Florida. *Proc. 1994 Natl. Conf. Environ. Eng.,* 436.

Mines, R.O., *et al.* (1992) Bionutrient Removal Using a Modified Pure Oxygen Process. Paper presented at 65th Annu. Conf. Water Environ. Fed., New Orleans, La.

Morales, L.M., *et al.* (1991) Capability Assessment of Biological Nutrient Removal Facilities. *Res. J. Water Pollut. Control Fed.,* **G3**, 900.

Parker, D.S., *et al.* (1975) *Process Design Manual for Nitrogen Control.* EPA 625/1-75-007, U.S. EPA, Washington, D.C.

Pickard, D.W., *et al.* (1985) Six Years of Successful Nitrogen Removal at Tampa, Florida. Paper presented at 58th Annu. Conf. Water Pollut. Control Fed., Kansas City, Mo.

Randall, C.W., *et al.* (1992) *Design and Retrofit of Wastewater Treatment Plants for Biological Nutrient Removal.* Technomic Publishing Co., Inc., Lancaster, Pa.

Reardon, R.O. (1993) Overview of Current Practice in Biological Nutrient Removal. Biological Nutrient Removal: The Florida Experience III and Beyond. March 10, 1994, Orlando, Fla.

Redd, K., *et al.* (1992) Pilot Testing and Evaluation of a Denitrifying Tertiary Filter. Paper presented at Calif. Water Pollut. Control Fed. Southern Reg. Training Conf., San Diego, Calif.

Savage, E.S. (1983) Biological Denitrification Deep-Bed Filter. Paper presented at Filtech Conf. Filtration Soc., London, Eng.

Savage, E.S., and Chen, J.J. (1972) Development Report for Nutrient Removal Project at Indian Lake Treatment Plant, North Huntington, Penn. Dravo Corporation, Pittsburgh, Pa.

Savage, E.S., and Chen, J.J. (1975) Operating Experience with Columnar Denitrification. *Water Res.* (G.B.), **9**, 751.

Soap and Detergent Association (1989) *Principles and Practice of Phosphorus and Nitrogen Removal from Municipal Wastewater.* New York, N.Y.

State of California (1978) Wastewater Reclamation Criteria. California Administrative Code, Title 22, Division 4, California Dept. Health Serv, Sanit. Eng. Sect., Berkeley, Calif.

Stensel, H.D., *et al.* (1973) Biological Kinetics of Suspended-Growth Denitrification. *J. Water Pollut. Control Fed.,* **45,** 249.

*Tampa Tribune* (1992) Ecosystem of Bay on Rebound. July 28.

Tetreault, M., and Parker, D.S. (1989) Separate Stage Denitrification—Key to Achieving a 3-3-2-0.6 AWT Effluent. Paper presented at 62nd Annu. Conf. Water Pollut. Control Fed., San Francisco, Calif.

TREEO (1988) *Design of Advanced Wastewater Treatment Systems*. Univ. Fla., Gainesville.

U.S. Environmental Protection Agency (1993) *Nitrogen Control Manual.* EPA 625-R/93-010, U.S. EPA, Washington, D.C.

Water Environment Federation (1992) *Design of Municipal Wastewater Treatment Plants*. Manual of Practice No. 8, Alexandria, Va.

# Chapter 8
# Design of Biological Phosphorus Removal Processes

The purpose of this chapter is to introduce the reader to fundamental concepts of enhanced biological phosphorus removal (EBPR) process design. These designs have been established by both laboratory and full-scale experience, enabling the reader to make the judgments and determinations necessary to develop a reliable design.

The advantages of biological phosphorus removal are that it

- Reduces or eliminates chemical addition for phosphorus (P) removal;
- Reduces or eliminates alkalinity addition to restore that consumed by chemical phosphorus precipitation;
- Selects for flocculating bacteria instead of filamentous forms, thereby improving sedimentation and clarification;
- Reduces aeration requirements and equipment;
- Allows retrofits that are simple and typically low in cost; and
- Improves nitrification rates in the aerobic zone by removing much of the soluble organic substrate in the anaerobic zone.

The primary disadvantages of EBPR are that it

- Requires additional capital costs, that is, baffles and mixers;
- Requires more careful design because it can be overdesigned as well as underdesigned;
- Is more sensitive to operate because it requires a limitation of nitrate and dissolved oxygen (DO) recycle to the anaerobic zone and prevention of phosphorus solubilization in the clarifier; and
- May make solids handling requirements more complex.

It is assumed that nitrification is neither required nor desired when implementing EBPR, and it is well known that biological nitrogen removal can be implemented without EBPR. While it is possible to implement either EBPR or biological nitrogen removal independently from the other, the processes are more efficient, stable, and economical when implemented together for most municipal wastewaters. For example, when the two processes are combined into a single-sludge system, the aeration energy requirements are reduced, chemical requirements are reduced, sludge production is reduced, and the rates of nitrification in the aerobic zone are increased (Randall *et al.*, 1992a), if the system is correctly designed and operated. However, EBPR sludge without any nitrification will settle better than sludge with combined phosphorus and nitrogen removal.

# DESIGN PRINCIPLES

The basic principle of EBPR design is simple. An anaerobic zone is placed ahead of an aerobic zone in a single-sludge activated-sludge system, and then the influent wastewater is fed directly to the anaerobic zone. Thus, the primary EBPR design criteria is that the bacterial biomass must cycle sequentially through anaerobic and aerobic environments, with most of the organic substrate initially available in the anaerobic zone. Any process configuration that satisfies these conditions will accomplish EBPR if properly operated. The purpose of the design is to select the desired bacterial population for the removal and intracellular storage of organic substrate in the anaerobic zone using energy from polyphosphate bonds, with subsequent use of the stored organic polymers in the aerobic zone to accomplish uptake and storage of phosphorus as polyphosphate. The anaerobic zone, if suitable organic substrate is available within it, acts as a microbial selector to enhance the population of phosphorus-storing (polyphosphate) bacteria and simultaneously suppresses the growth of most filamentous microorganisms.

The sizing of the anaerobic zone is also relatively simple because the process typically is not very sensitive to the anaerobic hydraulic retention time (HRT), although complicating factors such as wastewater characteristics, pH, and temperature effects must be considered. However, the design considerations required to maximize EBPR and to reduce the concentration of phosphorus in the effluent during operation can be fairly extensive. In particular, the design must ensure that the operators can maintain true anaerobic conditions in the anaerobic zone. For example, care must be taken to reduce DO in the process influent, to prevent significant entrainment of DO into the anaerobic mixed liquor during mixing, and to reduce the recycle of DO and nitrates with the return sludge. Also, resolubilization of phosphorus during clarification and recycle of phosphorus from sludge processing must be considered and reduced. The number one consideration, however, is that the designer incorporate sufficient flexibility for the operator to adjust the system to varying conditions. Given this flexibility, good performance and consistent results can be achieved reliably.

The treatment objectives of an EBPR activated-sludge system are to remove organic matter and phosphorus from the mixed liquor to desired effluent levels and to develop a biomass sludge that settles well enough to produce an effluent low in suspended solids after gravity sedimentation. The necessary process configuration is simple and requires only a small change from conventional activated-sludge design. The basic schematic for a continuous-flow system is shown in Figure 8.1. Separation of the anaerobic and aerobic zones can be accomplished in a plug-flow basin by the installation of a rigid baffle that prevents backflow. The anaerobic zone also needs sufficient mixing to maintain the mixed liquor suspended solids (MLSS) in suspension. Then, if the control of electron acceptor (DO and oxidized nitrogen [$NO_x$]) recycle and entrainment is enough to establish reduced conditions in the zone while available biochemical oxygen demand (BOD) is present, EBPR will occur.

The most precise EBPR design requires thorough wastewater characterization. Methods for characterization have been developed by Marais and others (Ekama and Marais, 1984a; Ekama *et al.*, 1982; and Siebritz, 1983), as recently described by Gibson and Dold (1991). The empirical design method they developed may then be used (Ekama and Marais, 1984b), or the designer may use the modification of the IAWPRC model proposed by Dold (1990) or the version recently published as International Association on Water Quality (London, England) Model No. 2. However, none of these models includes consideration of one factor that affects EBPR performance, namely, whether available substrate or phosphorus is limiting in the anaerobic zone. This will be discussed in more detail later.

The level of wastewater characterization desired by the above models is seldom available to the designer and may be impossible to obtain for many situations within the required time frame. When possible, a wastewater characterization study can be performed using procedures described by Gibson and Dold (1991), but one modification is suggested. The most important characterization is the determination of bioP available substrate in the process influent. This can be approximated by the method developed by Mamais *et al.* (1993) instead of the more laborious method developed at the University of Cape Town (Ekama and Marais, 1984b). Fortunately, experience has shown that acceptable EBPR designs can be developed without extensive wastewater characterization by incorporating operating flexibility into the treatment system or by providing prefermentation for volatile fatty acid (VFA) production. This is the approach presented in this chapter.

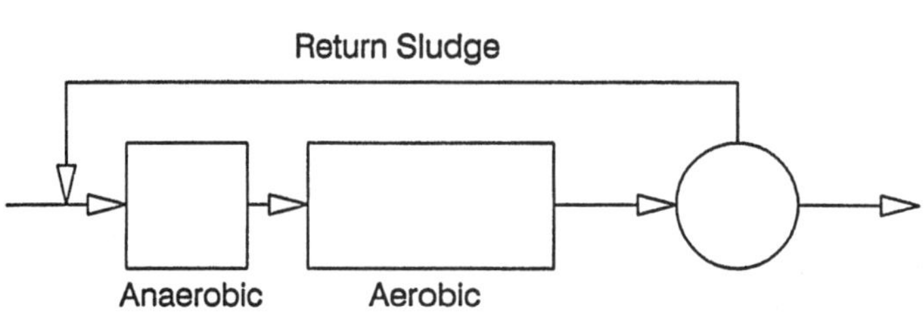


**Figure 8.1 Two-stage biological phosphorus removal system.**

**REQUIREMENTS FOR ENHANCED BIOLOGICAL PHOSPHORUS REMOVAL.** The following conditions, substrates, and nutrients are essential if EBPR is to be reliably achieved:

- Anaerobic–aerobic sequencing;
- Appropriate concentrations of readily available substrate in the anaerobic zone;
- Phosphorus in excess of cellular growth requirements;
- Adequate amounts of potassium and magnesium; and
- Appropriate pH range.

The previous list assumes that adequate amounts of viable poly-P microorganisms are in the anaerobic zone. Poly-P bacteria are defined as microorganisms capable of phosphorus uptake in excess of that required for cellular growth. Sometimes, short-term conditions occur during operation wherein the poly-P biomass in the anaerobic zone is inadequate for storage of sufficient organic matter to accomplish phosphorus removal to the desired level in the aerobic zone. This typically occurs when the return activated sludge (RAS) is not flow-paced with the influent wastewater. For near steady-state (that is, long-term) conditions, the poly-P biomass will always increase in the activated sludge until maximum phosphorus removal for the limiting substrate or nutrient is accomplished.

**Poly-P Microbial Population Dynamics.** An understanding of biological nutrient removal (BNR) systems sufficient for design requires an appreciation of microbial population dynamics. Anaerobic–aerobic sequencing is required to establish in the activated sludge a population of bacteria, known as poly-P bacteria, that uses phosphorus-storing mechanisms so that EBPR can be accomplished. Bacterial populations in activated sludge are dominated by those species that can grow the fastest under the operating conditions. Under fully aerobic conditions, the mechanisms of EBPR provide no growth advantage to the bacteria, resulting in the condition that the mechanisms are not used, the poly-P populations do not flourish, and EBPR is not accomplished. However, poly-P bacteria gain an advantage when the organic wastewater is initially introduced into an anaerobic zone because they can use energy stored in phosphate bonds (as adenosine triphosphate and polyphosphates) to sequester and polymerize a select number of organic compounds for intracellular storage, typically as polyhydroxybutyrate (PHB) or polyhydroxyvalerate, or, designated collectively, as polyhydroxyalkanoates. Thus, when the activated sludge exits the anaerobic zone, much or all of the organic substrate has been removed and stored by the poly-P bacteria, and the non-poly-P bacteria are unable to obtain large amounts of organic substrate for energy and growth in subsequent anoxic and aerobic zones. The fraction of poly-P bacteria in the activated sludge will be determined by the fraction of the substrate they obtain in the anaerobic zone, and the substrate fraction they obtain will be determined by which is limiting, available substrate or phosphorus, assuming sufficient nutrients. Thus, the fraction of the poly-P bacteria in the resulting activated sludge will be determined by the ratio of bioavailable substrate to available phosphorus in the anaerobic zone. Consequently, this ratio determines the effluent phosphorus concentration possible, the percent phosphorus in the mixed liquor volatile suspended solids (MLVSS), the required size of the anaerobic zone, and the effect of temperature on EBPR.

Fortunately, the designer can change the available substrate-to-phosphorus ratio by providing prefermentation or the addition of a suitable substrate.

**Readily Available Substrate in the Anaerobic Zone.** The key factor that determines the amount of phosphorus stored in the activated sludge is the amount of readily available organic matter in the anaerobic zone. There must be a large excess beyond that needed to deplete the electron acceptors (DO and $NO_x$) recycled or entrained to the anaerobic zone because the bacteria will preferentially metabolize the organic matter rather than store it if electron acceptors are available. Thus, the stored substrate, and subsequently the EBPR, will be reduced by the quantity of electron acceptors present.

The type of organic compound in the anaerobic zone is also important. It must be a soluble, relatively short-chain organic, such as a short-chain VFA, which the poly-P bacteria can take into the cell and polymerize. The VFAs from acetic (two carbons) through valeric (five carbons) are considered to be the prototype readily available organic compounds needed for EBPR. Although all of these VFAs are usable for EBPR, their relative effectiveness varies in accordance with the results of Abu-Ghararah and Randall (1991), shown in Table 8.1. The data show that acetic acid is the most efficient substrate for EBPR, and it is considerably more efficient than propionic acid, another typical product of wastewater fermentation. In fact, all of the VFAs listed in Table 8.1 are generated during wastewater fermentation, although the fractions vary with the nature of the wastewater organics. It is interesting to note that, of the VFAs listed, only valeric had a higher chemical oxygen demand (COD) used to phosphorus removal ratio than the 50:1 ratio typical of North American municipal wastewater. This indicates that a significant fraction of the organics in wastewater is not available for EBPR, which results in the high overall COD used:P removed ratio typically observed.

Also note that EBPR could not be accomplished with formic acid (one carbon). Prevailing theories teach that polymerization of one carbon compound is thermodynamically unfavorable for the bacteria, and the results confirm the theories.

It is important to realize that fermentation typically occurs in the wastewater collection system, assuming all the DO is depleted in the wastewater during its transport and nitrates are not available. This reduces both the need for prefermentation facilities at the plant and the required size of the anaerobic zone. It also is fortunate that acetic acid is the main product of this fermentation for most municipal wastewater. However, the amount of fermentation that occurs is a function of the flow time, the entrainment of DO into the wastewater during flow, and the temperature in the sewers.

**Table 8.1** **Effect of organic substrate on EBPR. (Reprinted from *Water Science and Technology,* Vol. 23, Abu-Ghararah, Z.H., and Randall, C.W., The Effect of Organic Compounds on Biological Phosphorus Removal, p. 585, Copyright 1991, with kind permission from Elsevier Science Ltd, The Boulevard, Langford Lane, Kidlington, OX5 1GB, UK.)**

| Substrate | Phosphorus uptake,[b] mg/L / COD[c] used, mg/L | COD[c] used mg[d] / Phosphorus removed, mg |
|---|---|---|
| Formic acid | 0 | ∞ |
| Acetic acid | 0.37 | 16.8 |
| Propionic acid | 0.10 | 24.4 |
| Butyric acid | 0.12 | 27.5 |
| Isobutyric acid | 0.14 | 29.1 |
| Valeric acid | 0.15 | 66.1 |
| Isovaleric acid | 0.24 | 18.8 |
| Municipal wastewater | 0.05 | 102[e] |

[a] Biological solids retention time = 13 days for all experiments.
[b] Total phosphorus uptake in aerobic zone.
[c] Chemical oxygen demand.
[d] COD used and phosphorus removed in total system.
[e] Value obtained with highly aerobic, low COD wastewater.

The amount of fermentation that has occurred in the collection system can be determined by directly measuring the VFA concentration in the raw wastewater and comparing that number to the total COD (TCOD). Experience has shown that fermentation also can be estimated by determining the $BOD_5$:COD ratio of the raw wastewater. What really determines performance, of course, is the amount of VFAs in the wastewater when it enters the anaerobic zone. If the plant has anaerobic digestion and there is recycle to the headworks from the digesters or from digested sludge processing, the amount of VFAs in the process influent is likely to be higher than that in the raw influent.

Figure 8.2 illustrates changes in the $BOD_5$:COD ratio of both raw wastewater and primary effluent that can occur at a municipal wastewater treatment plant (WWTP) over time. The data are from the Hampton Roads Sanitation District (HRSD) (Virginia) York River plant when it was being operated for EBPR with anaerobic digestion and recycle of belt filter press filtrate. The $BOD_5$:COD ratio of the raw influent varied from 0.37 to 0.73 during the 12-month period, and the ratio of the primary effluent varied even more, from 0.33 to 0.97, during the same period. This shows the fallacy of assuming a constant readily available substrate concentration in the influent for design purposes.

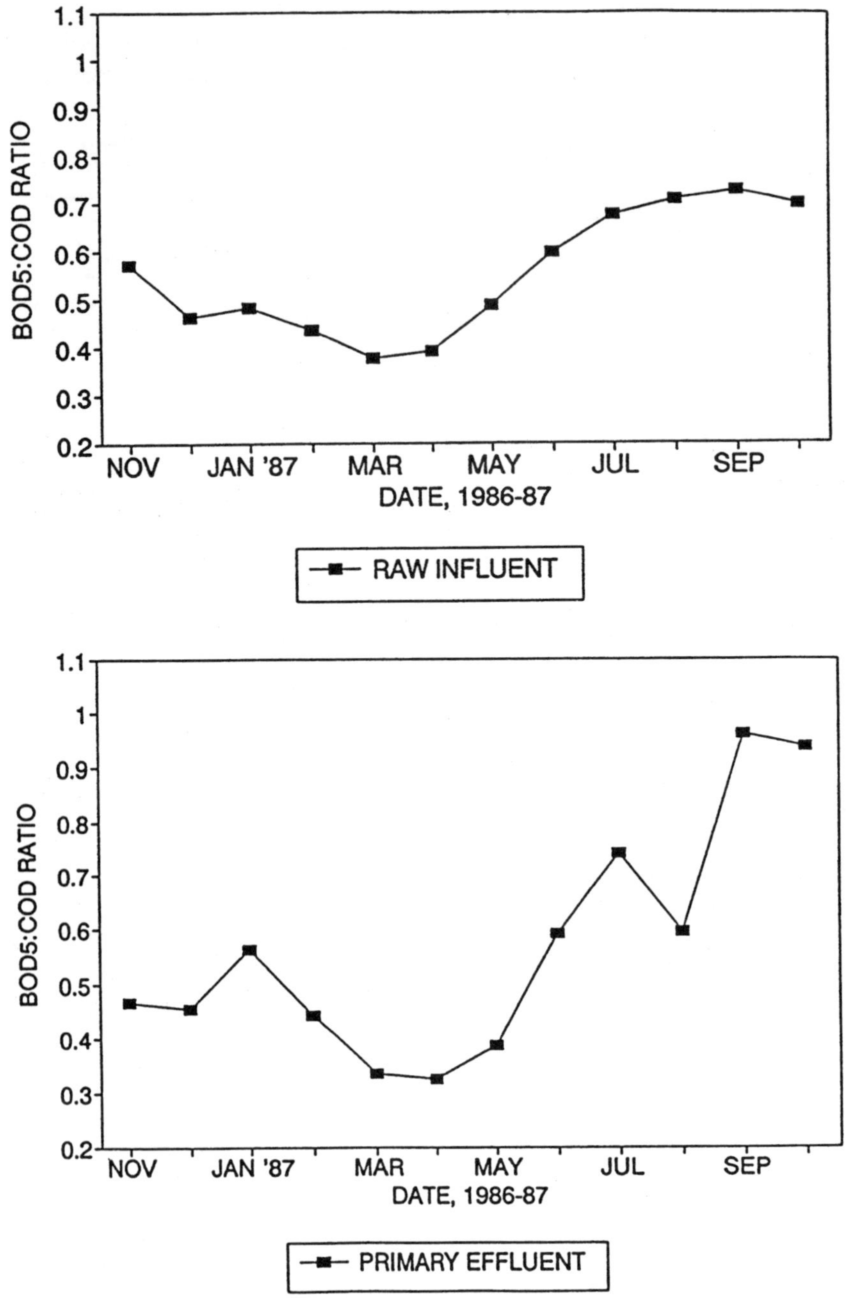


**Figure 8.2** **Seasonal variation in $BOD_5$:COD ratio at York River BNR plant (from Randall, C.W., *et al.* [1992b] *Design and Retrofit of Wastewater Treatment Plants for Biological Nutrient Removal.* Technomic Publishing Co., Inc., Lancaster, Pa., with permission).**

An accepted rule of thumb is that the readily available organic matter concentration within the initial anaerobic zone must be more than 25 mg/L, expressed as VFA, to accomplish significant EBPR (Ekama and Marais, 1984b). Increases of more than 25 mg/L will increase phosphorus release and organic storage in the anaerobic zone, and subsequent EBPR in the aerobic zone, up to some optimum VFA-to-phosphorus ratio, which will vary depending on which VFAs are present. Above that ratio, there will be no further increase in EBPR, and the BOD in the anaerobic zone effluent will increase. When this happens, the fraction of non-poly-P bacteria increases, which reduces the percent phosphorus in the MLSS, but also reduces the soluble phosphorus (SP) in the effluent. Conversely, studies at Virginia Tech have shown that high concentrations of acetic acid, for example more than 400 mg/L, also can cause failure of EBPR processes (Randall and Chapin, 1994).

**Phosphorus in Excess of Growth Needs.** Insufficient phosphorus can limit EBPR just as insufficient organic substrate can. In fact, one or the other will always be limiting. Obviously, there must be an excess of phosphorus compared to the quantity needed for growth before the mechanisms of EBPR can become established. Of course, if there is not an excess of phosphorus, then EBPR is not needed. However, it is important to understand which is limiting (available organic substrate or available phosphorus) because this fundamentally affects EBPR performance.

**Magnesium and Potassium.** Research performed by Pattarkine (1991) has demonstrated that both magnesium and potassium are essential nutrients for EBPR. Fortunately, the requirements are such that they are unlikely to be limiting in municipal wastewater, but they need to be considered when treating industrial wastewater or high fractions of industrial wastewater mixed with domestic wastewater. Calcium, however, is not an essential nutrient for EBPR if it is present in sufficient quantity for biomass growth. The results of a continuous-flow experiment by Pattarkine, which illustrates these relationships, are shown in Figure 8.3.

Pattarkine (1991) also established the typical ratios of magnesium and potassium taken up and released with phosphorus during EBPR. These results are shown in Figures 8.4 and 8.5. The scatter shows that there can be significant variation in the specific ratios. This is because the cations can substitute for each other to a limited extent, and calcium and some other cations also can substitute for either of them to a limited extent. However, EBPR will fail if either magnesium or potassium is absent from the wastewater or present in concentrations significantly below those indicated by the graphs.

**Required pH Range.** Studies by Tracy and Flammino (1985) have shown that EBPR mechanisms do not function at less than pH 5.4 (see Figure 8.6).

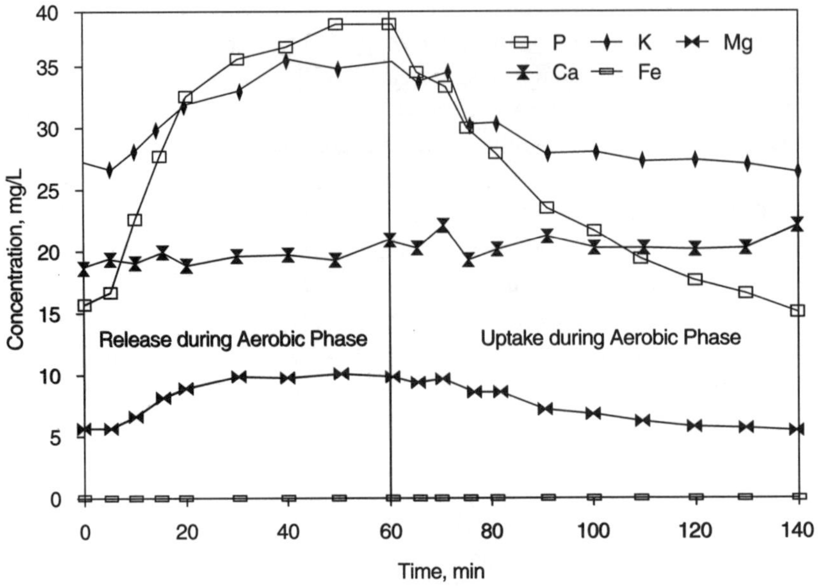


**Figure 8.3** **Typical profiles of phosphorus and metals: batch experiment (September 8, 1989) (from Randall, C.W., *et al.* [1992b] *Design and Retrofit of Wastewater Treatment Plants for Biological Nutrient Removal.* Technomic Publishing Co., Inc., Lancaster, Pa., with permission).**

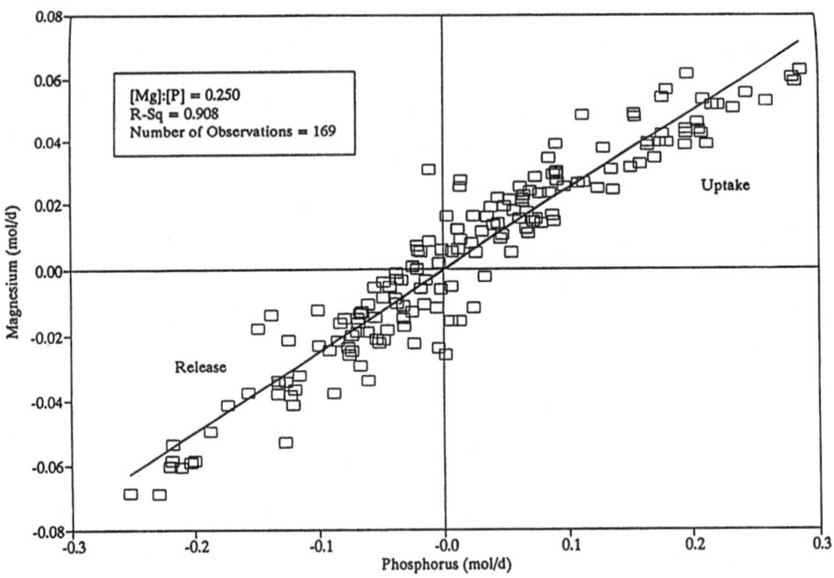


**Figure 8.4** **Release and uptake of magnesium with phosphorus: University of Cape Town (UCT) systems (COD source, municipal wastewater) (from Randall, C.W., *et al.* [1992b] *Design and Retrofit of Wastewater Treatment Plants for Biological Nutrient Removal.* Technomic Publishing Co., Inc., Lancaster, Pa., with permission).**

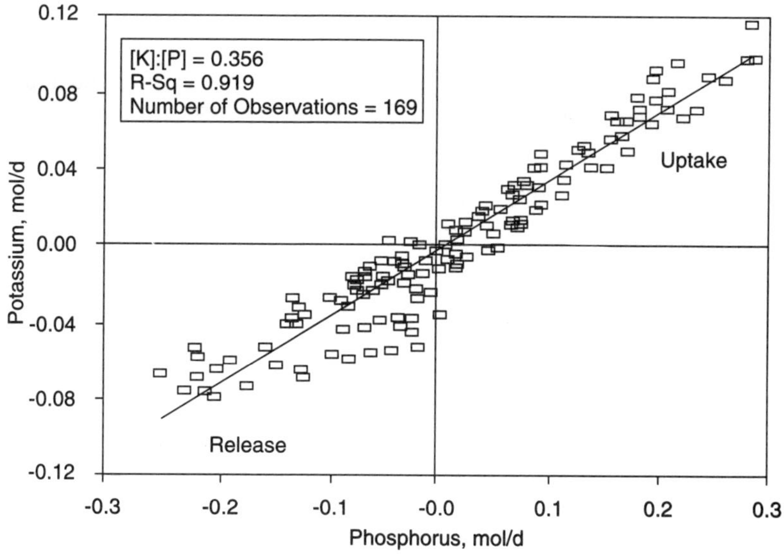


**Figure 8.5** **Release and uptake of potassium with phosphorus: UCT systems (COD source, municipal wastewater).**

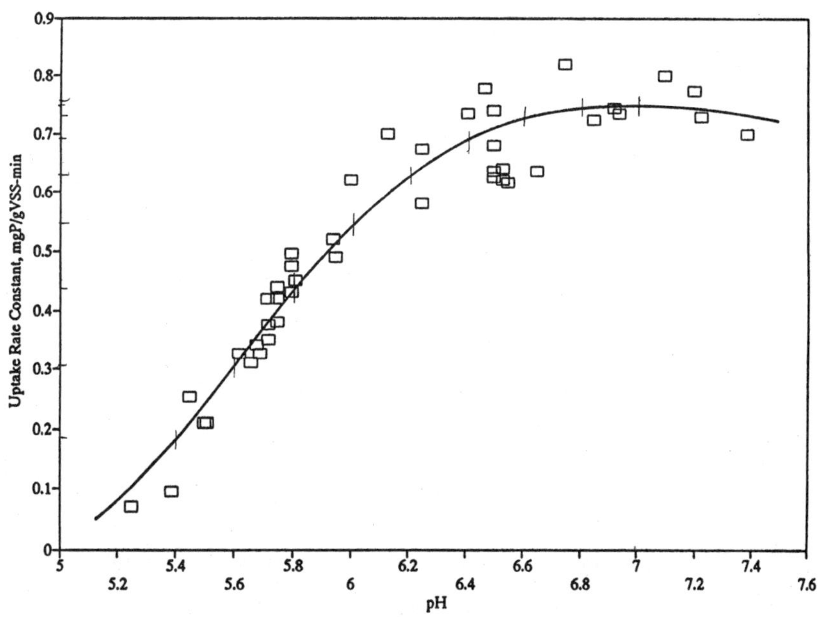


**Figure 8.6** **Effect of pH on the phosphate uptake rate constant (Tracy and Flammino, 1985).**

This result has been confirmed by Chapin (1993). Definitive studies have not been performed to study high pH values, but it is known that the EBPR mechanisms can operate in the pH range 8.5 to 9.0. Of course, chemical precipitation of phosphorus will probably be significant in this same range and may be dominant. At higher pH, it is doubtful that SP would be available in sufficient quantity for EBPR to occur.

**FACTORS AFFECTING PERFORMANCE.** It is obvious that EBPR cannot be accomplished without sufficient bio-P-available organic substrate, preferably measured as VFA, but also as approximated by either COD or BOD. The removal processes are accomplished by heterotrophic bacteria, and they require organic matter for their metabolism. It is possible, then, to state that the removal of 1 mg/L of phosphorus by EBPR will require a specific quantity of BOD, with the complication that EBPR metabolism is strongly affected by the specific organic compound available for assimilation. Therefore, the amount of phosphorus that can be removed by EBPR processes in any given system can be reasonably estimated if the influent organics have been fully characterized.

The COD consumption during biological phosphorus removal has been estimated by Ekama and Marais (1984a) to be 50 to 59 mg/L COD per mg/L P removed from South African municipal wastewater. Experiments at Virginia Tech indicate that 50 mg/L COD consumed per mg/L P removed is a conservative number for North American municipal wastewater, and it is recommended for design purposes.

**Influent Available Substrate:Available Phosphorus Ratio.** The most fundamental factor affecting the performance of an EBPR process is the available substrate:available P ratio in the influent to the biological process. This is because this ratio determines the food distribution, which determines the relative numbers of the different microorganisms in the activated sludge. Then, the performance of the biological system varies as the microbial fractions vary. This shift in heterotrophic populations between conventional and BNR systems has been demonstrated by Kavanaugh and Randall (1994).

The effluent phosphorus concentration possible with EBPR will vary with the biodegradable organic matter–to–TP (total phosphorus) ratio of the wastewater, as illustrated by Figures 8.7 and 8.8. The propensity could be best predicted using the VFA:available P ratio, but this information typically is not available. However, experience indicates that the $BOD_5$:TP ratio can be used to approximate this ratio. This is because the available substrate will be fully used in the $BOD_5$ test, and nearly all of the nonphosphate phosphorus in municipal wastewater is hydrolyzed to phosphate during biological treatment.

Figure 8.7 shows the relationship between the effluent TP concentration achieved during several full-scale and pilot-scale experiments as a function of the $BOD_5$:TP ratio of the wastewater influent to the anaerobic zone. These results indicate that a $BOD_5$:TP ratio of 20:1 or greater is needed to reliably

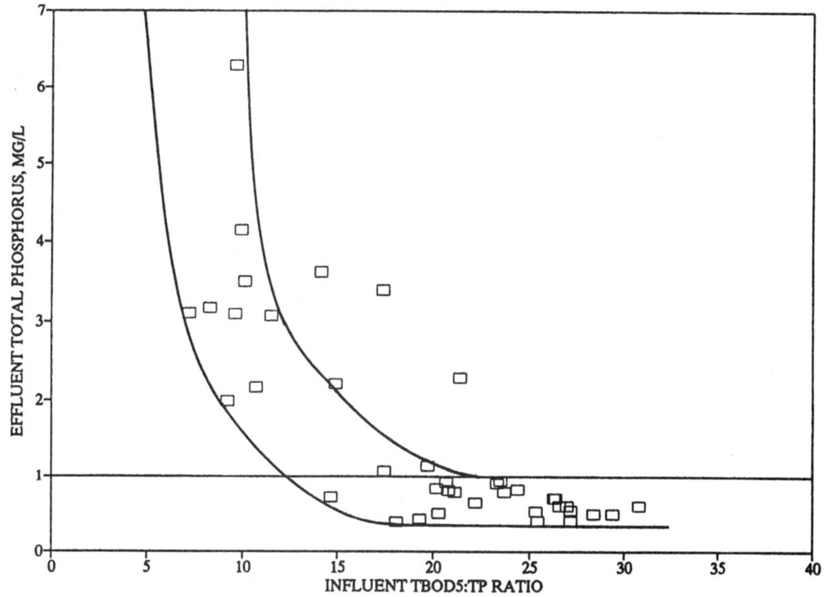


**Figure 8.7** **Effect of influent total BOD (TBOD):TP ratio on effluent TP (from Randall, C.W., *et al.* [1992b] *Design and Retrofit of Wastewater Treatment Plants for Biological Nutrient Removal.* Technomic Publishing Co., Inc., Lancaster, Pa., with permission).**

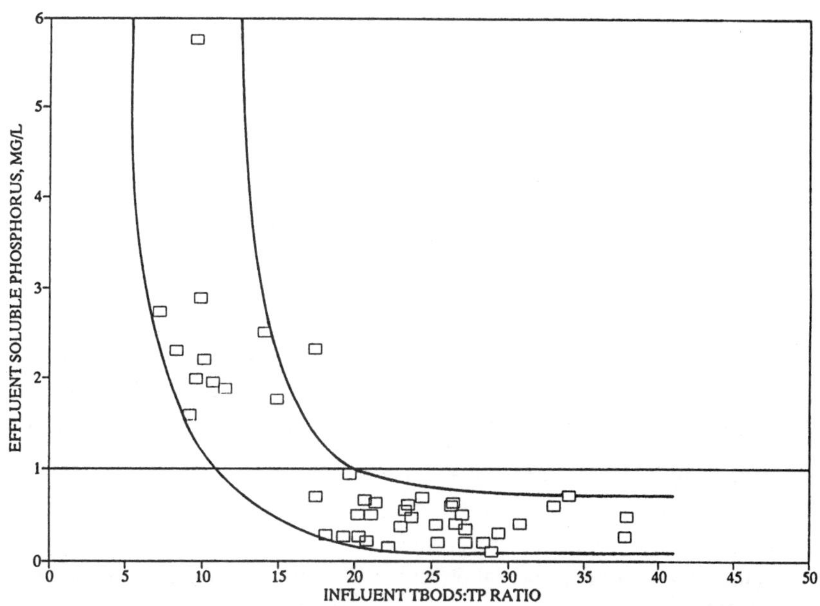


**Figure 8.8** **Effect of influent $TBOD_5$:TP ratio on effluent soluble phosphorus (from Randall, C.W., *et al.* [1992b] *Design and Retrofit of Wastewater Treatment Plants for Biological Nutrient Removal.* Technomic Publishing Co., Inc., Lancaster, Pa., with permission).**

achieve an effluent TP concentration of 1.0 mg/L or less, when gravity sedimentation is used as the last treatment step. Of 39 plotting points from nine full-scale plants and two pilot plants, with each point representing data averaged during time periods of from 1 month to 1 year, only one effluent TP concentration exceeded 1.0 mg/L when the $BOD_5$:TP ratio of the process influent wastewater was 20:1 or greater. Even lower effluent concentrations of phosphorus could have been achieved with chemically enhanced clarification or effluent filtration, as shown by the graph of effluent SP for the same data (Figure 8.8).

Whereas the $BOD_5$ data correlated reasonably well with effluent phosphorus, the data in Figures 8.7 and 8.8 clearly show that the influent ratio of $BOD_5$ to total organics (COD) varied widely, and, consequently, the $BOD_5$:TP ratio cannot be accepted as a reliable predictor of EBPR potential for all circumstances. Logically, the TCOD:TP ratio would be more reliable because COD is a more consistent measurement and a close approximation of ultimate BOD for most municipal wastewater. Graphs of the effluent soluble phosphorus and TP as a function of the process influent TCOD:TP ratio are given in Figure 8.9 for the same data set used to develop Figures 8.7 and 8.8. The graphs indicate that for TCOD:TP ratios of 45 or greater, EBPR can reliably produce an effluent quality of 1.0 mg/L TP or less. Of course, the actual effluent concentration would be a function of the mix of organic compounds in the influent, as indicated by Table 8.1, but the TCOD:TP and total BOD ($TBOD_5$):TP results shown here are conservative relative to what could be accomplished with good design and operation.

Although the TCOD:TP ratio is recommended for most raw wastewaters, the reader should note that aerobic biological pretreatment, or any set of circumstances that consumes the VFAs or compounds that can be fermented to VFAs before entering the anaerobic zone, will result in COD that cannot be used for EBPR. Figure 8.9 is not applicable for such circumstances. It should also be noted that the seasonal variation in the BOD:COD ratio reflects changes in the VFA content, which is the readily biodegradable COD, and this explains seasonal variation observed in the efficiency of some EBPR plants.

It is possible to calculate the effect of the influent BOD:TP ratio on the effluent concentration if the kinetic coefficients are known for activated-sludge treatment of the wastewater. Figure 8.10 shows calculated relationships between effluent SP and percent phosphorus in the waste sludge for different BOD:TP ratios. As the figure shows, the effluent phosphorus concentration cannot be predicted without specifying the kinetic and operating conditions.

Figure 8.11 illustrates percent phosphorus in the biomass as a function of the COD:TP ratio.

**Phosphorus Limitation Versus Biochemical Oxygen Demand Limitation.** The components of the wastewater influent to the anaerobic zone determine the rates of reaction in that zone and the performance of the

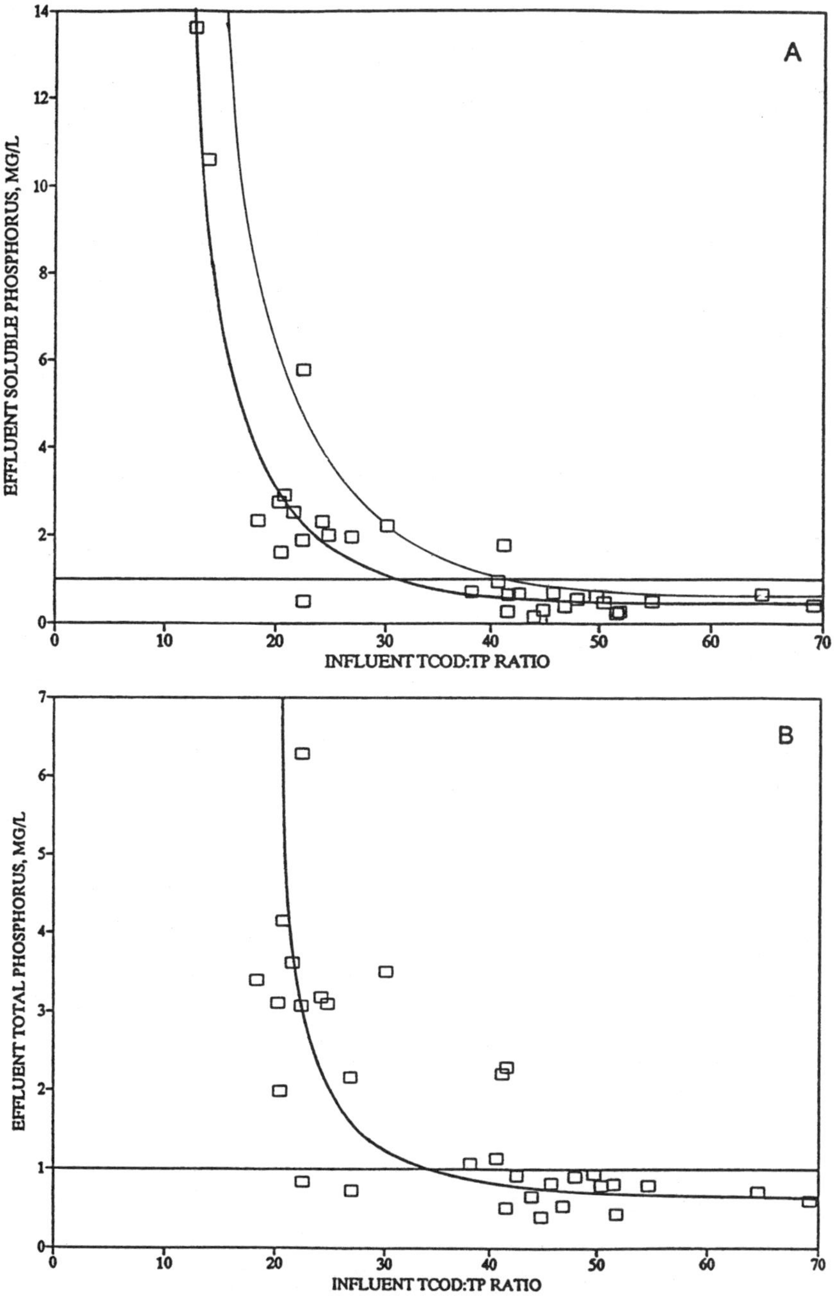


**Figure 8.9** **Effluent phosphorus as a function of the influent TCOD:TP ratio (from Randall, C.W., *et al.* [1992b] *Design and Retrofit of Wastewater Treatment Plants for Biological Nutrient Removal.* Technomic Publishing Co., Inc., Lancaster, Pa., with permission).**

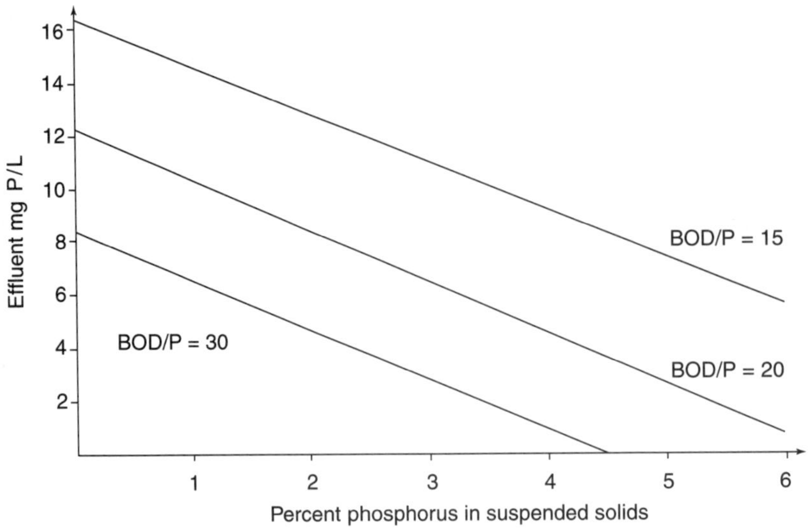


**Figure 8.10** **Calculated relationship between percent phosphorus in MLSS and effluent phosphorus (wastewater: 250 mg BOD/L, $Y_{BOD}$ = 0.75).**

EBPR process. If substantial concentrations of readily available organics such as acetic, propionic, and butyric acid are in the influent, the rates of reaction are very fast and little volume or HRT is needed for the poly-P bacteria to accomplish storage of organics with release of orthophosphates. However, the extent of this reaction will be limited by either the phosphorus available or the storable organics available. That is to say, all wastewater is either phosphorus limited or storable organics (BOD) limited with respect to EBPR. If phosphorus is limiting, the available organics will not be completely removed in the anaerobic stage and soluble organics will enter the aerobic stage. If available COD is limiting, the phosphorus removal will be limited and the desired effluent phosphorus concentration may not be achievable without either prefermentation or organic chemical addition to increase the available organics, or supplemental inorganic chemical addition to precipitate the excess phosphorus.

It should be recognized that available organics can be limiting for two reasons. They may be limiting because the influent wastewater does not contain enough total organics, as measured by BOD or COD, compared to the influent TP for removal of the phosphorus down to the desired concentration. This can be determined by the BOD:TP ratio or, preferably, by the COD:TP ratio. The available organics may also be limiting because the wastewater organic compounds have not undergone sufficient fermentation to generate the short-chain VFAs needed by the poly-P bacteria to form stored organics such as PHB. This can be determined for most municipal wastewater by comparing the biodegradable COD:TP ratio to the $BOD_5$:TP ratio. If the biodegrad-

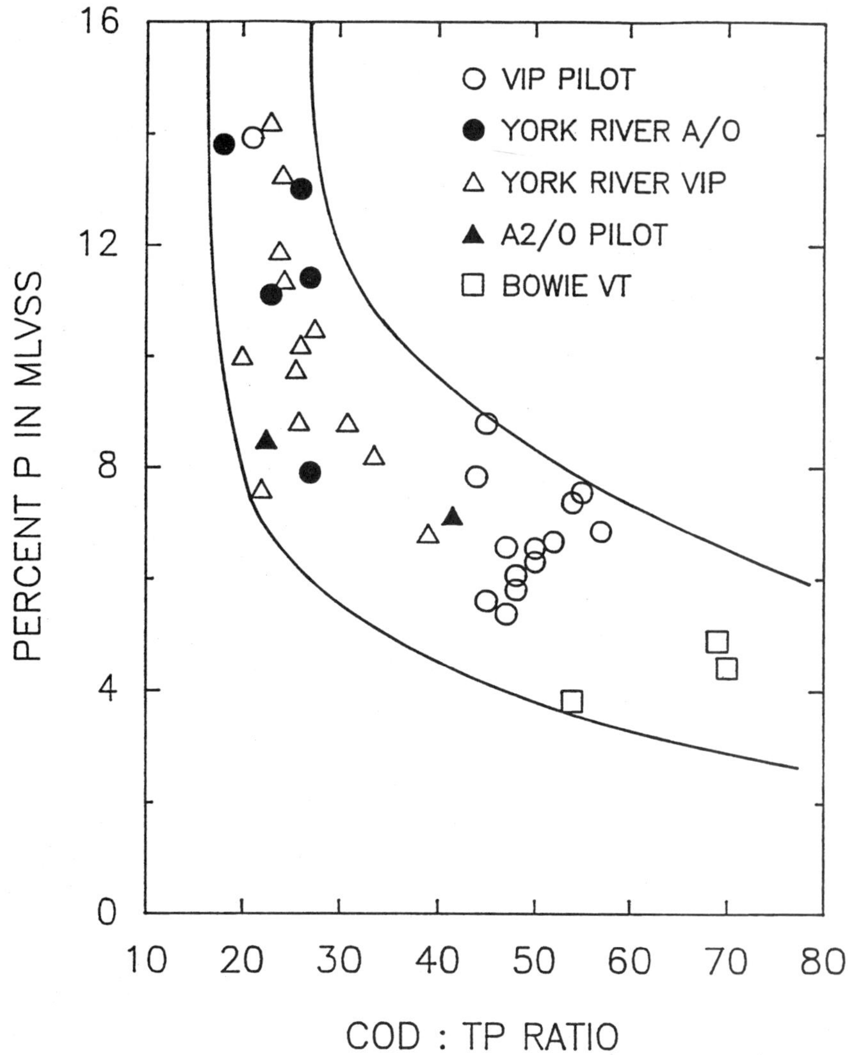


**Figure 8.11** **Effect of influent COD:TP ratio on percent phosphorus in MLVSS (from Randall, C.W., *et al.* [1992b] *Design and Retrofit of Wastewater Treatment Plants for Biological Nutrient Removal.* Technomic Publishing Co., Inc., Lancaster, Pa., with permission).**

able COD:TP ratio is considerably higher than 40:1, whereas the $BOD_5$:TP ratio is considerably lower than 20:1, then the wastewater has not undergone substantial fermentation, and design of the EBPR system should be modified to accomplish the desired level of fermentation. This can be achieved either by making the anaerobic stage larger, thus giving the fermenting bacteria a longer time to generate the desired VFAs, or by including a prefermentation unit.

**Percent Phosphorus.** Phosphates can be removed by the treatment system only through the waste sludge. Therefore, phosphorus removal is a function of the percent phosphorus in the MLVSS and the operating solids retention time (SRT). Because the growth advantage is given to the poly-P bacteria in EBPR systems by putting all of the wastewater into the anaerobic zone, they will grow to reach an equilibrium concentration with the limiting factor, that is, either phosphorus or available COD, for nearly all municipal wastewater. Thus, either phosphorus or COD will determine the fraction of poly-P bacteria in the activated-sludge population, and this will determine the percent phosphorus in the MLVSS. If COD is limiting, the activated sludge will be dominated by the poly-P bacteria and the percent phosphorus in the MLVSS will be high. However, the excess phosphorus remaining after the COD has been fully used will be discharged in the effluent, and the effluent phosphorus will tend to be greater than 1.5 mg/L. Conversely, if phosphorus is limiting, the SP in the effluent will be very low (less than 0.5 mg/L) unless secondary release occurs in either the activated-sludge basin or the secondary clarifier. The excess COD may be enough to permit significant growth of non-poly-P aerobic bacteria, however, and the poly-P bacteria may not be dominant. If so, the percent phosphorus in the MLVSS will be relatively low (4 to 6%).

Paradoxically, then, if the percent phosphorus in the MLVSS is high, the effluent phosphorus concentration may also be high. If the percent phosphorus is low, the effluent phosphorus will be low for a properly designed and operated EBPR system. Thus, the available COD:P ratio in the wastewater influent to the anaerobic zone determines both the percent phosphorus in the MLVSS and the effluent phosphorus concentration, and both parameters are inversely related to the COD:TP ratio. These relationships are shown in Figures 8.10, 8.11, and **8.12**.

Because they are both functions of the COD:P ratio, there also tends to be a correlation between MLVSS percent phosphorus and effluent SP, but it may be weak at times. For example, during the York River VIP™ (Virginia Initiative Plant) demonstration, the percent phosphorus varied from less than 7% to more than 14%, yet the effluent SP monthly averages varied only 0.49 to 1.16 mg/L when the process was not affected by excessively high flows or incomplete nitrification conditions. This occurred because the system was phosphorus rather than COD limited most of the time. In contrast, when the same plant was operated as a two-stage EBPR plant and with an excess of phosphorus, there was an increase of effluent SP from approximately 2 mg/L to almost 4 mg/L when the percent phosphorus in the MLVSS increased from 10 to 13%.

**Effect of Solids Retention Time.** Because the poly-P bacterial population changes as a function of the COD:P ratio, EBPR systems virtually are independent of biomass SRT in the range from 2 to 40 days (Barnard, 1988). Wentzel *et al.* (1988b) have noted that this phenomenon may be at least par-

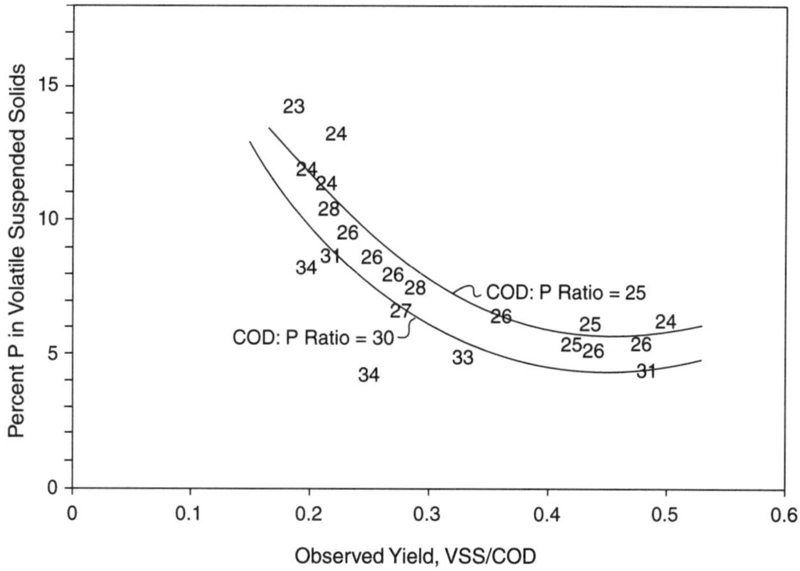


**Figure 8.12** **Percent phosphorus in the volatile suspended solids as a function of the observed yield (York River VIP™ process numbers = influent COD:P) (from Randall, C.W., *et al.* [1992b] *Design and Retrofit of Wastewater Treatment Plants for Biological Nutrient Removal.* Technomic Publishing Co., Inc., Lancaster, Pa., with permission).**

tially attributed to the low endogenous decay rate of the poly-P bacteria, measured as 0.05 $day^{-1}$ on a COD basis compared to 0.24 $day^{-1}$ for aerobic heterotrophic bacteria. This means that at longer SRTs, a proportionally larger part of the active biomass will consist of poly-P bacteria; consequently, the phosphorus content of the biomass increases with an increase in SRT.

The ability of the EBPR activated sludge to adjust its phosphorus content to the operating conditions is illustrated by the data plotted in Figure 8.12. The graph shows that the percent phosphorus in the MLVSS varied as a function of both the observed yield and the process influent COD:P ratio. Because the observed yield is determined by the operating SRT and is inversely related to it, the graph clearly shows that as the SRT increased, the percent phosphorus in the MLVSS increased.

A theoretical graph similar to Figure 8.13 has been presented by Reddy (1991) and compared with data from the VIP™ pilot-plant experiments (Daigger *et al.*, 1987b). However, comparison of that data with the York River data shows that the actual values of the relationships are not transferable between the two systems, indicating their site-specific nature. This is not surprising considering that the percent phosphorus in the MLVSS is a

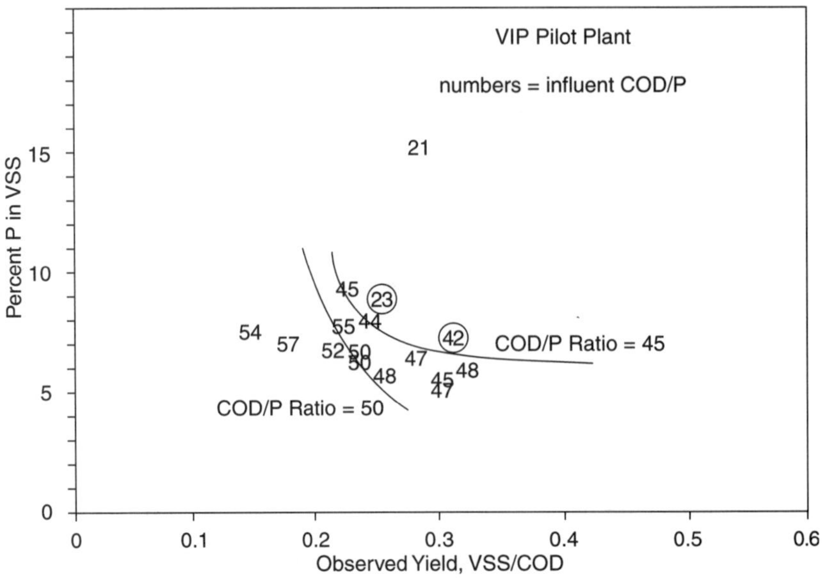


**Figure 8.13 Percent phosphorus in the volatile suspended solids as a function of the observed yield (from Randall, C.W., *et al.* [1992b] *Design and Retrofit of Wastewater Treatment Plants for Biological Nutrient Removal.* Technomic Publishing Co., Inc., Lancaster, Pa., with permission).**

function of the COD:P ratio, and these ratios differed markedly at the two plants.

It has been shown in practice that good phosphorus removal is possible at SRTs ranging from 3 to 68 days. Wentzel *et al.* (1988b) showed that although EBPR is possible at SRTs of less than 3 days, the process is not as stable and the effluent is not clear. Because the effluent suspended solids particles contain phosphorus, it is necessary to use chemical treatment or filtration to obtain a clear effluent so that a low phosphorus concentration can be achieved. The process becomes more stable at an SRT as low as 4 days, but at SRTs of 6 days and temperatures above 15°C, nitrification may interfere and an anoxic zone may be needed for process stability.

From studies with SRTs of 38 and 68 days, Kerdachi and Roberts (1982) concluded that the nonstoichiometric formation of crystals that contained phosphorus added to the stability of the EBPR process at high SRTs.

Reddy (1991) also proposed that at fixed influent organic loading and process operating conditions, an EBPR process will have a fixed maximum phosphorus storage capability. This concept is true for instantaneous conditions but is not important over the normal range of conditions for EBPR processes at equilibrium. That is to say, if the phosphorus loading is suddenly increased for a given COD loading condition, the phosphorus removal capacity of the poly-P population may be exceeded temporarily, but the poly-P population will increase with time and the phosphorus removal capacity will increase until either available phosphorus or available COD becomes limiting. Research performed by Wentzel *et al.* (1988a) showed that the actual maximum phosphorus storage capability is 32% of the MLVSS, which is much greater than that needed to remove phosphorus from municipal wastewater. For example, if the percent phosphorus of the MLVSS is 8% and the MLVSS concentration is 3 000 mg/L, operation at an SRT of 40 days would result in the removal of 6 mg/L TP. This is a typical amount that needs to be removed from North American wastewater to produce an effluent SP concentration of less than 1.0 mg/L.

The maximum phosphorus storage capacity can be used to explain why EBPR performance is independent of the operating SRT. When the SRT is increased (sludge wasting rate is decreased), the percent phosphorus increases up to the maximum value, and phosphorus removal changes little or not at all. The percent phosphorus increases because the fraction of the poly-P bacteria in the MLVSS increases, as shown by Wentzel *et al.* (1988a). Therefore, the operating SRT for two-stage EBPR processes typically will be selected to prevent or reduce nitrification, typically 4 to 5 days. Of course, SRT values this low increase waste sludge production, and this may not be desired. However, if SRT values high enough to permit partial or complete nitrification are selected, steps to prevent $NO_x$ recycle to the anaerobic zone must be taken. The University of Capetown (UCT)/VIP™ configuration (without nitrate recycle) is a potentially suitable selection, as is any configuration that removes the nitrates from the RAS before it enters the anaerobic zone. Another approach, for example, is to make the first compartment of the anaerobic zone into a denitrification tank for the return sludge. A fraction (5 to 20%) of the influent wastewater is sent to this compartment to provide organic carbon for denitrification, with the rest entering the second compartment. The approach has advantages over the UCT/VIP™ configuration in that a recirculation pump/pipe system is eliminated, and the denitrification is more efficient because of the high MLSS concentration within the compartment.

The text preceding the discussion of nitrate recycle indicates that the best way to size the anaerobic zone of an EBPR system is to make the size variable. That is, install sufficient baffling plus mixing and aeration equipment to expand or reduce the size of the anaerobic zone to obtain the desired effluent SP concentration over the likely range of influent wastewater characteristics and operating conditions. Of course, if prefermentation is used, then the influent wastewater characteristics can be controlled and a very small anaerobic zone (for example, 5%) of the total activated-sludge basin volume can be used. This is because the uptake and storage of VFAs is a rapid reaction when not limited by fermentation.

Further effects of the TCOD:TP ratio on the sizing of the anaerobic zone are explained in the following section.

**Anaerobic Hydraulic Retention.** The full-scale and pilot-scale data plotted in the two graphs of Figure 8.14 show that the sensitivity of EBPR performance to changes in the anaerobic nominal HRT is also a function of the available substrate:available P ratio in the anaerobic zone. The change in EBPR performance with a change in anaerobic HRT was relatively small when the TCOD:TP ratio varied from 42 to 68 (that is, phosphorus-limiting conditions). However, changes in anaerobic HRT between 0.5 and 2.7 hours had large effects on EBPR performance when the TCOD:TP ratio was between 20 and 43 (that is, COD-limiting conditions).

**Temperature Effects.** Research by McClintock *et al.* (1991 and 1993) revealed that the TCOD:TP ratio in the process influent also influences the temperature effects on EBPR processes. Figures 8.15 and 8.16 show percent phosphorus in the MLVSS plotted as a function of mixed liquor temperature for two sets of conditions, when phosphorus was limiting in the anaerobic zone and when COD was limiting in the anaerobic zone. It can be seen by comparing the results that temperature had a stronger effect on the system operated at a 5-day SRT when COD was limiting than when phosphorus was limiting. In fact, EBPR was completely lost at 10°C in the 5-day SRT system when COD was limiting, whereas this was not true when phosphorus was limiting.

**EFFECTS OF PRIMARY SETTLING AND SLUDGE PROCESSING.** Because the process influent available substrate:available P ratio is such a fundamental factor affecting performance of the EBPR process, it is important to understand how the operation of other wastewater processing steps affect this parameter.

The use of primary settling will substantially reduce the organic matter–to–phosphorus ratio in the wastewater and reduce the amount of phosphorus and nitrogen that can be removed by the WWTP. Recycle streams from sludge processing will have a similar effect. For some plants, the inclusion of primary settling or the recycle of sludge processing flows will result in effluent phosphorus concentrations greater than those desired. Results from the HRSD York River plant, given in Table 8.2, illustrate such changes. The raw wastewater had a $BOD_5$:TP ratio of 22.6:1. After primary settling, however, the ratio was 12.5:1. The corresponding COD:TP ratios were 41:1 and 22.4:1, respectively. The wastewater was changed from phosphorus limiting to COD limiting with respect to the anaerobic zone.

The ratio decreased so much because, in addition to primary settling, sludge processing flows were recycled back to the headworks. The recycle flows were from the primary sludge gravity thickener, the secondary sludge flotation thickener, and the filtrate from the belt filter press dewatering an anaerobically digested mixture of the two thickened sludges. The combined recycle had a $BOD_5$:TP ratio of only 5.8:1. However, only 30% of the phos-

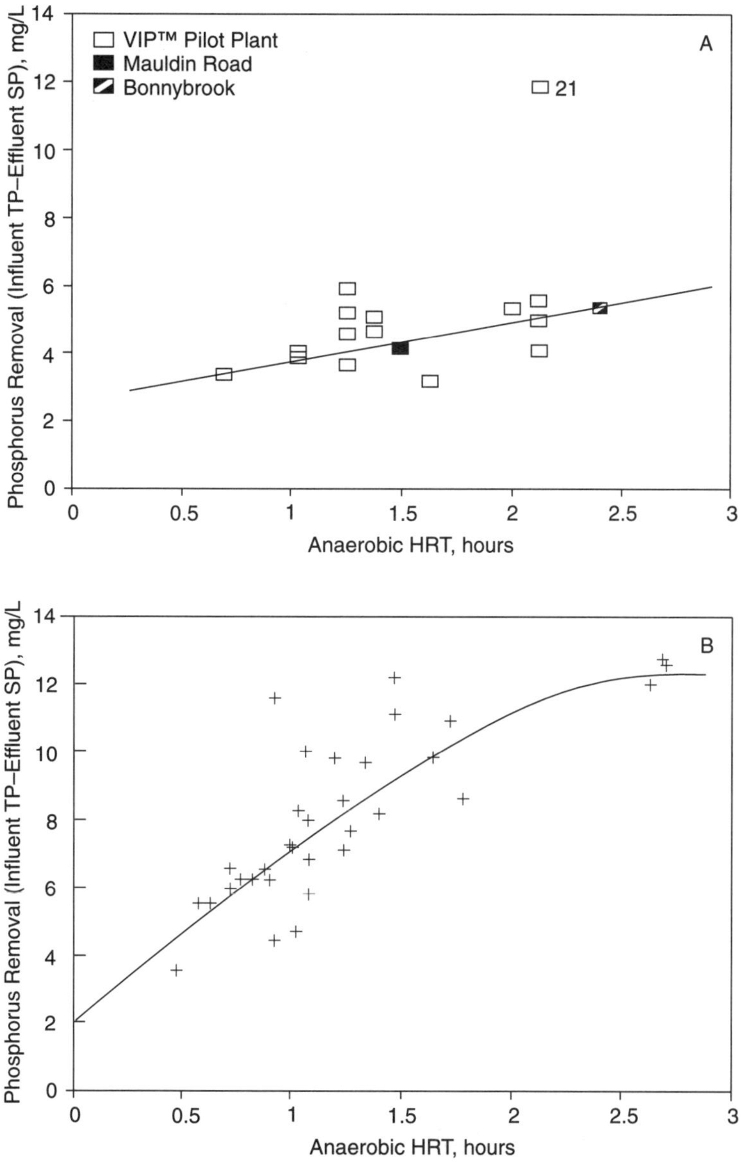


**Figure 8.14** **Effect of anaerobic HRT on biological phosphorus removal (A, TCOD:TP = 42–68; B, TCOD:TP = 20–43 [York River A/O™, $A^2$/O™, and VIP™) (from Randall, C.W., *et al.* [1992b] *Design and Retrofit of Wastewater Treatment Plants for Biological Nutrient Removal.* Technomic Publishing Co., Inc., Lancaster, Pa., with permission).**

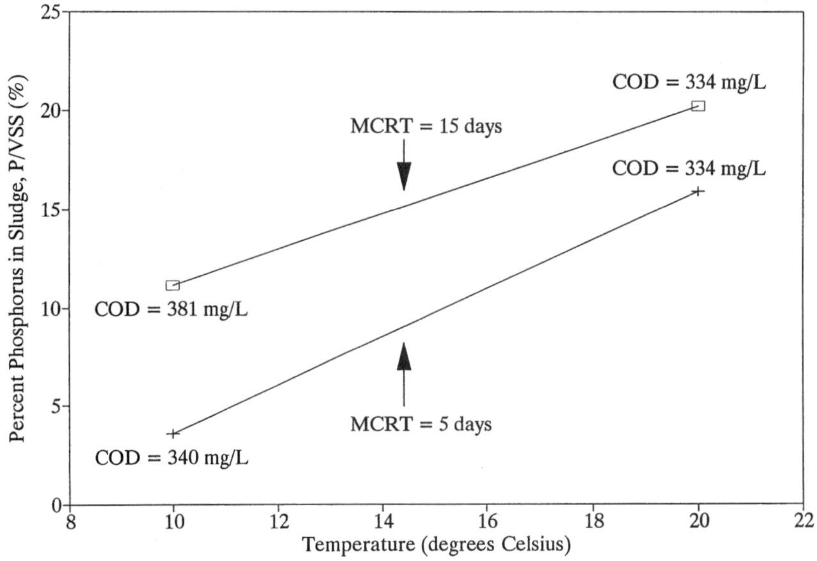


**Figure 8.15** **Effect of temperature and SRT on percent phosphorus in sludge: phosphorus-limiting conditions (VSS = volatile suspended solids).**

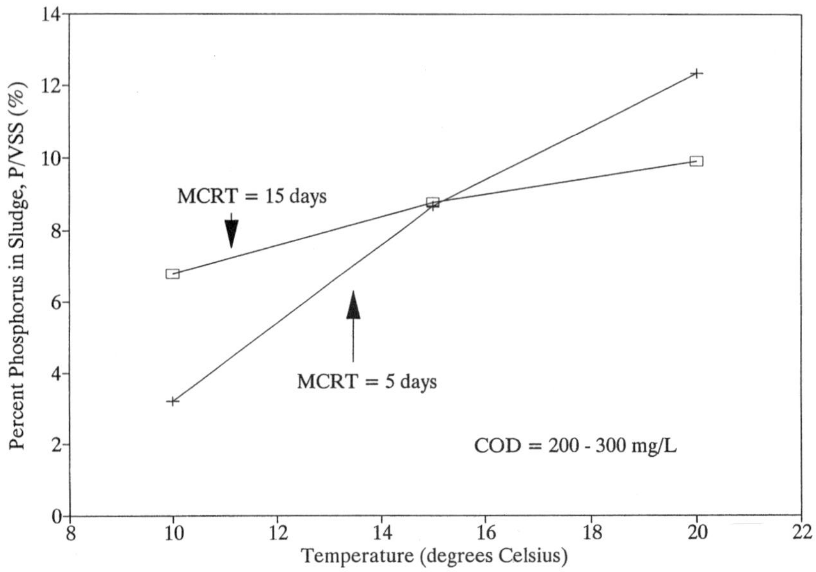


**Figure 8.16** **Effect of temperature and SRT on percent phosphorus in sludge: COD-limiting conditions (COD = 200–300 mg/L) (VSS = volatile suspended solids).**

**Table 8.2 Changes in organics-to-phosphorus ratios with recycle and primary treatment (from Randall, C.W., *et al.* [1992b] *Design and Retrofit of Wastewater Treatment Plants for Biological Nutrient Removal.* Technomic publishing Co., Inc., Lancaster, Pa., with permission).**

| Parameter | Raw influent | Combined influent | Change mg/L | Change % | Primary effluent | Decrease mg/L | Decrease % |
|---|---|---|---|---|---|---|---|
| $TBOD_5$, mg/L | 209 | 209 | 0 | 0 | 139 | 70 | 33 |
| TCOD, mg/L | 377 | 418 | 41 | 11 | 249 | 169 | 40 |
| TP, mg/L | 9.26 | 13.76 | 4.50 | 49 | 11.11 | 2.65 | 19 |
| TKN,[a] mg/L | 28.4 | 35.7 | 7.3 | 26 | 29.0 | 6.7 | 19 |
| $TBOD_5$:TP | 22.6 | 15.2 | — | −33 | 12.5 | — | 18 |
| TCOD:TP | 40.7 | 30.4 | — | −25 | 22.4 | — | 26 |

[a] Total Kjeldahl nitrogen.

phorus entering the digesters was recycled in the filter press filtrate because of the formation of struvite and other phosphorus-containing precipitates in the anaerobically digesting sludge; 70% of it was wasted with the dewatered sludge. There were no detrimental effects of the strovite formation over the 4-year demonstration period.

The primary clarifier removed a higher percentage of the organic matter from the combined flow than it did nutrients. It removed 26 and 34%, respectively, of the $BOD_5$ and COD, but only 18% each of the TP and total Kjeldahl nitrogen (TKN). The resulting nutrient concentrations to the biological process were TP 11.1 mg/L and total nitrogen (TN) 29.2 mg/L, whereas the $BOD_5$ and COD concentrations were 139 and 249 mg/L, respectively. It was also noted that primary settling reduced the organic matter:phosphorus ratios compared to those calculated for the combined flows, as shown in Table 8.2. Analysis indicated that calcium phosphate precipitates were forming to a limited extent and were being removed with the primary sludge. Similar changes should be incorporated when developing an EBPR design.

The EBPR process clearly could not produce an effluent TP concentration of 1.0 mg/L or less, given the low $BOD_5$:TP ratio, but it did remove an average of 9.4 mg/L phosphorus by biological mechanisms, except for high-flow or acclimation periods, and produced an average effluent SP concentration of 2.4 mg/L. Some MLVSS phosphorus contents as high as 15% on a dry weight basis were observed. Apparently the process was using approximately 26.5 mg/L COD to remove 1 mg/L of phosphorus. Furthermore, during the latter stages of the demonstration, it simultaneously removed TN down to a 2-month average concentration of 5.7 mg/L, while removing an average of 10.4 mg/L phosphorus and discharging an average SP concentration of 2.3

mg/L. The design factors given earlier would predict a COD requirement of 660 mg/L for these removals, yet the actual COD removal averaged only 234 mg/L. However, the $BOD_5$ removal averaged 218 mg/L, showing that all of the influent organics were in the readily biodegradable form, probably acetic acid, and much higher than predicted removals were possible. The actual $BOD_5$:P removal ratio was 21:1.

It appears that substrate-limited EBPR systems can remove considerably more nutrient than predicted but cannot produce an effluent TP concentration of less than 1.0 mg/L at the same time. The results also indicated that stored substrate can be used simultaneously for both nitrogen and phosphorus removal.

**Fundamentals of Prefermentation.** Given the importance of the available substrate:available P concentrations in the anaerobic zone to process performance, and the desirability of having phosphorus-limiting conditions so that low effluent phosphorus concentrations can be achieved, methods for supplementing the VFAs entering the anaerobic zone have been developed and are frequently used for design so that more predictable results can be obtained.

There are three different methods that can be used to hydrolyze primary sludge for VFA production—biological, chemical, and thermal hydrolysis. The biological method was pioneered by Barnard (1992) and has been in widespread use for several years. More recently, all three methods of hydrolysis were studied as part of the HYPRO project, which was funded by the European Economic Community and carried out by a number of Scandinavian companies and institutions (Henze and Harremoës, 1990).

In the biological hydrolysis process, anaerobic bacteria hydrolyze the organic particulate matter by means of extracellular enzymes. Biological hydrolysis is the initial, that is, acid fermentation, step of anaerobic digestion, and it can be segregated from the methane fermentation step by the selection of a suitable SRT. The products of biological prefermentation are rich in acetic acid and other VFAs and are ideal for EBPR. During the HYPRO studies, 10 to 15% of the TCOD was solubilized, and 75% of the soluble COD was found to consist of VFAs.

In the chemical hydrolysis process, either acid was added to adjust the pH to 2 or base was added to pH 14. A retention time of approximately 4 hours was used. Chemical hydrolysis with sodium hydroxide produced the same hydrolysis yield as biological hydrolysis, whereas hydrochloric acid and calcium hydroxide gave approximately half as much yield, but all chemical hydrolysis products contained very little VFA.

Thermal hydrolysis was studied in the HYPRO project using a temperature of 160°C and 6 bars of pressure, with a retention time of approximately 2 hours. Tests showed that thermal hydrolysis produced a hydrolysis yield approximately equal to biological yield, but the hydrolysis products contained very little VFA. The addition of sulfuric acid for pH adjustment along with thermal treatment resulted in hydrolysis of 35 to 40% of the TCOD, but, again, few VFAs were produced.

The preliminary HYPRO results showed that biological hydrolysis is superior to chemical and thermal hydrolysis for EBPR purposes.

**Biological Prefermentation.** The HYPRO project studied biological hydrolysis of both primary sludge and chemically pre-precipitated sludge at temperatures of 10, 20, and 30°C and retention times of 1, 2, and 3 days. The experiments were performed using a continuously operated pilot-scale 1-$m^3$ tank with mixing.

The results showed that COD hydrolysis increased with time for all retention times used. The optimum temperature was found to be between 20 and 30°C for the retention times of 1 to 3 days. At 10°C the hydrolysis rate was extremely slow and comparable to the rate observed for sludge stored at 1 to 2°C. The effect of precipitants on sludge hydrolysis rate was found to be very small and of no significance. It was concluded that the optimum operating conditions for the biological hydrolysis process would be a retention time of approximately 2 to 4 days and a process temperature of 20 to 30°C. Under these conditions, the pH will naturally adjust to a level between 5.3 and 6.0.

The fundamentals of biological prefermentation design and operation have been described in considerable detail by Barnard (1992), and his observations are summarized in the remainder of this section.

The alternative to the generation of VFA in the anaerobic zone of the activated-sludge plant is to use a prefermenter in which the fermentation step is independent of the activated-sludge system, and carbon available in the incoming raw wastewater is fermented to generate short-chain VFAs. In many instances, sufficient VFA is present in the raw wastewater to obviate any special process for their generation. This may have resulted from fermentation in slow-flowing outfall sewers or in pumping stations and force mains. Where a substantial portion of the waste is pumped to the plant through long force mains, excellent phosphorus removal may result without any real effort. This phenomenon has been observed at several plants in the U.S. and is particularly typical among plants in tidewater areas (where the sewers are long and the slopes are small). However, it varies with temperature and may be quite low in the winter, as observed at York River, Virginia.

Typically, it has been found that without any incidental fermentation in outfalls of force mains or deliberate fermentation in primary sedimentation tanks, it is difficult to produce sufficient VFA in the anaerobic zone of an enhanced BNR activated-sludge plant when the influent is weak (influent BOD less than 200 mg/L) and winter mixed liquor temperatures are below 17°C. Somewhere below this point, the additional secondary release (phosphorus release without substrate storage) is more than the uptake possible through the production of VFAs, and enlargement of the anaerobic zone becomes counterproductive.

Thus, while it has been found that with stronger wastes and high winter temperatures there may be merit in enlarging the anaerobic zone, for weaker wastes and lower winter temperatures, the smaller the anaerobic basin the better, provided sufficient VFA is produced in a prefermentation step. The anaerobic zone then becomes merely an anaerobic contact zone for the uptake and storage of VFAs with the release of phosphate. If sufficient VFAs are introduced, the total exclusion of air and $NO_x$ is not required and strict anaerobic conditions may be counterproductive. Gerber *et al.* (1986) have shown that when acetates were fed to sludge from a nutrient removal plant, primary release of phosphates commenced immediately even in the presence of nitrates and DO. At this point it is not yet clear which is the more harmful under these conditions—excessive anaerobiosis or some oxygen or nitrate. Because no more fermentation is required, oxygen or nitrate will allow only some organisms to utilize substrate, whereas secondary release under anaerobic conditions would require more VFAs for the uptake of such released phosphates.

**Design of Systems for Optimizing Volatile Fatty Acid Production.** The acetates and propionates produced are very soluble and can be elutriated from the sludge. The following methods have been proposed for achieving this:

1. Dewater the underflow from gravity thickeners—initial attempts for elutriating the VFAs from the thickener underflow at Kelowna, British Columbia (Canada), consisted of passing the underflow over a static screen for the removal of the coarser material (Barnard *et al.*, 1984). This will only work when spraying water on the screen to continuously wash the sludge, but this resulted in too much solid material passing through the screen. An alternative to this method could be the use of some dewatering device such as a centrifuge or belt filter press, with return of the liquid to the anaerobic basin. Note that this refers to the dewatering of thickened primary sludge and not of digested sludge.
2. Return the thickener overflow directly to the anaerobic basin—this would be normal practice in combination with the BNR plants but does not optimize the elutriation of VFA. In the thickening process, liquid containing VFA is expelled continuously, but as the sludge thickens and the optimal VFA production is reached, less liquid is expelled and much of the VFA is not passed to the activated-sludge unit unless the next step is sludge dewatering.
3. Recycle on the primary settling tank (PST) or gravity thickener—one of the simplest ways of producing VFAs is to allow a sludge blanket to form within the PST and to slowly recycle the sludge to the inlet. This concept is the basis of the "activated primary sedimentation tank" or a-PST (Barnard, 1984). Figure 8.17 is a schematic of an a-PST arrangement. The constant recycling process inoculates the in-

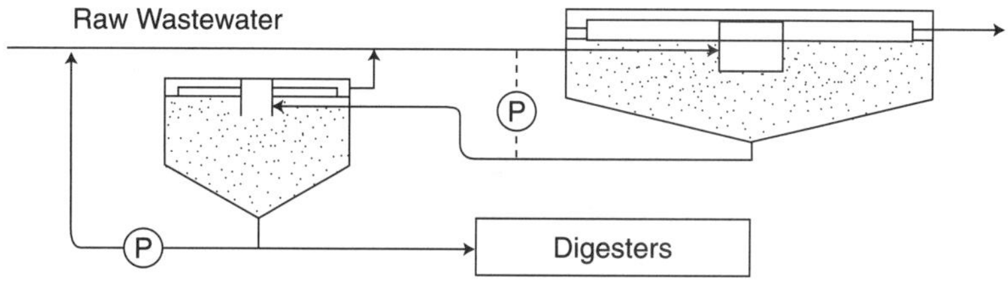


**Figure 8.17 The activated primary sedimentation tank (from Barnard, J.L. [1984] Activated Primary Tanks for Phosphate Removal. *Water SA*, 10, 121, with permission).**

coming solids with actively fermenting organisms, elutriates the VFA formed in the sludge blanket, and prevents the formation of methane and hydrogen sulphide through the constant exposure to air with every recycle. Such recycle may also be applied to the gravity thickener or from the thickener back to the PST. Further details regarding operation have been provided by Barnard (1984).

Because any continuous system will eventually lead to a slow buildup of methane organisms, the sludge should be wasted completely at regular intervals. These intervals vary from one location to the next and even from season to season and must be optimized for each individual plant.

4. Completely mixed fermenter—the sludge removed daily from the PST can be discharged to a completely mixed fermentation unit and from there to a thickener for removal of the liquid containing the VFAs. The SRT of the sludge can be controlled easily by the volume of sludge wasted per day from the PST. For example, if the minimum volume of sludge wasted per day from the PST is $x$ m$^3$ (that is, the maximum thickening possible in the PST), the volume of the fermenter may be 8$x$, which will result in a maximum SRT of 8 days. By wasting for longer periods from each PST and thus wasting a larger volume per day, the retention time in the fermenter is reduced and the SRT can be controlled. Even though this may mean that more liquid is wasted, the liquid will be removed and discharged to the BNR plant. Again, provision should be made for emptying the fermenter at regular intervals to prevent the development of methane organisms, which tend to develop and build up slowly in spite of the adverse conditions. Alternatively, one may provide for regular aeration of the fermenter during short periods to counteract the development of methane organisms. This has been successfully done at Johannesburg, South Africa, plants.
5. Batch systems—at Pietermaritzburg, Natal (South Africa), four small redundant digesters were used for batch treatment of the primary

sludge. The PST underflow is pumped for 2 days to one batch unit, the pumping automatically switched to the second, and so forth. At the other end of the batch units, pumps automatically empty the batch units to the thickener on a constant basis, with the thickened sludge being discharged to the digesters and the VFA-rich liquid discharged to the BNR plant. This has the advantage of full automation and achieves optimization of the process in an odorless situation.

6. Some experimental information—acid fermentation laboratory experiments performed at the University of Capetown by Wentzel *et al.* (1988b), with initial VSS concentrations ranging from 3 000 to 40 000 mg/L and runover periods of 9 to 19 days after the start of fermentation, found that the lag phase before acid starts varied from 0 to 7 days, that the peak acid concentration was obtained 6 to 9 days after the start of fermentation, and that production followed a first-order reaction.

   A maximum yield of approximately 0.125 kg VFA as COD was obtained per kilogram initial VSS as COD. For an SRT of 3 to 4 days, the yield was 0.075 kg/kg. The VFAs were produced in proportions of 1:1:0.08:0.07 for acetic, propionic, butyric, and valeric acids, respectively.

   When four reactors in series were operated semicontinuously using primary sludge at VSS concentrations between 37 000 and 57 000 mg/L, the following observations were made: There was no lag period for acid production, between 6 and 10% of the VSS (as COD) was solubilized to VFA (as COD), and approximately 50 to 70% of the 0.45-$\mu$m-filtered COD was VFA. The acetic:propionic:butyric: valeric acid ratio of the VFA was 1:1:0.3:0.1, and of the total VFA generated, 43% was acetic, 41% propionic, and 16% butyric and valeric.

   The series system had no significant advantage over a single reactor with equal retention time.

**Return of Organic Carbon.** When designing a plant with EBPR, cognizance should be taken of the return of organic carbon in the form of BOD or COD to the biological process. This is not more than would typically be expected when thickening primary sludge, but if extensive fermentation is required, the return of COD may be substantial and must be reflected in the calculations for the BNR process.

**Monitoring of the Process.** Ideally, acid fermentation will provide sufficient acetic acid to obtain approximately 50 mg/L acetic acid as COD in the primary effluent to the anaerobic zone. It has been demonstrated in practice that it is possible to remove phosphates biologically to levels below 0.1 mg/L as phosphorus (after filtration). However, if this level of effluent phosphorus is desired, a more scientific method for the monitoring of the prefermenter is required, and investment in a gas chromatograph is recommended. With the assistance of good information, a seasonal history of the performance of the plant can be built up. If this is not possible, a check should be run on the performance of the fermenter by a commercial laboratory.

The literature is confusing about the ideal SRT for optimal results with a prefermenter. This is probably because the type and condition of the sludge arriving at different plants vary greatly. Some sludges may contain easily degradable solids; others may contain more complex proteins. The solids or the liquid arriving at the plant may be fresh or may have undergone a long period of prefermentation. The wastewater may be strong or quite weak, and the wastewater temperature may vary more than 20°C. Attempts have been made to produce models, but in the end the matter needs to be solved by experiments at the plant.

**Practical Points about Prefermenters.** The less mechanical the equipment used, the better, especially because the unit must be enclosed for odor control. This creates a corrosive atmosphere that attacks not only metal but also the concrete used for the cover. Because a high degree of stirring is not desired, external pumps may be considered. The cover may be aluminum or may be in touch with the liquid, which would prevent concrete corrosion. Provision should be made for venting the air space to an odor control unit. The effluent from a prefermenter should be handled with care to prevent odor problems.

The recycling of sludge in the fermenter leads to the buildup of a fibrous gooey substance that is difficult to handle, blocks pumps, and will accumulate in the digesters. Though a fine screen in the sludge line to the fermenter will help, the best location for this screen is on the recycle line because most of the fibrous mass seems to reform in the fermenter. Because the substance is odorous and difficult to handle, a press is required to squeeze the liquid out and return it to the fermenter or the PST. The remainder of the fibers could be wasted to the screenings. Because these fibers would mostly end up in the digesters, the cost of removing them is worthwhile because of the additional benefit.

The importance of the fermenter for reliable biological phosphorus removal would require that sufficient redundancy be provided to ensure that the production and supply of VFA is not interrupted. Two or more fermenters and at least two thickeners would be required. Alternatively, provision could be made for recycling around the primary sedimentation tanks, of which there should be more than one.

There are still a number of unknowns and contradictions in the art of prefermenting. For example, the recycle of digester supernatant, even though one can measure what appears to be adequate VFAs, seems to inhibit rather than help the process. This has been attributed to the presence of hydrogen sulfide, but one of the most reliable plants for phosphate removal had an influent with a very high hydrogen sulfide concentration. Rabinowitz and Oldham (1985) observed that while the VFAs kept increasing as the SRT of the fermenter increased up to 10 days, optimal phosphate removal was observed at an SRT in the fermenter of between 6 and 8 days.

When using two primary sedimentation tanks as shown in Figure 8.18, there are many possible ways in which the WWTP could be operated. In this case, the sludge is separately recycled back to the influent of each tank, the pumps are connected directly to the underflow of each tank, and the lines are interconnected by means of two-way valves. Thus, any of approximately 12 operating combinations is possible. For example, the underflow of tank 1 can be pumped to tank 2 while the underflow of tank 2 is pumped to the digesters or recycled on itself and then pumped to the digesters. It is thus possible to keep fermentation going in one tank while cleaning the sludge out of the other.

*FACTORS AFFECTING SUBSTRATE STORAGE.* Because of the interrelationships described previously, the available COD:available P ratio of the wastewater influent to the biological process is the most important factor affecting the performance of EBPR systems and should be used for design. As Figures 8.9, 8.11, and 8.12 show, the TCOD:TP ratio approximates the available COD:available P ratio sufficiently for preliminary design purposes. The phosphorus removal actually obtained, however, is determined by the amount of stored substrate contained by the poly-P bacteria when they enter

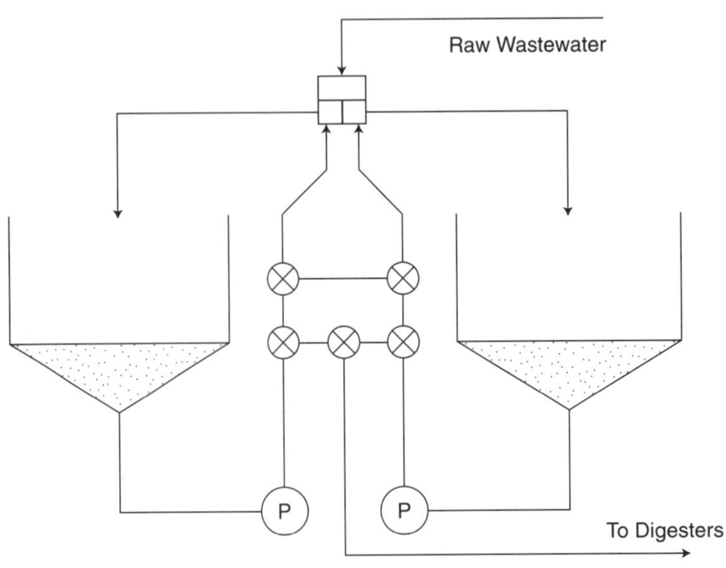


**Figure 8.18 Arrangement of two activated primary tanks (from Barnard, J.L. [1984] Activated Primary Tanks for Phosphate Removal. *Water SA*, 10, 121, with permission).**

the aerobic zone, and other factors affect the efficiency of storage in the anaerobic zone.

The amount of substrate stored by poly-P bacteria in the anaerobic zone is a function of

- The quantity of VFAs in the anaerobic zone,
- The type(s) of VFAs in the anaerobic zone,
- The mass of poly-P bacteria in the anaerobic zone,
- The type(s) of poly-P bacteria in the anaerobic zone,
- The quantity of stored phosphorus contained by the poly-P bacteria when they enter the anaerobic zone,
- The mass of the electron acceptors, DO and $NO_x$, entering the anaerobic zone per unit time,
- The cations available in the anaerobic zone,
- The HRT of the anaerobic zone,
- The temperature of the anaerobic zone mixed liquor, and
- The pH of the anaerobic zone mixed liquor.

The last four factors may prevent completion of the organic storage reaction for any given operating conditions and, consequently, limit phosphorus removal.

The reactions in the aerobic zone also determine system phosphorus removal. There must be sufficient DO for the poly-P bacteria to completely metabolize the stored organics, and the DO concentration should not limit the rate of oxygen transfer to the cells. There should also be enough DO in the bioreactor effluent to prevent phosphorus release in the secondary clarifier, but the amount contained in the RAS should be near zero. Dissolved oxygen concentrations ranging from 0.7 mg/L at the influent end of the aerobic zone to 3.0 mg/L at the effluent end appear to be satisfactory for high-rate plug-flow systems (Barnard, 1988). Experience with high internal recycle rate, high-SRT oxidation ditch systems has shown that performance is determined by the mass of oxygen transferred rather than the concentration maintained, and it may be necessary to maintain the DO concentration below 0.3 mg/L to prevent recycle of excessive oxygen (Sen *et al.*, 1990a). When this is the case, redox monitoring should be used instead of DO monitoring for aeration control purposes, or some other surrogate parameter such as effluent turbidity or alkalinity can be used (Sen *et al.*, 1990b).

There is also evidence that phosphorus removal in the aerobic zone does not begin until all of the VFAs have been removed from solution (Pattarkine, 1991). Therefore, if the VFAs are not completely assimilated in the anaerobic zone, the size of the aerobic zone should be enlarged to ensure completeness of the phosphorus removal reaction after soluble substrate removal. The breakthrough of readily available organics to the aerobic zone is also likely to stimulate filamentous growth, particularly if the aerobic zone is completely mixed rather than plug flow. Organic breakthrough is unlikely to be significant when treating North American municipal wastewater and can easily be avoided by including an anoxic zone ahead of the aerobic zone if conditions are such that it would otherwise occur.

**Electron Acceptor Recycle/Entrainment.** The efficiency of both nitrogen removal and phosphorus removal can be reduced by conditions that result in the metabolism of usable substrate by other biochemical pathways. For example, design or operating conditions can result in the recycle or entrainment of DO and nitrates in the flows to the anaerobic zone, or DO to the anoxic zone. Bacteria that accomplish BNR will use both DO and $NO_x$ forms as electron acceptors and metabolize the available organics in the anaerobic zone to carbon dioxide and water, before they can be stored for phosphorus removal, to the extent that DO and $NO_x$ are available. The EBPR bacteria will use DO and $NO_x$ preferentially, when available, because they obtain more energy for growth through use of these electron acceptors than through anaerobic substrate storage. Similarly, denitrifying bacteria will use DO instead of $NO_x$ in the anoxic zone if DO is available because they obtain more energy for growth using DO. These reactions reduce the organic matter available for nitrogen and phosphorus removal and, consequently, may limit the effluent quality that can be achieved. Therefore, good design and operation will reduce the quantities of the unwanted electron acceptors available to the bacteria in the anaerobic zone.

One milligram per litre of nitrate as nitrogen is equivalent to 2.86 mg/L of DO for BOD and COD removal purposes. This equivalency can be used to calculate the reduction in available substrate for EBPR, if the amount of DO and $NO_x$ entering the anaerobic zone is known.

**Temperature Effects.** There has been considerable controversy regarding the effects of temperature on EBPR processes. While it is agreed that EBPR processes are relatively insensitive to temperature changes compared to nitrification, early research reports concluded that poly-P bacteria were less sensitive to temperature than typical activated-sludge bacteria. However, these claims have proven to be false. Sell (1981) concluded that the bacteria responsible for EBPR are psychrophiles that grow better at lower temperatures than the typical activated-sludge BOD-removing bacteria, which tend to have their maximum growth rate at 20°C. Moreover, in subsequent laboratory-scale automated fill-and-draw studies conducted over the temperature range of 5 to 15°C, Sell (1981) found that there was no difference in EBPR efficiency at temperatures of 15 and 10°C, but there was a substantial improvement in performance when the temperature was lowered to 5°C. Similar results were obtained when the experiments were repeated. Ekama *et al.* (1984), also based on laboratory-scale research, reported greater EBPR at 14°C than at 22°C, which seemingly confirmed the Air Products and Chemicals results. It should be noted that both research teams used small-scale anaerobic units that were sealed to reduce DO entrainment.

More recent full-scale, pilot-scale, and laboratory-scale results have shown that EBPR is reduced as temperature decreases, but different mechanisms apply for different conditions. Waltrip (1991) reported a temperature dependency for the EBPR process during the HRSD York River plant VIP™ demonstration (Figure 8.19). The correlation is weak and the decrease is relatively small, but a trend is discernible. It has been concluded that the decrease in EBPR was caused by an increase in the entrainment of DO to the process influent flow and across the anaerobic zone surface as a result of the higher DO transfer driving force resulting from the lower temperatures. In contrast to the laboratory-scale studies, the anaerobic zone was open to the atmosphere. The data in his report can be fit by a $\theta$ of 1.045 for temperatures below 20°C (13 to 20°C) and by a $\theta$ of 1.037 5 for temperatures from 20 to 27°C. The reference phosphorus removal value at 20°C was 4.8 mg/L, and the average SRT for all of the data was 5.8 days. For comparison, the reference phosphorus removal value at 15°C in the Air Products and Chemicals study was 3.30 mg/L.

While the York River results indicate secondary effects that can result in a reduction in EBPR with lower temperatures, recent laboratory- and pilot-scale research has shown conclusively that poly-P bacteria, or at least the EBPR functions, wash out of activated sludge before typical heterotrophic bacteria. Pilot-scale studies by McClintock *et al.* (1991) at 10°C with an SRT of 5 days resulted in complete failure of the EBPR process, while heterotrophic COD removal continued with high efficiency.

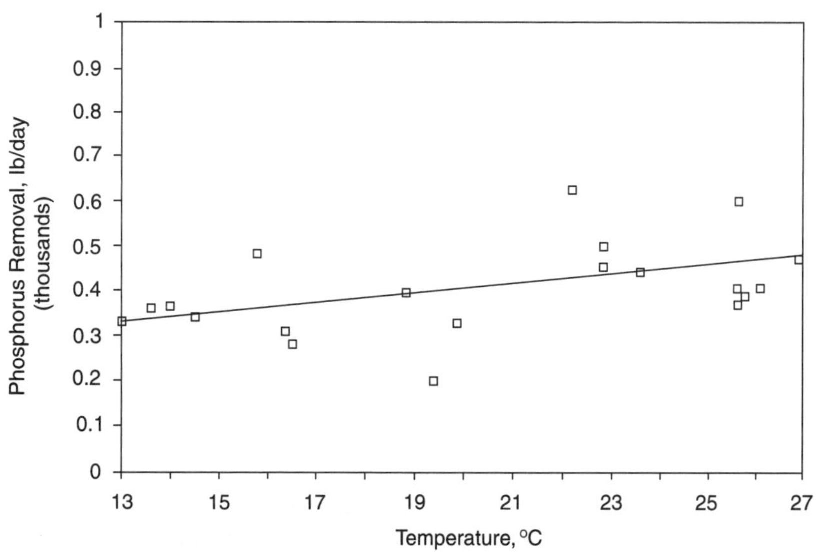


**Figure 8.19** **Temperature effects on mass phosphorus removal (VIP™ configuration, HRSD York River plant; SRT = 5.8 days; phosphorus removal = 4.8 mg/L at 20°C) (lb/d × 5.250 = mg/s).**

Mamais and Jenkins (1992) performed temperature and SRT studies of EBPR systems in the laboratory and found that EBPR functions can be washed out of activated-sludge systems without losing BOD removal efficiency. They also found that the EBPR aerobic SRT washout value was clearly a function of the mixed liquor temperature. Their results, combined with a washout value observed at the Los Angeles Hyperion Plant and the washout value reported by McClintock *et al.* (1991), are shown in Figure 8.20. This graph should be used to determine the minimum SRT for EBPR design and operation. However, although this graph can be used to determine washout SRTs, it does not provide information regarding temperature effects when the EBPR system is operated at SRTs greater than the washout value.

The pilot-scale studies by McClintock *et al.* (1991) in the temperature range from 10 to 20°C showed that EBPR was affected by temperature above washout, but the temperature coefficient was a function of both the influent COD:P ratio and the operating SRT, as previously shown in Figures 8.15 and 8.16. A temperature-adjustment $\theta$ of 1.038 was calculated for a 15-day SRT, whereas a value of 1.137 was determined for a 5-day SRT. However, the EBPR process completely failed at 10°C and 5-day SRT. The reference phosphorus removal values at 20°C were 6.7 and 8.1 mg/L, respectively, at the 15- and 5-day SRTs, somewhat higher than for the studies previously mentioned. More research on the effects of temperature on EBPR is needed, but

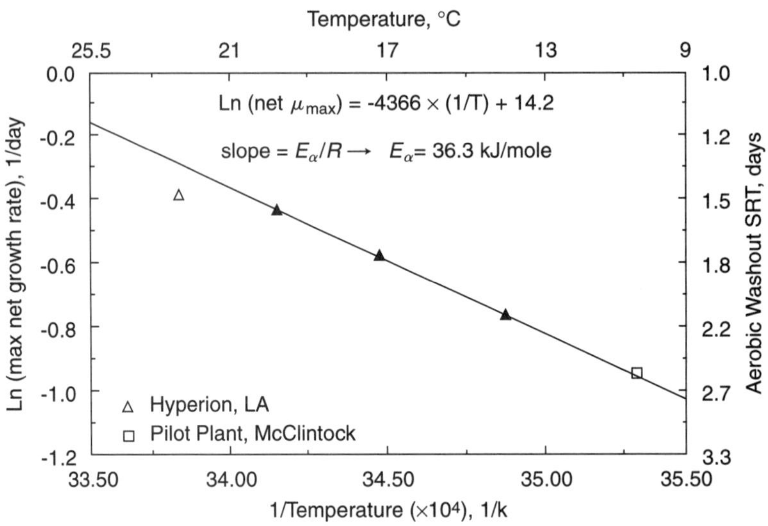


**Figure 8.20 Enhanced biological phosphorus removal washout as a function of temperature. (Reprinted from *Water Science and Technology,* Vol. 26, Mamais, D., and Jenkins, D., The Effects of MCRT and Temperature on Enhanced Biological Phosphorus Removal, p. 955, Copyright 1992, with kind permission from Elsevier Science Ltd, The Boulevard, Langford Lane, Kidlington OX5 1GB, UK.)**

the indications are that the effects are small when the system is operated above washout conditions. Based on the limited information presented, it is recommended that a temperature-correction $\theta$ value of 1.045 be used.

**DETERMINATION OF DESIGN REQUIREMENTS. Sizing the Anaerobic Zone.** The uptake of VFAs in the anaerobic zone is quite rapid at temperatures of 14°C and above, yet follows first-order kinetics (Pattarkine, 1991). Because it is rapid, the size (nominal HRT) of the zone tends to be less significant than the operating SRT (the SRT in the anaerobic zone) when there is an excess of VFAs in the process influent, relative to the available phosphorus (high TCOD:TP ratio). The low sensitivity between process performance and anaerobic zone HRT under these conditions was previously shown in graph A of Figure 8.14. This low sensitivity also can be shown to exist between EBPR performance and the anaerobic mass fraction for steady-state conditions because the poly-P fraction of the population will increase as the anaerobic mass fraction decreases, within the typical limits of operation. Consequently, the anaerobic zone can be designed with a short HRT (<1.0 hour), and as a small fraction (5%) of the total activated-sludge basin volume, if the zone influent contains an excess of VFAs. The best way to ensure an excess of VFAs is to provide them by prefermentation or some other form of VFA addition.

Because VFA removal is first order, it is possible to size the anaerobic zone based on the VFA removal rate; however, Marickovich (1995) has shown that the VFA removal rate in the zone is a function of the zone's mixed liquor temperature. This temperature dependency is incompletely known at present, however, so it is not possible to accurately base an anaerobic zone design on the VFA removal rate without a pilot study. Therefore, it is recommended that the anaerobic zones of EBPR systems be designed as a fraction of the total basin volume (5 to 10%) and that the relationship of Mamais and Jenkins (1992) (see Figure 8.20) be used to determine the operating SRT of the process for the design temperature. Of course, a suitable safety factor (2 to 3) should be applied to the calculated minimum SRT for design purposes.

The size of the anaerobic zone has a much greater effect on EBPR performance when the VFA concentration in the zone is low relative to the available phosphorus concentration, as shown by graph B of Figure 8.14. This is because EBPR is a function of the rate of VFA production by fermentation under those conditions. Then, the fermentation rate is a function of the types of organics available for fermentation, as well as of the organic concentrations, and of the mixed liquor temperature. However, the rate of fermentation in the anaerobic zones of EBPR systems has not been sufficiently defined for design purposes. Fortunately, there are indications that the fermentation rate is relatively insignificant to the overall EBPR process when the anaerobic zone is 25% of the total activated-sludge basin volume. These indications are illustrated by Figure 8.21.

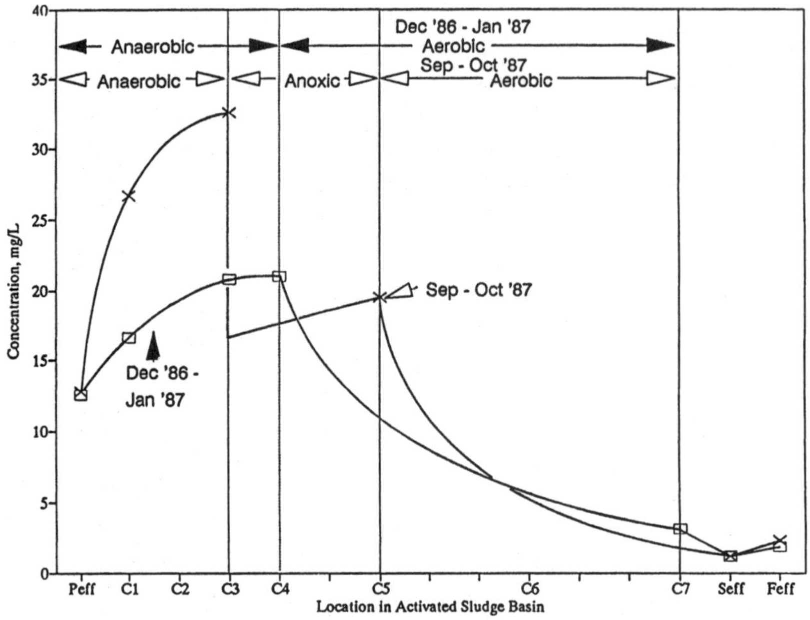


**Figure 8.21 Phosphorus profiles at York River during two-stage biological phosphorus removal.**

Although the extent of fermentation of the wastewater entering the anaerobic zone differed markedly for the two profiles shown in Figure 8.21, the net phosphorus removals and the effluent phosphorus concentrations were identical. The data in the lower graph of Figure 8.21 show that the $BOD_5$ was equal to the COD from September to October, 1987, while it averaged only 51% of the COD during December 1986 to January 1987. The phosphorus profile for December 1986 to January 1987 is a 6-week average profile for daily observations during A/O™ operation, and it shows that the release of phosphorus was virtually complete in section C3 of the anaerobic zone, which was equivalent to an anaerobic mass fraction of 25% and represented an anaerobic HRT of 1.5 hours. The total nominal HRT of the biological process was 6.0 hours, and the anaerobic zone was 33% of the total volume. The anaerobic mass fraction was also 33% for this configuration. The influent TP concentration of 11.76 mg/L was reduced to an effluent SP concentration of 2.74 mg/L, for a reduction of 9 mg/L. Clearly, the anaerobic HRT was long enough for near-complete fermentation within the zone. Furthermore, phosphorus release, which occurs simultaneously with substrate storage, was complete after only 90 minutes even though extensive fermentation was required in the anaerobic zone. The average raw influent temperature during the profile period was 13.6°C.

During the 2-month period of September to October, 1987, the more complete fermentation before the anaerobic zone improved phosphorus removal, but only by 1.58 mg/L, or approximately 18%. The process was being operated in the three-stage ($A^2/O^{TM}$) configuration, and the anaerobic zone was 25% of the total volume. The average COD:TP ratio during this period was 25.6:1, whereas the $BOD_5$:TP ratio was 23.5:1. The raw influent temperature averaged 20°C during the period. A lot more phosphorus was released and at a faster rate during September and October compared to December and January; nevertheless, phosphorus removal increased by only a small fraction and the effluent quality did not improve at all. The secondary effluent SP concentrations averaged 1.16 mg/L and 1.17 mg/L, respectively, for the two time periods. Clearly, the EBPR process was not very sensitive to either prefermentation or temperature for the two periods compared when the anaerobic volume was 25% of the total basin volume.

The relationship between the anaerobic mass fraction and the effluent SP for the same plant, which includes the data used in Figure 8.21, is shown in Figure 8.22. The graph shows that there was no significant improvement in effluent quality when the anaerobic mass fraction exceeded 25%. The anaerobic volume fractions and the anaerobic mass fractions were identical for the configurations used; therefore, the same conclusion applies for the anaerobic HRT.

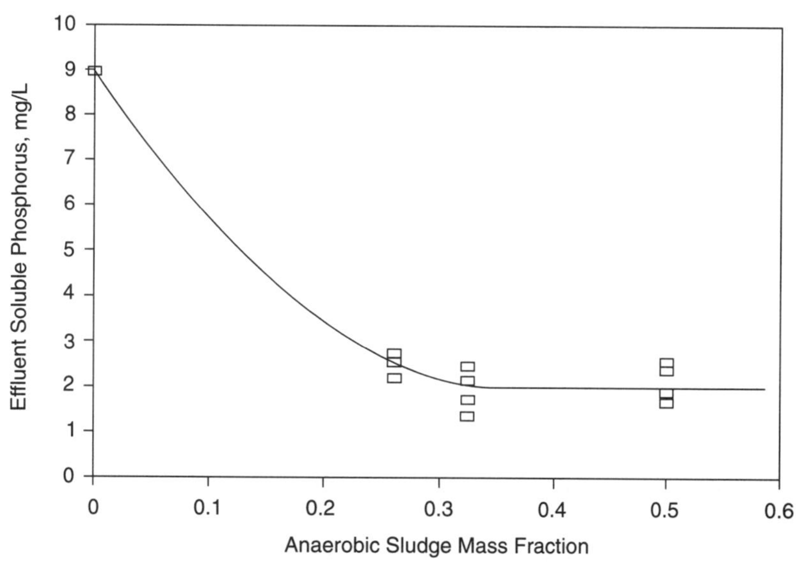


**Figure 8.22** **Effect of anaerobic sludge mass fraction on effluent soluble phosphorus (York River plant, two-stage and three-stage operation; influent flows of 19–26 ML/d; COD-limiting conditions) (from Randall, C.W., *et al.* [1992b] *Design and Retrofit of Wastewater Treatment Plants for Biological Nutrient Removal.* Technomic Publishing Co., Inc., Lancaster, Pa., with permission).**

Excellent performance has been obtained in the laboratory with a UCT pilot-plant system operated at 20°C with an SRT of 15 days and a total HRT of 2.37 hours (Marickovich, 1995). The anaerobic volume was 31% of the total volume, and the system had an anaerobic mass fraction of 16%. Thus, the anaerobic HRT was only 44 minutes and the mass fraction was relatively small, yet the system removed 21 mg/L of phosphorus from an influent concentration of 25 mg/L. The process influent COD concentration was 400 mg/L, and the effluent COD was 30 mg/L. The wastewater was typical municipal wastewater except that phosphorus had been added to make certain that the system would not be phosphorus limited.

It appears, based on the wide variation of observed results, that the dominant factor which determines the appropriate size of the anaerobic zone is the composition of the wastewater entering the anaerobic zone. If the wastewater is highly fermented (that is, if a large fraction of the organics consists of short-chain VFAs), substrate storage occurs very rapidly and a small anaerobic zone, on the order of 5% of the total basin volume, is sufficient for EBPR. Such a volume is appropriate for systems that include prefermenters. Conversely, if the wastewater is very fresh (if it contains virtually no VFAs), then a much larger anaerobic zone should be used to promote fermentation. Full-scale data obtained by the author indicate that an anaerobic mass fraction of 15% is sufficient for most municipal wastewater applications and that 25% is sufficient even if extensive fermentation is required. It was demonstrated that an anaerobic HRT of 90 minutes was sufficient for extensive fermentation at a temperature of 13.6°C.

After the anaerobic zone fraction is selected, the designer should determine the washout SRT using the relationship of Mamais and Jenkins (1992) and then calculate the required volume of the aerobic zone using the desired MLSS concentration.

**Aeration Requirements.** The use of an anaerobic zone as the first treatment stage results in a reduction of the oxygen requirements for the treatment of wastewater by EBPR processes (McClintock *et al.*, 1991, and Randall *et al.*, 1992a). This phenomenon has been termed anaerobic stabilization, but the biochemical mechanisms involved are incompletely understood. However, it has been repeatedly documented, and data from a pilot-scale study designed to investigate the phenomenon are shown in Table 8.3 to illustrate its magnitude and consistency.

During the study by McClintock *et al.* (1991), mass-balance techniques were used to determine and compare the oxygen requirements of a conventional, fully aerobic activated-sludge system and a UCT/VIP™ system placed side by side and treating the same wastewater. Both units were operated at three temperatures and two SRTs. The results in Table 8.3 show that the EBPR system always required less oxygen than the conventional system. The oxygen mass balance for the conventional system was within 4.5% of the theoretical amount for all experiments except IB, illustrating that the pro-

**Table 8.3 Comparison of oxygen requirements for conventional aerobic activated-sludge treatment and biological phosphorus removal (BPR) treatment (McClintock *et al.*, 1993).**

| Phase | System[a] | $TOR_{Predicted}$,[b] g/d | TOR,[c] g/d | AnS/$COD_s$, % | Total SRT | Temp, °C |
|---|---|---|---|---|---|---|
| IA | Conv. | 23.88 | 24.95 | | 5 | 10 |
| | BPR | 20.16 | 16.47 | 18.3 | 5 | 10 |
| IB | Conv. | 31.93 | 27.41 | | 5 | 15 |
| | BPR | 19.91 | 18.23 | 8.4 | 5 | 15 |
| IC | Conv. | 43.27 | 42.41 | | 5 | 20 |
| | BPR | 36.19 | 32.56 | 10.0 | 5 | 20 |
| IIA | Conv. | 26.12 | 25.63 | | 15 | 10 |
| | BPR | 21.96 | 19.42 | 11.6 | 15 | 10 |
| IIB | Conv. | 34.91 | 34.07 | | 15 | 15 |
| | BPR | 28.71 | 21.71 | 24.6 | 15 | 15 |
| IIC | Conv. | 33.82 | 34.20 | | 15 | 20 |
| | BPR | 29.48 | 21.44 | 27.3 | 15 | 20 |
| | Average Reductor | | | 16.7 | | |

[a] Conv. represents a conventional activated-sludge system with nitrification.
[b] Calculated from COD and N mass balance.
[c] Determined by direct measurement; oxygen uptake rate measurements during Phase I were by BOD bottle method (APHA *et al.*, 1995); oxygen uptake rate measurements during Phase II were made *in situ*; conventional HRT was 8 hours, and BNR HRTs were 2.75 hours (anaerobic, anoxic, aerobic).

cedures used were reliable and accurate. The oxygen mass balance for the EBPR system was always less than the theoretical amount, by an average of 16.7%. Data from several other studies that support the existence of anaerobic stabilization are reported in the paper by Randall *et al.* (1992a) and have been previously published (Brannan, 1986; Daigger *et al.*, 1987a; Randall *et al.*, 1985; and Randall *et al.*, 1987).

An additional observation is that the total oxygen used by the BNR system, which included denitrification as well as EBPR, was an average of 31% less than the total oxygen used by the fully aerobic conventional system.

Although a strong theoretical explanation for anaerobic stabilization has not been developed as yet, the existence of the phenomenon is considered to be conclusive enough to reduce the design oxygen requirements of activated-sludge systems incorporating EBPR processes. It is recommended that the theoretical requirements be reduced by 10% for design purposes if the phosphorus removal achieved is 4 mg/L or more.

**Mixing Requirements.** The purpose of mixing in anaerobic and anoxic zones is simply to keep the MLSS suspended while reducing surface turbulence to reduce the transfer of oxygen from the atmosphere. Therefore, the minimum power input necessary to keep the solids suspended should be used, and the stirring mechanism should be designed to avoid vortexing. Vortexing is most easily avoided by using submerged, side-mounted mixers that can be raised and lowered as desired. The power input should be sufficient to maintain a velocity of 0.3 m/s (1.0 fps) throughout the zone. Because horizontal velocities can be difficult to quantify in a reactor with a single-point mixing device, a more practical criterion is complete turnover of the cell contents every 20 minutes, based on the primary pumping rate of the mixer.

An appropriate formula for determining the mixing requirements of activated sludge as a function of the MLSS concentration has been developed from information presented by Reynolds (1982) and is as follows:

$$P/V = 0.000\,94\,(\mu)^{0.3}\,(\text{MLSS})^{0.298} \quad (8.1)$$

Where

| | | |
|---|---|---|
| $\mu$ | = | absolute viscosity in centipoises = 1.008 7 at 20°C; |
| $P/V$ | = | power units per unit volume, kW/1 000 L; and |
| MLSS | = | mixed liquor suspended solids concentration, mg/L. |

Of course, this formula has limited usefulness and should be used only for nearly square zones with a depth of no more than 6 m (20 ft).

**Baffle Design.** It is important that the baffles between zones be designed and constructed to prevent backflow because backflow from either an anoxic or an aerobic zone will contain electron acceptors, either nitrate or DO, which will reduce the available substrate, plus stimulate filamentous growths and cause *Norcardia* foam formation. It is particularly important to remember that the water surface in the aerobic zone will be considerably higher than the water surface in the nonaerated zones, and backflow from the higher surface must be prevented. The baffle openings need to be small enough to ensure a forward flow of 0.3 m/s (1 fps) at minimum flow, and the openings between zones should be several feet below the water surface.

It is also desirable to include a means to remove scum from the nonaerated zones in the design. This can be done by including slots in the baffle walls, making certain the slots are located slightly above (20 mm at peak flow) the higher of the water surfaces separated by the baffle. Also, the scum could be easily handled by including skimmers in each zone, if desired.

**Secondary Clarifier Design and Operation.** Clarifier design is crucial to the successful operation of an EBPR system. The clarifier must provide effective clarification because the biomass suspended solids contain phosphorus and must be removed to low levels to meet effluent requirements. For example, to achieve an effluent phosphorus concentration of 1.0 mg/L when the biomass suspended solids contain 4.5% phosphorus, the effluent suspended solids cannot exceed 15 mg/L, assuming an effluent SP concentration of 0.3 mg/L. Consequently, a design overflow rate of 12.2 to 16.3 m/d (300 to 400 gpd/sq ft) is recommended. The resolubilization of phosphorus in the sludge blanket can also be a problem, but the effects can be reduced by increasing the side water depth (SWD) of the clarifier so that it can be operated without upflow through the sludge blanket. That is to say, the resolubilized phosphorus will exit the clarifier in either the RAS or the waste activated sludge (WAS), instead of the effluent overflow. A suitable SWD range is 4.3 to 5.5 m, and it should be possible to vary the RAS rate from 0.25 to 1.25 $Q$.

The need to prevent upflow through the sludge blanket also makes the rimflow clarifier design undesirable for EBPR systems, and it should not be included in the design. It is possible to successfully operate an EBPR system with a rimflow clarifier, but it severely restricts the operating flexibility.

It is further recommended that clarifiers be equipped with full-width scum skimmers to provide more efficient scum removal.

**Supplemental Chemical Addition.** Biological phosphorus removal can be enhanced by the addition of supplemental phosphorus-precipitating chemicals, such as alum, ferric chloride, and ferrous sulfate, to the activated sludge without inhibiting the EBPR mechanisms, if the dosages are appropriately controlled. The simultaneous use of the two processes reduces the amount of chemical needed to reliably achieve the required effluent phosphorus concentration. However, there is a tendency to steadily increase the chemical addition rather than trust the EBPR process to reliably remove phosphorus when the two processes are combined, and this eventually starves the EBPR process, washing the poly-P function out of the system. Care must be taken to avoid chemical overfeeding to take full advantage of the EBPR operation.

The chemical can be added at several locations within the EBPR system, including all zones. The coagulating chemical typically is added to the influent flow to the secondary clarifier because this reduces interference with EBPR and alkalinity consumption. However, experiments with a Bio-Denipho pilot plant showed that a more stable effluent quality could be obtained by adding ferrous sulfate to the anaerobic zone rather than to the aerobic zone (HYPRO, 1991).

**DESIGN STEPS FOR ENHANCED BIOLOGICAL PHOSPHORUS REMOVAL.** The primary design considerations for EBPR include how big the anaerobic zone should be relative to the aerobic zone. Also, determine if prefermentation is required or desired and, if so, what operating SRT should be used.

Secondary considerations include the following: Should the anaerobic zone be subdivided by baffles, how much mixing is required, and what must be done to minimize DO and $NO_x$ inputs to the anaerobic zone? Additionally, how should the secondary clarifiers and waste sludge processing train be designed to minimize solubilization and recycle of phosphorus to the biological reactor?

The following steps are recommended for EBPR design using the graphs and techniques of this section:

1. Select the desired effluent phosphorus concentration.
2. Determine the TCOD:P and $TBOD_5$:P ratios of the process influent.
3. Use the data from Figure 8.9 to estimate the effluent SP concentration possible with the TCOD:P ratio of the process influent. The $BOD_5$:P ratio in either Figure 8.7 or 8.8 may be used as an alternate if COD data are not available.

    a. If the TCOD:P ratio is greater than 40:1, the effluent phosphorus concentration possible with optimum design and operation will be less than 1.0 mg/L.
    b. If the TCOD:P ratio is less than 40:1 ($BOD_5$:P < 20:1), an effluent phosphorus concentration of less than 1.0 mg/L will be difficult to achieve and maintain. If the required effluent concentration cannot be obtained by EBPR with the process influent, then either the influent VFAs need to be supplemented by prefermentation or other methods, or effluent polishing by chemical precipitation is necessary.

4. Compare the process influent $BOD_5$ and COD values to estimate the degree of fermentation of the process influent. Recognizing that the $BOD_5$:COD ratio varies throughout the year, the limiting ratio should be selected for design. These EBPR processes can maintain efficiency throughout short periods of a week or more of inadequate influent VFAs, the period being related to the operating SRT, but reduction of phosphorus removal will occur if the condition persists for a month or more. It is recommended that the lowest monthly average $BOD_5$:COD ratio be used for design decisions if year-round compliance is required.

    a. If the $BOD_5$:COD ratio is less than 0.5 and the TCOD:P ratio is greater than 40, it can be concluded that an effluent SP concentration of less than 1.0 mg/L can be achieved by EBPR treatment without VFA supplementation if the anaerobic zone is adequately sized for fermentation. The appropriate size can be estimated, based on the phosphorus removal desired, using the data from Figure 8.14a, if the removal required is less than 6 mg/L. If the removal required is greater than 6 mg/L, use an anaerobic mass fraction of 25% or greater.
    b. If the $BOD_5$:COD ratio is less than 0.5 and the TCOD:P ratio is less than 40, use the data from Figure 8.14b to estimate the required anaerobic HRT, if it is possible to achieve the required effluent phosphorus concentration. If the effluent phosphorus con-

centration cannot be achieved, select either prefermentation or chemical polishing. If chemical polishing is selected, use the anaerobic HRT obtained from Figure 8.14b. If prefermentation is selected, use an anaerobic mass fraction of 5 to 10%.

   c. If the $BOD_5$:COD ratio is greater than 0.5 and the TCOD:P ratio is greater than 40, use a minimum anaerobic HRT of 90 minutes or a minimum anaerobic mass fraction of 15%, whichever is more appropriate for the system being designed.
   d. If the $BOD_5$:COD ratio is greater than 0.5, and the TCOD:P ratio is less than 40, use a minimum anaerobic HRT of 90 minutes or a minimum anaerobic mass fraction of 15%, but use the data from Figure 8.14b to estimate whether or not enough phosphorus removal can be obtained with an anaerobic HRT of 2 hours or less to achieve the desired effluent quality. If sufficient removal cannot be predicted, then prefermentation or chemical polishing should be used to supplement EBPR.
      — (NOTE: Chemical coprecipitation with either aluminum or iron salts, that is, addition of the chemical directly to the activated sludge, can be used to obtain effluent phosphorus concentrations as low as 0.2 mg/L simultaneously with EBPR. If lower concentrations are required, then effluent filtration and/or tertiary precipitation will probably be required.)

5. Adjust the selected design anaerobic HRT for temperature using a $\theta$ coefficient of 1.045 for temperatures less than 20°C, if the system is COD limited (TCOD:TP $< 40$). Assume that the design HRT times from Figure 8.14 are for 20°C. No adjustment is necessary if the system is phosphorus limited.
6. Adjust the temperature-adjusted anaerobic HRT by dividing by the pH correction factor, obtained from data in Figure 8.6.
7. Determine the minimum aerobic SRT for EBPR using the Mamais–Jenkins relationship given in Figure 8.20. Multiply the calculated SRT by a safety factor of 2.5, then calculate the volume needed for the desired design MLVSS concentration in the aerobic zone.
8. For a two-stage EBPR system, design the activated-sludge basin as a plug-flow unit and divide the anaerobic zone into at least three subzones using baffles. Equip the last anaerobic subzone with both mixing and aeration equipment so that it can be operated as either an anaerobic or an aerobic section. If designing a three-stage BNR system, the three anaerobic subzones should be equipped with mixing equipment only, with the third subzone piped for anoxic operation, but part or all of the anoxic zone should be equipped for either anoxic or aerobic operation. The purpose is to provide the operator with as

much flexibility as possible for unforeseen problems and changing circumstances.

**DESIGN EXAMPLE.** For this example, assume a municipal flow of 10 ML/d with a COD of 380 and a $BOD_5$ of 160 mg/L, after primary settling. Other primary effluent concentrations are 6.0 mg/L TP, 24 mg/L TKN, 140 mg/L alkalinity, and 150 mg/L TSS, with a pH of 7.2. Design a two-stage EBPR activated-sludge plant to meet effluent concentrations of 20 mg/L $BOD_5$, 20 mg/L TSS, and 1.0 mg/L TP. Assume no effluent ammonium limitations for the primary design, but also develop an alternate design with an effluent ammonium limitation of 2.0 mg/L. The effluent pH must be between 6 and 9. Design for a temperature of 12°C.

1. The TCOD:TP ratio of the primary effluent is 380/6.0 = 63. An effluent concentration of less than 1.0 mg/L can be easily accomplished with EBPR treatment of this wastewater.
2. The $BOD_5$:COD ratio of 0.42 indicates that the wastewater needs substantial fermentation to increase EBPR. The data predicts that an anaerobic HRT of 2.0 hours is needed to remove the required 5.0 mg/L of phosphorus.
3. The system can be designed with an operating SRT of 4 days to prevent nitrification under most conditions. For a total nominal HRT of 6 hours in the activated-sludge basin, the aerobic SRT would be only $4/6 \times 4 = 2.7$ days. Accepted nitrification kinetics predict that little or no nitrification would occur at this SRT up to mixed liquor temperatures in excess of 20°C. The Mamais–Jenkins graph shows that EBPR will not wash out at an aerobic SRT of 2.7 days until the temperature is below 9°C. Because this is less than the design temperature, the value is acceptable for design.

   Appropriate design and operation of the secondary clarifier, or use of a small anoxic zone near the basin effluent, would prevent the recycle of nitrates formed during partial nitrification back to the anaerobic zone, even at temperatures above 20°C. Alternatively , the RAS recycle could be returned to an anoxic zone before entering the anaerobic zone, such as with a UCT-type modification or inclusion of a preanoxic section without wastewater addition.
4. An appropriate design MLSS concentration is 4 000 mg/L or greater to ensure that the clarifiers are sized sufficiently to provide operational flexibility. Assuming a percent MLVSS fraction of 85%, the design MLVSS concentration would be 3 400 mg/L. However, operation at an SRT of 4 days would result in an MLVSS concentration of only 2 000 mg/L, based on a COD yield coefficient of 0.4 and an endogenous decay coefficient of 0.1 $day^{-1}$, adjusted by a temperature coefficient ($\theta$) of 1.05. The SRT at design MLSS concentration and temperature would be 8 days, which would provide sufficient operat-

ing flexibility to maintain EBPR at low temperatures. Because the system being designed is phosphorus limited as indicated by the high TCOD:P ratio (>40:1), a temperature adjustment for phosphorus removal is not necessary, assuming washout conditions are avoided.

5. There is an insignificant effect of EBPR on alkalinity; in fact, it increases alkalinity slightly, and therefore there is no need for a pH adjustment of any of the design parameters for this example.
6. Oxygen requirements for this example can be calculated by the following equation:

$$\text{Oxygen required} = Q(1-1.45Y)(S_o - S_e) + 1.45K_dXV_a$$

Where

$V_a$ = aerobic volume, ML;
$X$ = MLVSS concentration, mg/L; and
1.45 = the COD equivalence of the MLVSS.

Therefore (assume effluent residual COD = 20 mg/L):

$$\text{kg } O_2/\text{d} = 10(1 - 1.45 \times 0.4)(380 - 20) + 1.45 \times 0.1 \times 3\,400 \times 4/6 \times 2.5 = 2\,334$$

before the 10% credit is taken.

$$\text{Net oxygen required} = 2334 - 0.1(2\,334) = 2\,100 \text{ kg/d}$$

7. The mixing requirements for the anaerobic zone are calculated from Equation 8.1:

$$P/V = 0.000\,94(1.008\,7)^{0.3}(4\,000)^{0.238} = 0.006\,78 \text{ kW/1 000 L}$$

$$\text{Total power required} = 0.006\,78 \times 2/6 \times 2.5 \times 10^6 = 5\,650 \text{ kW.}$$

**Alternate Design.** The alternate design would require an HRT sufficient to maintain nitrification. Nitrification would also require increased amounts of oxygen and would consume considerable alkalinity. At least partial denitrification would be required to protect the anaerobic zone from electron acceptors, but if the nitrates were recycled to use influent BOD for denitrification, the increased oxygen requirements for nitrification would be substantially reduced, and a significant fraction of the alkalinity would be recovered.

An EBPR system operated at an SRT sufficiently low to prevent nitrification will produce a biomass sludge that has a greater specific gravity and settles somewhat better than a sludge produced by a conventional fully aerobic activated-sludge system. Sludge volume indexes (SVIs) between 40 and 80

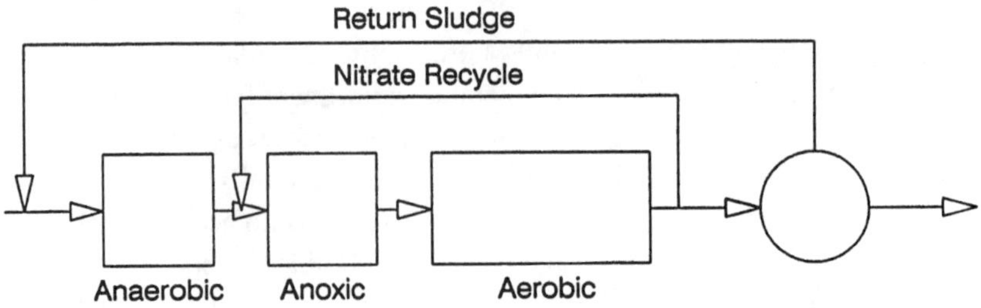


**Figure 8.23 Three-stage biological nutrient removal system.**

are typical. This is partly because the bacterial cells contain polyphosphate granules and partly because the mode of operation virtually eliminates filamentous microorganisms. However, sludge settleability will lessen considerably if the SRT is increased to the extent that partial or complete nitrification occurs unless all nitrates and DO are removed from the RAS flow before it enters the anaerobic zone. The best way to reduce the nitrates to acceptable levels is to insert an anoxic zone between the anaerobic and aerobic zones. The basic schematic for this type of continuous-flow system is shown in Figure 8.23. The anoxic zone with internal recycle ensures that the quantity of nitrates entering the clarifier is low, and this reduces or eliminates the nitrates entering the anaerobic zone.

# *EHHANCED BIOLOGICAL PHOSPHORUS REMOVAL CONFIGURATIONS*

A variety of configurations have been developed for activated-sludge systems that accomplish EBPR, and several have been patented. Most design considerations are the same for the different configurations, but there are special considerations for some.

**SIDESTREAM CONTINUOUS-FLOW SYSTEM.** A process configuration for the utilization of EBPR was first developed by Levin and Shapiro (1965). They showed that enhanced phosphorus storage by bacteria was possible and included an anaerobic reactor in the system they developed to accomplish phosphorus release. However, they did not understand the linkage between organic substrate storage and phosphorus release or fully appreciate the detrimental effects of nitrates in the anaerobic reactor. Consequently, they used a sidestream anaerobic reactor to accomplish phosphorus release from the biomass and then removed the phosphorus from the supernatant using chemical precipitation. The configuration developed, which was termed

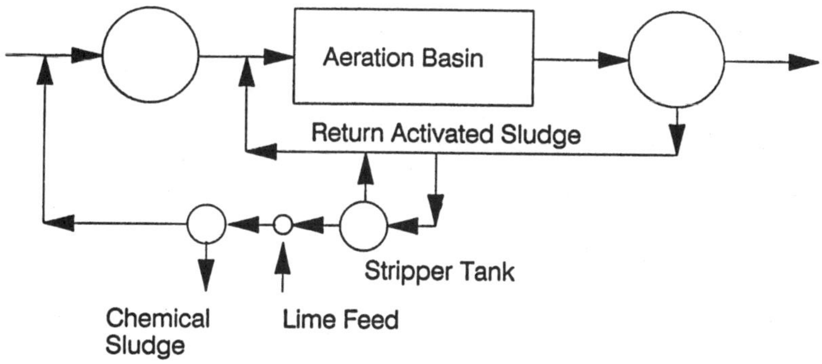


**Figure 8.24** **Schematic of PhoStrip™ process (from Randall, C.W., *et al.* [1992b] *Design and Retrofit of Wastewater Treatment Plants for Biological Nutrient Removal.* Technomic Publishing Co., Inc., Lancaster, Pa., with permission).**

the PhoStrip™ process, is shown in Figure 8.24. It is the only sidestream configuration currently in use for BNR.

Reliable operation of this configuration requires diversion of a portion of the wastewater to the anaerobic reactor to accomplish rapid denitrication and subsequent phosphate release with substrate storage.

**MAINSTREAM CONTINUOUS-FLOW SYSTEM.** The design considerations previously discussed apply fully to the well-known mainstream continuous-flow systems initially conceptualized by Barnard. Barnard (1974 and 1975) first understood the significance of the anaerobic zone for EBPR, and he developed the first mainstream systems accomplishing EBPR by placing an anaerobic reactor, first in front of an aerobic zone, then in front of the modified Ludzack–Ettinger (MLE) (Ludzack and Ettinger, 1962) configuration, and then in front of the Bardenpho™ configuration. He realized the importance of an anaerobic zone to stimulate phosphorus removal and proposed placing an anaerobic stage ahead of the aerobic stage so that the influent organics could be used to create anaerobic conditions instead of relying on endogenous respiration as in the PhoStrip™ process (Barnard, 1975). The resulting configuration, called Phoredox™, is the basic one shown in Figure 8.1. It was first applied on a full-scale basis at Germiston, South Africa, in 1974. This same configuration was later patented by Air Products and Chemicals, Inc., in the U.S. under the trademark of Anaerobic/Oxic, or A/O™ (Spector, 1979).

After pilot-plant and full-scale experimentation, Barnard combined the Phoredox™ concept with the MLE process at Klerksdorp, South Africa, in 1974. This resulted in the three-stage process shown in Figure 8.23, which also was later patented in the U.S. by Air Products and Chemicals, Inc., and trademarked as Anaerobic/Anoxic/Oxic, or $A^2/O$™. The Phoredox™ concept was next combined with the Bardenpho™ process for the new Goud-

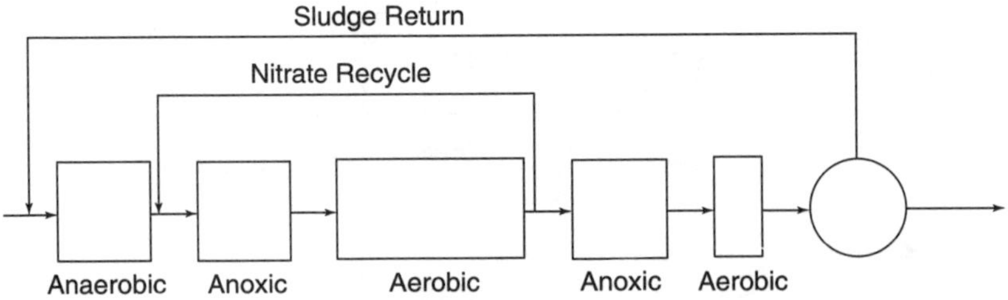


**Figure 8.25 Five-stage biological nutrient removal system (modified Bardenpho™).**

koppies plant in Johannesburg, South Africa, which went on line in 1978, to produce the five-stage Modified Bardenpho™ configuration shown in Figure 8.25. This system was designed for effluent phosphorus and nitrogen concentrations of less than 1.0 and 3.0 mg/L, respectively.

Some configurations were developed to correct perceived shortcomings of the original ones and provide more reliable performance. Marais and his coworkers (Ekama *et al.,* 1984) determined the detrimental effects of nitrates entering the anaerobic zone on the performance of EBPR processes. They developed a modification of the three-stage Phoredox™ process that eliminated the recycle of nitrates in the RAS to the anaerobic zone. This process, dubbed the UCT process, is shown schematically in Figure 8.26. They later modified the UCT process to provide even better protection of the anaerobic zone from nitrate recycle, as illustrated by Figure 8.27.

Randall (1984) recommended use of the UCT configuration as a high-rate BNR system for economical treatment of municipal wastewater in the U.S. This concept was piloted by CH2M/HILL for HRSD (Daigger *et al.*, 1987b). The pilot study was successful, and the process was patented as a public domain patent by CH2M/HILL and HRSD as the VIP™ process. Differences between the UCT and VIP™ processes include a staged configuration and design as a high-rate process (design based on nitrification rather than sludge stabilization) for the VIP™ process in comparison to UCT. These features result in increased phosphorus removal and a significant reduction in reactor volume.

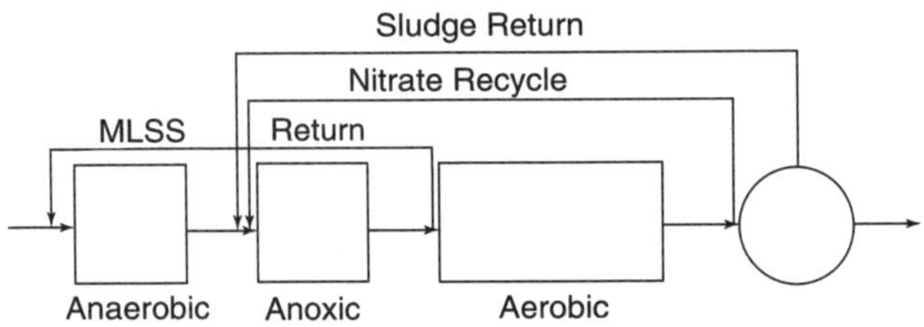


**Figure 8.26 University of Cape Town type biological nutrient removal system.**

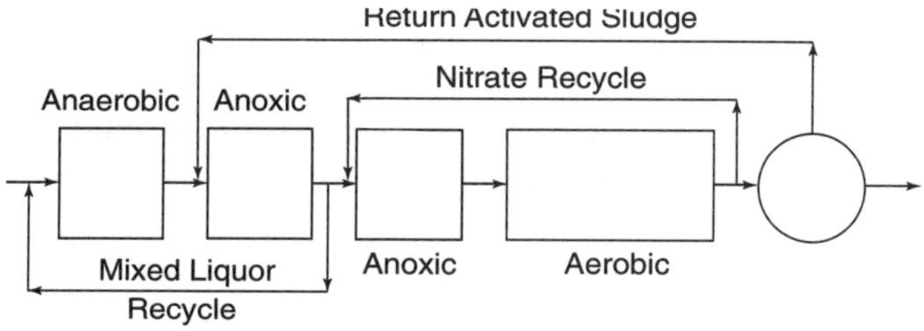


**Figure 8.27 Schematic of the modified UCT process.**

Several other investigators demonstrated during the 1980s that EBPR could be accomplished with a wide variety of process configurations.

**Four-Stage Design.** Wanner *et al.* (1989) of Czechoslovakia developed the reaeration–anaerobic–denitrification–nitrification (R–AN–D–N) configuration to achieve improved nitrification in a BNR system without a large increase in the process volume. After a series of experiments comparing four different flow schematics, R-AN-D-N was selected for a new plant at Pilsen, Czechoslovakia. The schematic diagram for this configuration is shown in Figure 8.28.

The primary advantage of this flow diagram is the reaeration of the RAS before it is mixed with the influent wastewater. Because the RAS concentration is typically two to three times greater than the MLSS, the HRT for the same volume is considerably longer than it would be for the MLSS. This significantly increases the aerobic SRT for a minimal volume increase and makes the nitrification process more stable. Also, the influent flow is divided

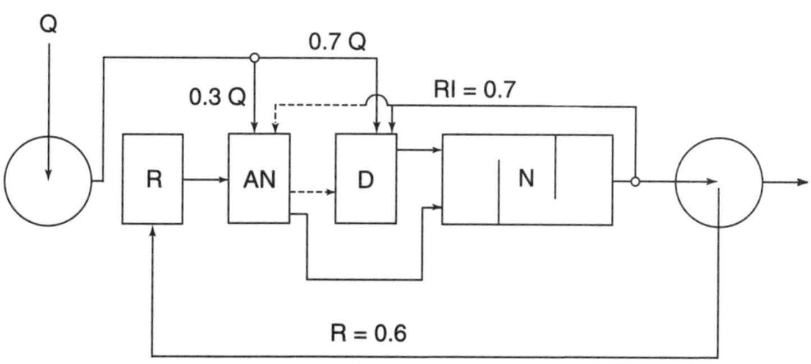


**Figure 8.28 The technological line of the R–AN–D–N process designed for the new WWTP for the town of Pilsen, Czechoslovakia (from Wanner, J., *et al.* [1989] An Innovative Technology for Upgrading Nutrient Removal Activated Sludge Plants. *Proc. IAWPRC Int. Spec. Conf. Upgrading Wastewater Treat. Plants,* Munich, Ger., 9, with permission).**

between the anaerobic zone and the denitrification zone. Actually, considerable denitrification must take place in the anaerobic zone before true anaerobic conditions can be achieved because of the nitrate concentration in the flow from the reaeration zone to the anaerobic zone. The flow of nitrates can be reduced, but not eliminated, by a high recycle rate of flow from the nitrification to the denitrification zone. To achieve EBPR, sufficient wastewater must be added to the anaerobic zone to both eliminate the nitrates from the reaeration zone and establish anaerobic conditions suitable for substrate storage by the poly-P bacteria. This can be accomplished most efficiently by dividing the anaerobic zone into two or more sections in series so that the anaerobic conditions of the second and subsequent sections are not disturbed by nitrate inflow.

**Five-Stage Design.** According to Barnard (1992), for high levels of nitrogen and phosphate removal, the five-stage process is required even though the removal of nitrates in the second anoxic zone is low. When a final effluent TN of less than 3 mg/L is required, the effluent nitrate must not exceed 1.5 mg/L to allow for small amounts of ammonia plus nondegradable soluble organic nitrogen that cannot be removed biologically.

Thus far, most instances of consistently good removals of both nitrogen and phosphorus has been performed by five-stage, Bardenpho™-type plants having point source aerators. In some plants, initially good performances were recorded with subsequent disappointing results. This may have stemmed from a lack of understanding by the operators as to what was actually happening in the process. Actually, better than predicted removals are possible. The Kelowna plant in British Columbia, Canada, will be used as an example. When the UCT model was applied to the plant, it predicted that the effluent nitrate concentration should be approximately 7 mg/L (Barnard *et al.*, 1984). However, during actual performance, less than 1 mg/L ammonia and nitrate could be maintained over months. This difference between theory and practice typically has been ignored, but if not taken into account one would have to come to a final conclusion that it is impossible to obtain a TN value of less than 3 mg/L, despite the fact that it has been obtained consistently by a number of plants as can be seen from the plot in Figure 8.29.

All the plants that achieved such low TN values used either surface or turbine aerators. Surface aerators were mostly used in Carrousel plants forming the aeration portion of five-stage Bardenpho™ plants. The effects of denitrification in the aeration basin will be demonstrated by using the following example.

*DESIGN EXAMPLE.* Assume that the plant is to be designed using the following information before primary sedimentation. Consider two situations, the first in which total effluent nitrogen must be less than 5 mg/L and the second in which the effluent TN must be less than 3 mg/L. Consider the flow

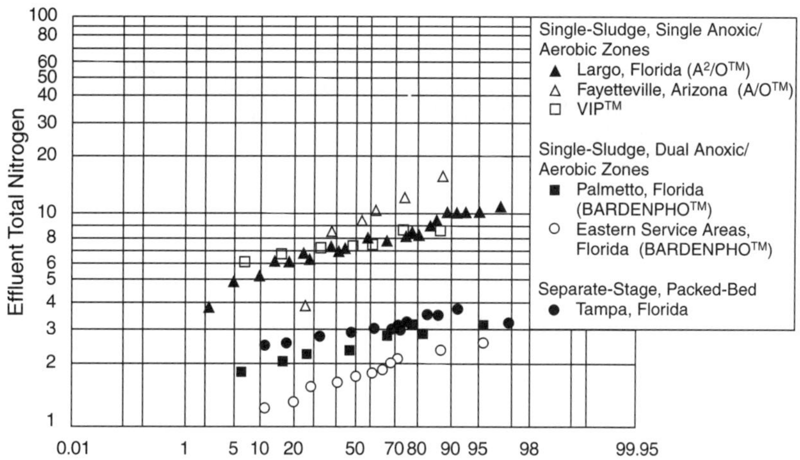


**Figure 8.29** **Performance of plants with respect to nitrogen removal (from Randall, C.W., *et al.* [1992b] *Design and Retrofit of Wastewater Treatment Plants for Biological Nutrient Removal.* Technomic Publishing Co., Inc., Lancaster, Pa., with permission).**

sheet in Figure 8.30 with a mixed liquor recycle of $aQ$ and an RAS recycle rate of $bQ$, where $Q$ is the average influent wastewater flow rate.

- Flow, $Q$ = 80 ML/d.
- Influent COD, $S_o$ = 400 mg/L.
- Influent suspended solids, $X_o$ = 200 mg/L.
- Nondegradable COD, $S_{nd}$ = 30 mg/L.
- Total influent nitrogen (TKN), $N_o$ = 40 mg/L.
- Total influent phosphorus (TP), $P$ = 6 mg/L
- Effluent suspended solids, $X_e$ = 10 mg/L.
- Specific growth rate of *Nitrosomonas*, $\mu m$ = 0.35/d.
- Minimum mixed liquor temperature, $T$ = 12°C.
- Denitrification rates, $D_1(T)$

$$= 0.72 \times X_a \times \theta_1^{(T-20)} \text{ mg N/mg active MLSS/d} \quad (8.2)$$

$$D_2(T) = 0.16 \times X_a \times \theta_2^{(T-20)} \text{ mg N/mg active MLSS/d} \quad (8.3)$$

$$D_3(T) = 0.072 \times X_a \times \theta_3^{(T-20)} \text{ mg N/mg active MLSS/d} \quad (8.4)$$

- Fraction of active mass nondegradable, $f$ = 0.2.
- Yield constant for carbonaceous compounds, $Y$ = 0.48 g/g $COD_r$.
- Endogenous decay rate, $K_b(T) = [0.24 \times (1.029)^{(T-20)}]/d$.

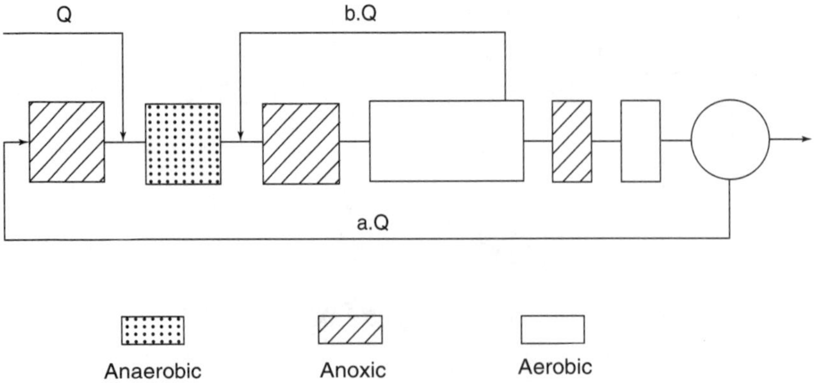


**Figure 8.30** **Flow diagram for design example (from Randall, C.W., *et al.* [1992b] *Design and Retrofit of Wastewater Treatment Plants for Biological Nutrient Removal.* Technomic Publishing Co., Inc., Lancaster, Pa., with permission).**

- $\theta_1 = 1.2$.
- $\theta_2 = 1.03$.
- $\theta_3 = 1.03$.

Calculate the values of the denitrification rates:

$$\begin{aligned} D_1(12) &= 0.72 \times (1.2)(12-20) \\ &= 0.167 \text{ mg/mg active MLSS/d} \end{aligned}$$

$$\begin{aligned} D_2(12) &= 0.160 \times 1.03^{(12-20)} \\ &= 0.126 \text{ mg/mg active MLSS/d} \end{aligned}$$

$$\begin{aligned} D_3(12) &= 0.072 \times 1.03^{(12-20)} \\ &= 0.057 \text{ mg/mg active MLSS/d} \end{aligned}$$

Determine the minimum mean cell residence time (MCRT) for nitrification:

$$\mu_m = \mu_m \times e^{0.098(T-15)} \cdot \frac{\text{DO}}{K_o + \text{DO}} \cdot [1 - 0.833(7.2 \text{ pH})] \qquad (8.5)$$

Where

$\mu_m$ = Actual growth rate at specific conditions;
DO = Dissolved oxygen in the nitrification stage, mg/L;
$K_o$ = half saturation constant = 0.3; and
pH = pH value of the mixed liquor.

Thus,

$$\mu_m = 0.35\ e^{0.098(12-15)} \cdot \frac{2}{0.3 + 2} \cdot [1 - 0.833(7.2 - 7.0)]$$

$$= 0.189/\text{d}$$

For the decay constant of the nitrifiers where $K_d$ (20) is assumed at 0.05 $\text{d}^{-1}$,

$$K_d(T) = K_d(20)(1.022)^{(T-20)}$$

$$= 0.05 \times 1.022^{(-8)}$$

$$= 0.042/\text{d}$$

For the minimum MCRT for nitrification under aerobic conditions,

$$\theta_m = \frac{1}{0.189 - 0.042}$$

$$= 6.8 \text{ days}$$

Assume that approximately 40% of the basin volume will be dedicated to anaerobic and anoxic volumes and apply a safety factor of 1.6. Then, for the minimum safe MCRT for nitrification,

$$\theta_m = 6.8 \times 1.6/(1 - 0.4)$$

$$= 18.1 \text{ days}$$

Determine the COD, TKN, and suspended solids removal in the primary sedimentation tanks, taking into account any feedback from sludge treatment units. If actual figures are unavailable, assume the following removals:

- COD = 35%,
- TKN = 10%, and
- Suspended solids, SS = 70%.

Assume a 5% COD feedback from the activated primary or gravity thickener. The masses of these parameters entering the BNR plant are as follows:

$$\text{COD}:MS_o = 0.7 \times 400 \times 80 = 22\ 400 \text{ kg/d}$$

$$\text{TKN}:MN_o = 0.9 \times 40 \times 80 = 2\ 880 \text{ kg/d}$$

$$\text{SS}:MX_o = 0.3 \times 200 \times 80 = 4\ 800 \text{ kg/d}$$

Where

$MS_o$ = mass of substrate,
$MN_o$ = mass of nitrogen, and
$MX_o$ = mass of biosolids (biomass).

Because not all the COD is removed in the BNR plant, the effluent COD must be determined from a laboratory-scale unit. In this case, assume 30 mg/L is not degradable. Thus, the final effluent COD may be on the order of 35 mg/L, which would correspond to less than 5 mg/L BOD.

The COD removed in the BNR plant is

$$MS_{bi} = 22\,400 - 35 \times 80 = 19\,600 \text{ kg/d}$$

Using the IAWPRC model, calculate the active mass, $MX_a$:

$$MX_a = Y \times MS_{bi} \times \theta/[1 + k_b(T) \times \theta]$$

$$= 0.48 \times 19\,600 \times 18.1/(1 + 0.191 \times 18.1) \quad (8.6)$$

$$MX_a = 38\,205 \text{ kg}$$

$$MX_v = YMS_{bi} \times \theta \times [1 + f \times K_b(T) \times \theta]/[1 + K_b(T) \times \theta]$$

$$= 0.48 \times 19\,600 \times 18.1 \times$$

$$(1 + 0.2 \times 0.191 \times 18.1)/(1 + 0.191 \times 18.1)$$

$$MX_v = 64\,620 \text{ kg} \quad (8.7)$$

$$MX_t = Mx_v/0.83 \text{ where MLVSS} = 0.83 \text{ MLSS}$$

$$= 64\,620/0.83 \quad (8.8)$$

$$MX_t = 77\,860 \text{ kg}$$

Calculate the nitrogen in the sludge assuming that the nitrogen content of the volatile solids is 10%. Thus, the nitrogen loss (that which will not need to be denitrified) will be

- In primary sludge: $0.10 \times 40 \times 80 = 320$ kg/d.
- In WAS: $0.10 \times 64\,620/18.1 = 357$ kg/d.
- In effluent suspended solids: $0.10 \times 80 \times 10 = 80$ kg/d.
- As effluent ammonia: $0.5 \times 80 = 40$ kg/d.
- Effluent undegradable TKN: $0.4 \times 80 = 32$ kg/d.
- Remaining to be nitrified/denitrified: 2 371 kg/d.
- In terms of concentration: $2\,371/80 = 29.6$ mg/L.

Use any convenient model to calculate the sludge recycle (RAS) rate required if

- Maximum MLSS = 3 000 mg/L,
- SVI = 120 mL/g, and
- RAS rate = $0.6 \times Q$.

Conservatively, assume that 15% of the TN nitrified will be lost through simultaneous denitrification in the aeration basin. This will leave $N_n = 25.2$ mg/L to be denitrified in the anoxic zone.

Assume that a mixed liquor recycle rate of $5Q$ is used and that full denitrification takes place in the first anoxic zone. Then, from a mass balance around the first anoxic zone, taking the nitrates in the effluent of the nitrification zone as $N_i(1)$,

$$N_n \times Q = N_i(1) \times (0.6 + 5 + 1) \times Q$$

$$N_i(1) = N_n/(0.6 + 5 + 1)$$

$$= 25.2/6.6$$

$$= 3.8 \text{ mg nitrate-N/L}$$

Assume first that this is sufficient for discharge under the requirement of less than 5 mg/L TN and that no second anoxic zone will be needed. For this design, it would be necessary to provide some means for removal of nitrate from the RAS before discharge into the anaerobic basin. Calculate the size of the final clarifiers and assume that there will be some sludge blanket formed at the tank bottom. If the peak flow equals $2.5Q$ and the MLSS equals 3 000 mg/L, the sludge flux to the final clarifiers will be

$$\text{Sludge flux} = (2.5 + 0.6) \times (80\,000/24)(\text{m}^3/\text{h}) \times 3(\text{kg/m}^3)$$

$$= 31\,000 \text{ kg/h}$$

$$\text{Limiting sludge flux} = 6 \text{ kg/m}^2\cdot\text{h}$$

$$\text{Surface area of final clarifiers} = 31\,000/6 = 5\,166 \text{ m}^2$$

Use four clarifiers, each with a diameter of 40 m (5 026 $m^2$). If an anoxic sludge blanket of 200 mm formed in each clarifier, then

$$\text{Anoxic volume} = 5\,026 \times 0.2$$

$$= 1\,005 \text{ m}^3$$

Active mass contained in this volume can be calculated from Equations 8.6 and 8.7:

$$\text{Active mass} = 1\,005 \times 3 \times 38.21/77.86$$

$$= 1\,479 \text{ kg}$$

Nitrates that can be removed through endogenous respiration in the final clarifiers can be calculated using Equation 8.4:

$$= D3 \times MX_a$$

$$= 0.057 \times 1\,479 \text{ kg/d}$$

$$= 84.3 \text{ kg/d}$$

$$= 1.1 \text{ mg N/L}$$

Nitrates remaining to be removed in a preanoxic zone

$$= 3.8 - 1.1$$

$$= 2.7 \text{ mg/L or } 216 \text{ kg/d}$$

For active mass required for endogenous respiration denitrification in the preanoxic zone,

$$MX_a = 216/0.057$$

$$x_a = 3\,789 \text{ kg}$$

$$\text{Required } MX_t = 3\,789 \times 77.86/38.21$$

$$X_t = 7\,721 \text{ kg}$$

At a recycle rate of $0.6Q$, the suspended solids concentration of the underflow of the final clarifiers will be 8 kg/m$^3$. Thus, volume of preanoxic basin will be 7 721/8, or 965 m$^{3.}$

Determine the volume of the first anoxic basin. Assume that a phosphorus release of five times the phosphorus in the influent will be required. Thus, a phosphorus release of up to 30 mg/L will be required in the anaerobic basin. This will take up approximately 30 mg/L of the easily degradable COD. If the easily degradable COD is 30% of the influent COD (120 mg/L), then 90 mg/L will be left for denitrification using the high-rate form of the nitrification Equation 8.3; thereafter, the second rate will be used. Assuming that approximately 7 mg COD is required for the removal of 1 mg nitrate-nitrogen,

Mass of nitrogen removed per day by the high rate of denitrification
= 90/7 = 12.8 mg/L, or 1 029 kg/d
To be removed at the second rate = 5 × 80 × 3.8–1 028 = 492 kg/d
The active fraction of the MLSS = 3 × 38.21/77.86 = 1.47 kg/m$^3$
Thus, the high rate of denitrification = 0.167 × 1.47 = 0.246 kg/m$^3$/d
Volume required for high rate = 1 029 kg/d/0.246 = 4 183 m$^3$
Volume for removing 492 kg by the second denitrification rate
= 492/(0.126 × 1.47) = 2 656 m$^3$
Total first anoxic volume = 4 183 + 2 656 = 6 839 m$^3$
Total basin volume = $MX_t$/MLSS = 77 860/3 = 25 953 m$^3$

The anaerobic basin does not form part of the aerobic–anoxic reactor. Allow 1 hour retention in the anaerobic basin. This will give the following volumes:

- Preanoxic zone = 965,
- Anaerobic = 3 333,
- First anoxic = 6 839,
- Aerobic = 18 149, and
- Total volume = 29 286 m$^3$.

The anaerobic/anoxic fraction of the reactor is 38%, which is close to the assumed value.

If it is desired that the effluent TN be reduced to a value of less than 3 mg/L, the exercise could be repeated using a second anoxic zone with endogenous denitrification instead of a preanoxic zone. However, do not try to reduce the effluent nitrates to below 1.5 mg/L because that would make the second anoxic zone unreasonably large. The TKN of the effluent should be below 1.2 mg/L, giving a TN value of less than 3.

The mass of nitrates to be removed in the second anoxic zone will be

$$1.6 \times 80 \times (3.8 - 1.5) = 294 \text{ kg/d}$$

$$\text{Required active mass} = 294/0.057 = 5\ 158 \text{ kg}$$

$$\text{Required total mass} = 5\ 158 \times 77.86/38.21$$

$$= 10\ 510 \text{ kg}$$

$$\text{Volume at 3 kg/m}^3 = 3\ 503 \text{ m}^3$$

The oxygen requirements could be determined using any reliable model.

**Cyclic Process Designs.** Cyclic processes such as single and multiple oxidation ditches, rotating aeration systems, and sequencing batch reactors can be easily and precisely controlled by automatic instrumentation interfaced with a computer. The controls are simpler than those for conventional continuous-flow configurations because the length of the anaerobic, anoxic, and aerobic cycles/zones can be varied to match the transient loading conditions if the monitoring instruments have been appropriately selected and are both accurate and reliable. For example, the oxygen can be turned on and off to prolong or shorten the aerobic period as the organic and nitrogen loading rates vary each day, using either ammonia or nitrate (or both) monitoring. Dissolved oxygen, alkalinity, and turbidity can also be successfully used for accurate, rapid monitoring to control BNR processes, and redox potential has been successfully demonstrated in France. Control methodology of this type ensures that the process is operating at optimum conditions at all times, and this can be used to reduce the design volumes.

The more typical cyclic processes are described in the following sections.

*OXIDATION DITCH CONFIGURATIONS.* Experience has also shown that a simple racetrack oxidation ditch can be operated to accomplish both nitrogen and phosphorus removal without chemical addition by simply controlling the amount of oxygen transferred to the mixed liquor (Sen *et al.*, 1990a). The resulting process can be visualized as diagrammed in Figure 8.31, though the distinct zones do not exist because of the very high internal recirculation rate. Better phosphorus removal can be obtained by adding an anaerobic reactor upstream of the oxidation ditch, as shown in Figure 8.32. This was also demonstrated by Sen *et al.* (1990a), and they further improved performance by putting two oxidation ditches in series to form the VT2 process shown in Figure 8.33. This system consistently produces effluent TN and TP concentrations of less than 4.0 and 0.7 mg/L, respectively, all months of the year, without chemical addition or effluent filtration.

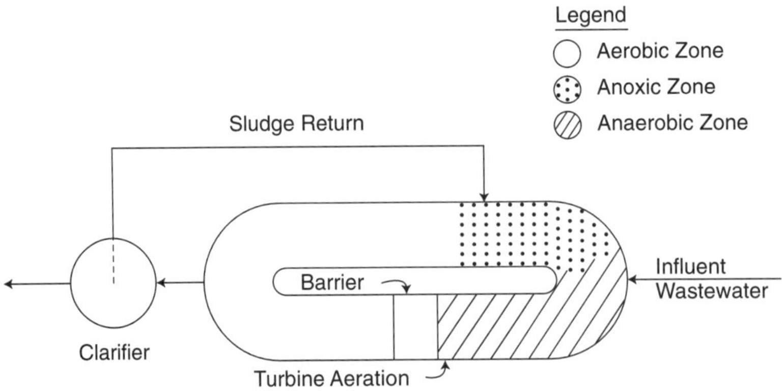


**Figure 8.31 Oxidation ditch design for nitrogen and phosphorus removal.**

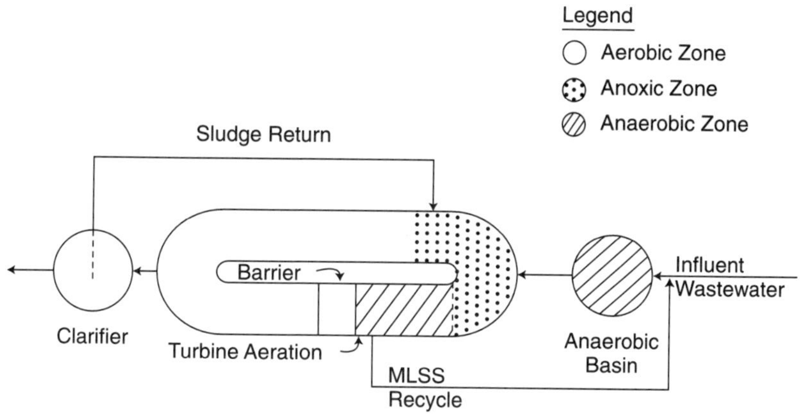


**Figure 8.32** **Oxidation ditch design for nitrogen and phosphorus removal.**

*BIO-DENITRO™/BIO-DENIPHO™ PROCESSES.* Configurations with multiple oxidation ditches have also been developed in Denmark for the biological removal of nitrogen alone or of both nitrogen and phosphorus (Bungaard, 1988). Labeled Bio-Denitro™ and Bio-Denipho™, respectively, these configurations are given in Figure 8.34.

In principle, the Bio-Denipho™ process includes the same biological mechanisms as other EBPR processes, comprising an anaerobic step succeeded by a nitrogen removal process step. Denitrification, however, is achieved according to the Bio-Denitro™ method.

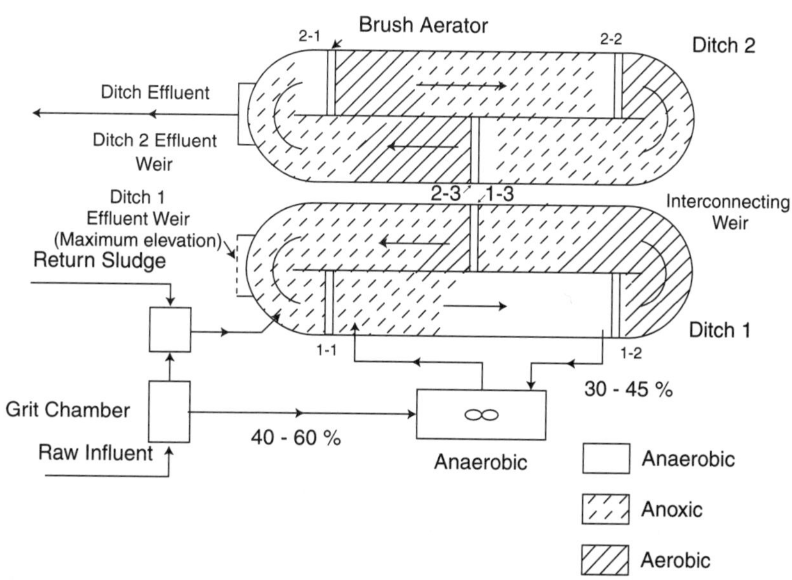


**Figure 8.33** **Schematic of the VT2 BNR process.**

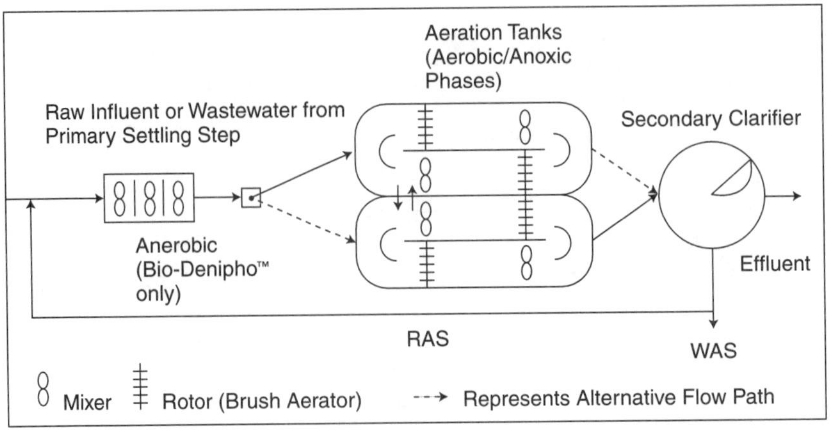


**Figure 8.34** **Schematic of the Bio-Denitro™ and Bio-Denipho™ processes (from Bungaard, E. [1988] Nitrogen and Phosphorus Removal by the Bio-Denitro & Bio-Denipho Processes—Experience and Development. *Proc. Int. Workshop Wastewater Treat. Technol.*, Danish Assoc. Consult. Eng., Copenhagen, Den., with permission).**

The Bio-Denitro™ system comprises two tanks working in an alternating mode of operation. The operation cycle consists of two main phases (A and B) and two intermediate phases (C1 and C2), as shown in Figure 8.35.

In each of the main phases, one tank is operating at anoxic conditions and fed with mixed liquor from the anaerobic tank to ensure denitrification, while the other tank is fed with effluent from the anoxic tank and operated at aerobic conditions to achieve nitrification. In the next main phase, the function of the two tanks is interchanged, and the direction of flow through the system is reversed.

By introducing intermediate phases in which both tanks are operated at either anoxic or aerobic conditions, the relative denitrification and nitrification capacity of the plant can be varied easily. Further, the duration of each individual phase can be changed without any problems. This makes it possible to adjust the operation to actual load variation (daily, weekly, or seasonally) or change the load (new industry).

In the alternating mode of operation, no recycling of the wastewater is necessary.

In cases of strict effluent TN requirements and/or unfavorable C:N ratio in the influent, the Bio-Denitro™/Bio-Denipho™ system can be extended with another aeration tank after the alternating system. This system is called Trio N&P, as shown in Figure 8.36. Typically, this third tank increases the flexibility of the plant and makes it possible for the other two tanks to operate under anoxic conditions for a period. This increases the possible denitrification time to more than 50%.

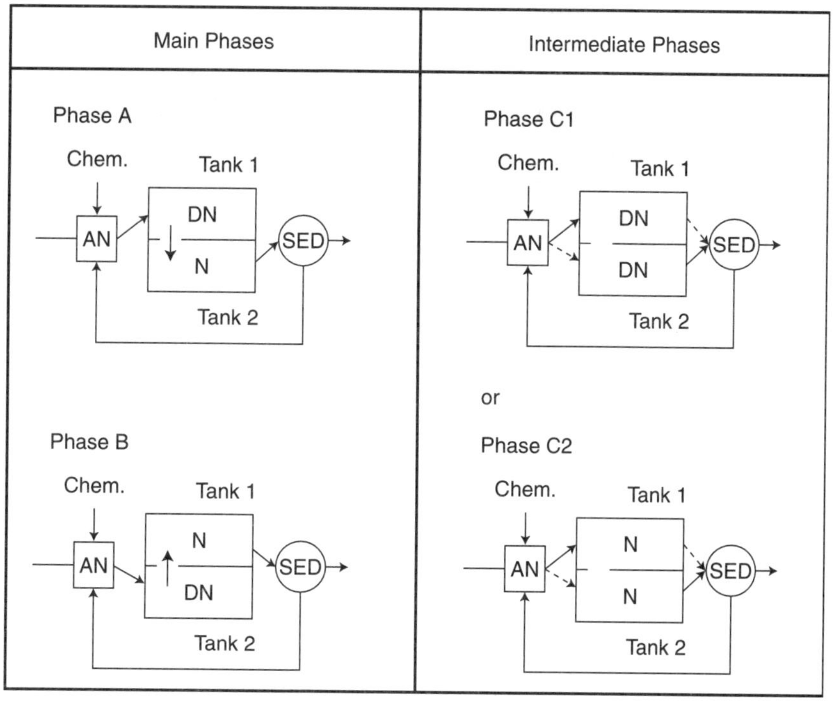


AN: Anaerobic Zone DN: Denitrification N: Nitrification SED: Sedimentation

**Figure 8.35** **The Bio-Denipho™ system (AN = anaerobic zone; DN = denitrification; N = nitrification; SED = sedimentation) (from Bungaard, E. [1988] Nitrogen and Phosphorus Removal by the Bio-Denitro & Bio-Denipho Processes—Experience and Development. *Proc. Int. Workshop Wastewater Treat. Technol.*, Danish Assoc. Consult. Eng., Copenhagen, Den., with permission).**

*SCHREIBER PROCESSES.* The Schreiber process consists of an aeration tank equipped with a rotating aeration bridge that provides both mixing and oxygen supply (Schreiber Wastewater, 1991). The process is termed countercurrent aeration because the air bubbles are released from a bridge traveling faster than the overlying water layers, with the result that the bubbles travel to the water surface at a significant angle rather than directly upward. This lengthens considerably their path through the water and increases oxygen transfer per unit horsepower. The BNR processes are controlled by turning the air on and off as dictated by DO, turbidity, nitrate, and ammonia monitoring. Although the process has been designed with an anaerobic anoxic center zone separate from the outer zone containing the traveling bridge, the system most suitable for nutrient removal with automatic controls is the design with only one zone. Termed "Simultech," the process schematic is given in Figure 8.37.

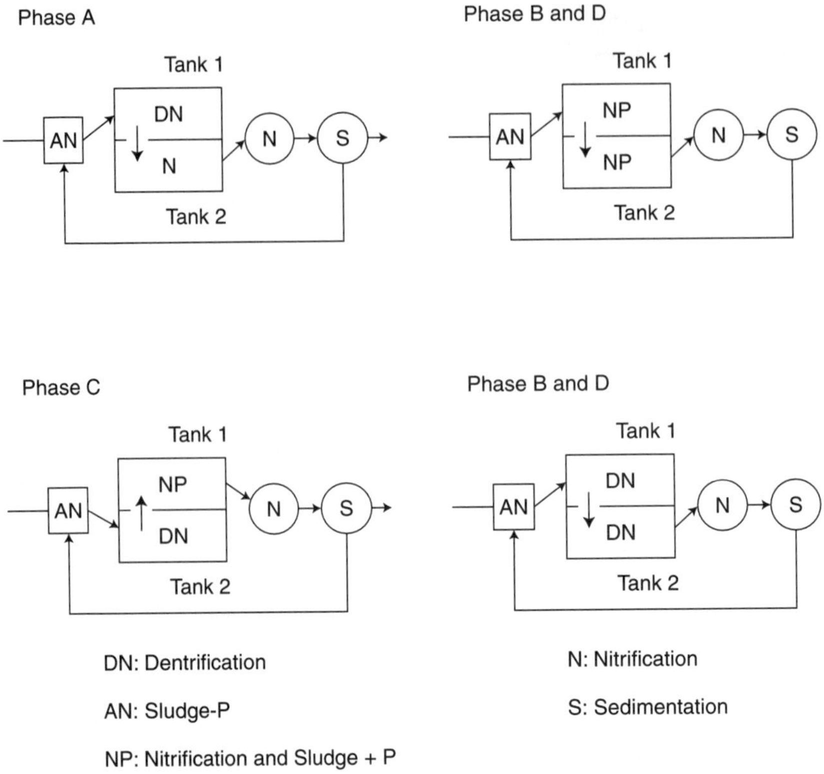


**Figure 8.36** **Bio-Denipho™ trio operation cycle (DN = denitrification; AN = sludge phosphorus; NP = nitrification and sludge + phosphorus; N = nitrification; S = sedimentation) (from Bungaard, E. [1988] Nitrogen and Phosphorus Removal by the Bio-Denitro & Bio-Denipho Processes—Experience and Development. *Proc. Int. Workshop Wastewater Treat. Technol.*, Danish Assoc Consult. Eng., Copenhagen, Den., with permission).**

**SEQUENCING BATCH REACTOR PROCESS.** Sequencing batch reactors (SBRs) are single reactors that cycle through anoxic, anaerobic, and aerobic conditions to accomplish BNR. Settling is also accomplished in the same reactor, as is supernatant and sludge wasting. While all processing steps occur in the same reactor, two or more reactors are essential for wastewater treatment unless a very large holding basin is available. The operating sequence is shown in Figure 8.38. Process operation and performance have been thoroughly described by Irvine *et al.* (1983, 1985, and 1987). Three tanks are recommended for operating flexibility.

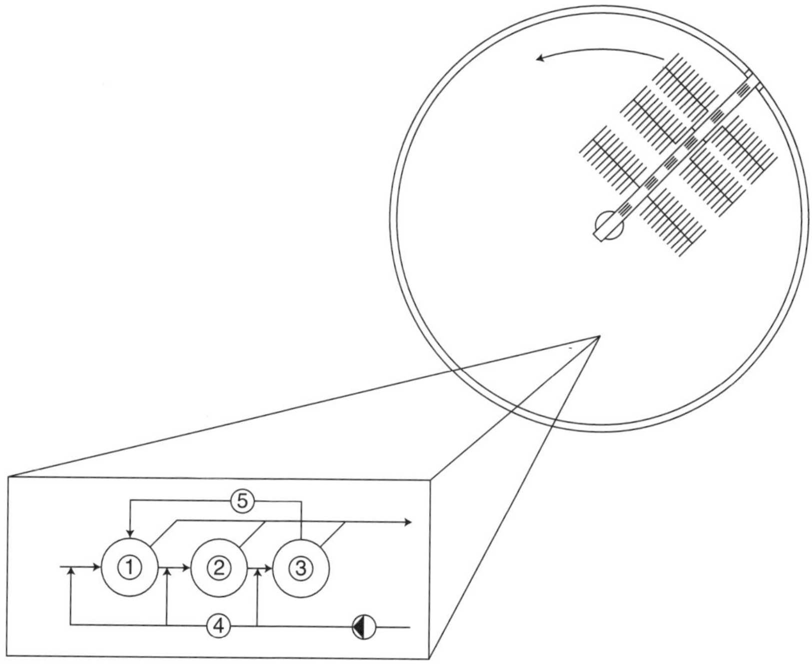


**Figure 8.37** "Simultech" process (1 = oxic process phase; 2 = anoxic process phase; 3 = anaerobic process phase; 4 = RAS; 5 = continuously repeating process cycle [represented as internal circulation]) (Schreiber Wastewater, 1991).

# *Effluent limitations*

There has been some confusion concerning the lower limits of phosphorus and nitrogen concentrations possible with BNR processes. In other words, how low can the final concentrations be if the microorganisms have an abundance of substrate available compared to the nutrient concentrations (that is, when the system is nitrogen and/or phosphorus limited rather than organic substrate limited)?

Effluent concentrations show that soluble effluent phosphorus concentrations averaging less than 0.35 mg/L have been achieved over long periods of time and that effluent TP concentrations of less than 0.50 mg/L are possible with only secondary clarification for suspended solids removal. Barnard (1988) has stated that effluent SP concentrations of less than 0.1 mg/L can be reliably obtained by phosphorus-limited EBPR processes. When complete nitrification is being achieved, effluent ammonia concentrations as low as 0.05 mg/L are possible, and concentrations less than 0.50 mg/L are typical. However, BNR process effluents typically will contain soluble organic nitro-

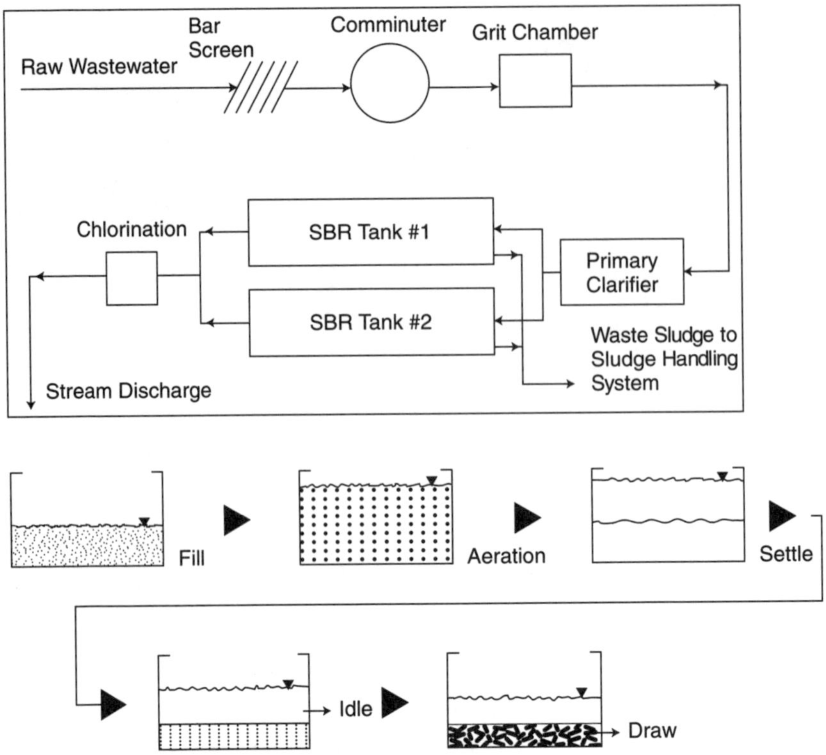


**Figure 8.38** **Schematic and operating sequence for two-tank SBR (Irvine *et al.*, 1983).**

gen concentrations of 1.0 to 1.5 mg/L, which cannot be removed without activated carbon treatment or an equivalent separation technique. Nevertheless, effluent TKN concentrations of less than 1.5 mg/L are possible.

Those BNR systems using postdenitrification units can reduce effluent $NO_x$ concentrations to approximately 1 mg/L, if desired. When combined with an effluent TKN concentration of less than 1.5 mg/L, effluent TN concentrations of less than 3.0 mg/L are possible. It should be recognized, however, that effluent TN concentrations of approximately 6 mg/L can be achieved much more economically than concentrations of less than 3 mg/L. This is because the reactor volumes for the lower concentration must be considerably larger to ensure complete nitrification at low temperatures. Also, the energy reduction benefits of denitrification are maximized when the nitrate recycle rate is approximately equal to the influent wastewater flow (Brannan and Randall, 1987), and this rate means the effluent nitrate-nitrogen concentration will typically be 4 to 5 mg/L.

Effluent nutrient standards for WWTPs should be set only after full consideration of the seasonal nature of nonpoint nutrient inputs, the magnitude of atmospheric inputs, and the economics of different levels of wastewater treatment.

# References

Abu-Ghararah, Z.H., and Randall, C.W. (1991) The Effect of Organic Compounds on Biological Phosphorus Removal. *Water Sci. Technol.*, **23**, 585.

American Public Health Association, American Water Works Association, and Water Environment Federation (1995) *Standard Methods for the Examination of Water and Wastewater.* 19th Ed., Washington, D.C.

Barnard, J.L. (1974) Cup P and N Without Chemicals. *Water Wastes Eng.*, **11,** 7, 33.

Barnard, J.L. (1975) Biological Nutrient Removal without the Addition of Chemicals. *Water. Res.* (G.B.), **9**, 485.

Barnard, J.L. (1984) Activated Primary Tanks for Phosphate Removal. *Water SA*, **10**, 121.

Barnard, J.L. (1988) Case Studies in Phosphate Removal. *Proc. Int. Workshop Wastewater Treat. Technol.*, Danish Assoc. Consult. Eng., Copenhagen, Den.

Barnard, J.L. (1992) Design of Prefermentation Processes. In *Design and Retrofit of Wastewater Treatment Plants for Biological Nutrient Removal.* C.W. Randall *et al.* (Eds.), Technomic Publishing Co., Lancaster, Pa.

Barnard, J.L., *et al.* (1984) Design Strategies for Nutrient Removal Plant. *Water Sci. Technol.*, **11/12,** 133.

Brannan, K.P. (1986) *Substrate Stabilization in the Anaerobic Stage of a Biological Phosphorus Removal System.* Ph.D. diss., Va. Poly. Inst. Univ., Blacksburg.

Brannan, K.P., and Randall, C.W. (1987) The Economics of Nitrogen Removal in a Single-Sludge Activated Sludge System. *Proc. Conf. Impact Control Nutrients Surf. Waters*, Environ. Syst. Eng. Dep., Clemson Univ., Charleston, S.C.

Bungaard, E. (1988) Nitrogen and Phosphorus Removal by the Bio-Denitro & Bio-Denipho Processes—Experience and Development. *Proc. Int. Workshop Wastewater Treat. Technol.*, Danish Assoc Consult. Eng., Copenhagen, Den.

Chapin, R.W. (1993) The Inhibition of Biological Nutrient Removal by Industrial Wastewater Components and High Acetic Acid Loadings. M.S. thes., Va. Poly. Inst. Univ., Blacksburg.

Daigger, G.T., *et al.* (1987a) Factors Affecting Biological Phosphorus Removal for a University of Cape Town Type Process Applied in a High Rate Mode. *Proc. IAWPRC Int. Spec. Conf., Biol. Phosphate Removal Wastewater*, Rome, It.

Daigger, G.T., *et al.* (1987b) Factors Affecting Biological Phosphorus Removal for the VIP Process, a High-Rate University of Cape Town Process. In *Biological Phosphate Removal Wastewaters.* R. Ramadori (Ed.), Pergamon Press, Oxford, Eng.

Dold, P.L. (1990) Incorporation of Biological Excess Phosphorus Removal in a General Activated Sludge Model. *Proc. 13th Int. Symp. Wastewater Treat.*, Montreal, Que., Can.

Ekama, G.A., and Marais, G.v.R. (1984a) Biological Nitrogen Removal. In *Theory, Design and Operation of Nutrient Removal Activated Sludge Processes.* Water Res. Comm., Pretoria, S. Afr.

Ekama, G.A., and Marais, G.v.R. (1984b) The Influence of Wastewater Characteristics on Process Design. In *Theory, Design and Operation of Nutrient Removal Activated Sludge Processes.* Water Res. Comm., Pretoria, S. Afr.

Ekama, G.A., *et al.* (1982) Considerations in the Process Design of Nutrient Removal Activated Sludge Processes. *Proc. IAWPR Post-Conf. Seminar Phosphorus Removal*, Pretoria, S. Afr.

Ekama, G.A., *et al.* (1984) Biological Excess Phosphorus Removal. In *Theory, Design and Operation of Nutrient Removal Activated Sludge Processes.* Water Res. Comm., Pretoria, S. Afr.

Gerber, A., *et al.* (1986) The Effect of Acetate and Other Short Chain Carbon Compounds on the Kinetics of Biological Nutrient Removal. *Water SA*, **12**, 7.

Gibson, J., and Dold, P.L. (1991) Wastewater Characterization for Activated Sludge Modelling. *Proc. 26th Can. Symp. Water Pollut. Res.,* Burlington, Ont., Can.

Henze, M., and Harremoës, P. (1990) Chemical–Biological Nutrient Removal—The HYPRO Concept. In *Chemical Water and Wastewater Treatment.* H.H. Hahn and R. Klute (Eds.), Springer-Verlag, Berlin, Ger., 499.

HYPRO (1991) Hydrolysis Project, A Nordic Collaborative Project. I Kruger AS, Danisco Biotechnology AS, Dep. Environ. Eng., Water Qual. Inst., Tech. Univ. Den.

Irvine, R.L., *et al.* (1983) Municipal Application of Sequencing Batch Treatment. *J. Water Pollut. Control Fed.*, **55**, 984.

Irvine, R.L., *et al.* (1985) An Organic Loading Study of Full-Scale Sequencing Batch Reactors. *J. Water Pollut. Control Fed.*, **57**, 847.

Irvine, R.L., *et al.* (1987) Analysis of Full-Scale SBR Operation at Grundy Center, Iowa. *J. Water Pollut. Control Fed.*, **59**, 132.

Kavanaugh, R.G., and Randall, C.W. (1994) Bacterial Populations in a Biological Nutrient Removal Plant. *Water Sci. Technol.*, **29**, 7, 25.

Kerdachi, D., and Roberts, M. (1982) Full Scale Phosphate Removal Experiences in the Umhlatuzana Works at Different Sludge Ages. *Proc. Post Conf. Seminar Phosphate Removal Biol. Treat. Proc., IAWPRC*, **2,** Pretoria, S. Afr.

Levin, G.V., and Shapiro, J. (1965) Metabolic Uptake of Phosphorus by Wastewater Organisms. *J. Water Pollut. Control Fed.*, **36**, 800.

Ludzack, F.L., and Ettinger, M. B. (1962) Controlling Operation to Minimize Activated Sludge Effluent Nitrogen. *J. Water Pollut. Control Fed.*, **34,** 920.

Mamais, D., and Jenkins, D. (1992) The Effects of MCRT and Temperature on Enhanced Biological Phosphorus Removal. *Water Sci. Technol.*, **26**, 5–6, 955.

Mamais, D., *et al.* (1993) A Rapid Physical-Chemical Method for the Determination of Readily Biodegradable Soluble COD in Municipal Wastewater. *Water Res.* (G.B.), **27**, 195.

Marickovich, D. (1995) *The Mechanism of Temperature Inhibition of Biogical Phosphorus Removal.* Ph.D. diss., Va. Poly. Inst. State Univ., Blacksburg.

McClintock, S.A., *et al.* (1991) The Effects of Temperature and Mean Cell Residence Time on Enhanced Biological Phosphorus Removal. *Proc. 1991 Spec. Conf. Environ. Eng.*, Am. Soc. Civ. Eng., New York, N.Y.

McClintock, S.A., *et al.* (1993) Effects of Temperature and Mean Cell Residence Time on Biological Nutrient Removal Processes. *Water Environ. Res.*, **65,** 110

Pattarkine, V.M. (1991) The Role of Metals in Enhanced Biological Phosphorus Removal from Wastewater. Ph.D. diss., Va. Poly. Inst. State Univ., Blacksburg.

Rabinowitz, B., and Oldham, W.K. (1985) The Use of Primary Sludge Fermentation in the Enhanced Biological Phosphorus Removal Process. *Proc. Univ. Br. Columbia Conf. New Dir. Res. Waste Treat. Manage.*, Vancouver, B.C., Can.

Randall, C.W. (1984) The Development of Energy Efficient Activated Sludge Systems Through Utilization of Nutrient Removal Processes. In *Toxic and Hazardous Wastes. Proc. 16th Mid-Atlantic Ind. Waste Conf.*, M.D. LaGrega and D. A. Long (Eds.), Technomic Publishing Company, Inc., Lancaster, Pa.

Randall, C.W., and Chapin, R.W. (1994) Acetic Acid Inhibition of Biological Phosphorus Removal. Paper presented at 67th Annu. Conf. Water Environ. Fed., Chicago, Ill.

Randall, C.W., *et al.* (1985) The Oxygen Requirements of Biological Nutrient Removal Processes. In *New Directions and Research in Waste Treatment and Residuals Management.* Vol. 1, Univ. Br. Columbia, Vancouver, Can.

Randall, C.W., *et al.* (1987) Factors Affecting Anaerobic Stabilization During Biological Phosphorus Removal. In *Biological Phosphage Removal From Wastewaters.* R. Ramadori (Ed.), Pergamon Press, Oxford, Eng.

Randall, C.W., *et al.* (1992a) Nitrification Kinetics in Single-Sludge Biological Nutrient Removal Activated Sludge Systems. *Water Sci. Technol.*, **25**, 195.

Randall, C.W., *et al.* (1992b) *Design and Retrofit of Wastewater Treatment Plants for Biological Nutrient Removal.* Technomic Publishing Co., Inc., Lancaster, Pa.

Reddy, M. (1991) The Concept of Phosphorus Storage Capability and Its Implications for Design of Systems for Enhanced Biological Uptake of Phosphate. *Water Sci. Technol.*, **23**, 577.

Reynolds, T.D. (1982) *Unit Operations and Processes in Environmental Engineering*. Brooks/Cole Engineering Division, Wadsworth, Inc., Monterey, Calif.

Schreiber Wastewater Treatment Technologies (1991) *Simultech Nutrient Removal System.* Schreiber Corporation, Inc., Trussville, Ala.

Sell, R.L. (1981) Low Temperature Biological Phosphorus Removal. Paper presented at 54th Annu. Conf. Water Pollut. Control Fed., Detroit, Mich.

Sen, D., *et al.* (1990a) *Biological Nitrogen and Phosphorus Removal in Oxidation Ditch and High Nitrate Recycle Systems.* CBP/TRS 47/90, Chesapeake Bay Program, U.S. EPA, Washington, D.C.

Sen, D., *et al.* (1990b) Alkalinity and Turbidity Measurements in Feedback Controls to Improve Performance of a Biological Nitrogen and Phosphorus Removal System. In *Instrumentation, Control and Automation of Water and Wastewater Treatment and Transport Systems.* R. Briggs (Ed.), Pergamon Press, Oxford, Eng.

Siebritz, I.P. (1983) *Biological Excess Phosphorus Removal in the Activated Sludge Process*, Ph.D. diss., Univ. Cape Town, S. Afr.

Spector, M.L. (1979) U. S. patent 4 162 153, July 24.

Tracy, K.D., and Flammino, A. (1985) Kinetics of Biological Phosphorus Removal. Paper presented at 58th Annu. Conf. Water Pollut. Control Fed., Kansas City, Mo.

Waltrip, G.D. (1991) Full-Scale Nutrient Removal Demonstration Project at the York River Treatment Plant with the VIP Process. Paper presented at 45th Annu. Conf. Virginia Water Pollut. Control Assoc., Williamsburg, Va.

Wanner, J., *et al.* (1989) An Innovative Technology for Upgrading Nutrient Removal Activated Sludge Plants. *Proc. IAWPRC Int. Spec. Conf. Upgrading Wastewater Treat. Plants,* Munich, Ger., 9.

Wentzel, M.C., *et al.* (1988a) Enhanced Polyphosphate Organism Cultures in Activated Sludge Systems. Part 1: Enhanced Culture Development. *Water SA*, **14**, 81.

Wentzel, M.C., *et al.* (1988b) Enhanced Polyphosphate Organism Cultures in Activated Sludge Systems. Part 2: Experimental Behaviour. *Water SA*, **15**, 71.

# Chapter 9
# Structured Process Models for Biological Nutrient Control

Activated-sludge systems become more complex as their function is expanded from carbonaceous removal alone to include nitrification, denitrification, and enhanced biological phosphorus removal (EBPR). Typically, a nutrient removal system involves multiple reactors, some aerated and some not, with interreactor recirculation of mixed liquor. The number of biological reactions and the number of compounds involved in the process also increase correspondingly. This is because a nitrification–denitrification enhanced biological phosphorus removal (NDEBPR) system involves three separate groups of microorganisms (poly-P heterotrophs, non-poly-P heterotrophs, and nitrifying autotrophs) operating on a large number of chemical components in three distinct environmental regimes (aerobic zones, anoxic zones where nitrate but not oxygen is present, and anaerobic zones where nitrate and oxygen are excluded as far as possible). These features make for complex behavior, which has increased the level of difficulty in design, operation, and control.

Given this complexity, the performance of any proposed system design can be determined only by experimentation (for example, in pilot plants) or by a mathematical model that simulates the behavior accurately. Comprehensive experimentation to evaluate the influence of a wide range of parameters is costly and time consuming. Increasingly, mathematical modeling of system behavior is being used as a tool to facilitate the tasks of designers and operators.

Currently, the approach in modeling activated-sludge systems is to use mechanistic models. The best known model of this kind is the International Association on Water Pollution Research and Control (IAWPRC) (now International Association on Water Quality [IAWQ]) Activated Sludge Model No. 1 (ASM1). These mechanistic models incorporate mathematical expressions that represent the biological interactions, based on hypotheses proposed for the biological processes occurring within the system. The model should quantify, for each process, both the kinetics (reaction rate, concentration dependence) and the stoichiometry (effect on the masses of compounds involved). For example, consider the biological process of growth of heterotrophs on a soluble organic substrate. The kinetics may be formulated as a Monod expression. The rate will be determined by the concentrations of substrate and heterotrophs within the bioreactor and by the magnitudes of the kinetic parameters (maximum specific growth rate, half-saturation coefficient). The mass of heterotrophs formed in the process will be determined by the magnitude of the stoichiometric yield parameter. Other stoichiometric parameters may quantify the associated effect on other compound masses (for example, the amounts of nitrogen, phosphorus, and oxygen consumed).

This type of mechanistic model describes the biochemical reactions within the activated-sludge system, that is, the set of distinct biological processes and the manner in which these act on the various model compounds. The biological model itself should not be confused with a simulation model. Simulation of the response of a given activated-sludge system involves two steps after the biological model has been specified. First, the reactor configuration and the flow scheme must be specified, including the reactor sizes, influent characteristics, recycle flow rates, wastage rate, and other specifics. After this information is fixed, it is possible to perform mass balances over each reactor (or reactor zone) for each model compound. These mass balances quantify the behavior of each compound in the particular system being simulated. The mass balances constitute the state equations that relate the dependent variables (compound concentrations) to the independent variables such as reactor volume. The mass balances form a set of simultaneous equations which, when solved, characterize the system behavior. (The equations typically are nonlinear because the kinetic expressions for the biological processes typically are nonlinear.) The simultaneous solution provides values of the state variables (concentrations) at points in space (different reactors) and time (where there is a time-varying input to the system). In this way, the change of state of the system is related to the transport (input and output) and conversion (reaction) processes occurring within the system.

Simulating the response of a nutrient removal system based on a comprehensive biological model is mathematically complex and typically achieved with a computer program. A simulation program is useful for a number of reasons:

- System analysis and optimization—if a system model provides accurate predictions of response behavior, then these predictions can be compared to observed responses in analyzing the operation of existing systems. Any discrepancies can be useful in identifying problems in operation. An accurate model also can be used to optimize the performance of existing systems. Various operating strategies can be proposed and tested rapidly without having to resort to potentially difficult practical evaluation.
- System design—a simulation program does not design a system directly. However, a simulation program can be a useful tool for the design engineer to evaluate proposed system designs rapidly. In addition, a dynamic model can provide valuable design information that has often only been available through empirical estimates. For example, a parameter such as peak oxygen demand can be obtained directly from the simulation program run under time-varying input patterns. This means that peak aeration capacity can be quantified accurately.

Table 9.1 lists names and contact addresses for five current software products that can be used for simulation of NDEBPR systems. The listing may well exclude other products about which the authors are not aware. Providing this information is not an endorsement of these products. Rather, the aim is to provide impetus to simulation as an engineering tool for assistance in system analysis, optimization, and design.

Simulation models are not the only types of model that can assist the designer and operator of nutrient removal systems. In particular, the design engineer has a need for tools to assist in process design. This chapter presents the following two models in an attempt to address these differing needs:

- A general dynamic model—this mechanistic activated-sludge system kinetic model incorporates carbon oxidation, nitrification, denitrification, and EBPR behavior. The model is complex and suited to implementation in a computer simulation program.
- A simplified steady-state model—this model, in essence, is a simplification of the general model. The model allows a "pencil and calculator" approach, which is of use to the design engineer.

**Table 9.1 Examples of biological nutrient removal computer software.**

| Product name | Contact |
|---|---|
| ASIM™ | Abt. Ingenieurwissen<br>Schaften<br>Uberlandstrasse 13<br>EAWAG<br>CH-8600 Dubendorf<br>Switzerland |
| BioWin™ | EnviroSim Associates Ltd.<br>482 Anthony Drive<br>Oakville<br>Ontario L6J 2K5<br>Canada |
| EFOR™ | EFOR<br>I. Kruger AS<br>Gladsaxevej 363<br>DX-2860 Soborg<br>Denmark |
| GPS-X™ and SimWorks™ | Hydromantis Inc.<br>1685 Main St. West<br>Suite 302<br>Hamilton<br>Ontario L8S 1G5<br>Canada |

At the outset it should be recognized that models are not perfect. This is particularly true for nutrient removal systems, in which there are many unresolved aspects that are the subject of ongoing research. Therefore, before presenting the models it is pertinent to discuss some of these issues. This information should provide the reader with background on the current status of nutrient removal system modeling and some guidance as to constraints on the use of the models. The chapter starts with a brief review of the development of models for biological nutrient removal systems.

# DEVELOPMENT OF MODELS FOR BIOLOGICAL NUTRIENT REMOVAL

In 1982, the IAWPRC (now IAWQ) appointed a task group to review modeling of activated-sludge systems incorporating carbonaceous energy removal, nitrification, and denitrification. The initial deliberations of the group resulted in a preliminary version of the IAWPRC Model (Grady *et al.*, 1986). Dold and Marais (1986) conducted a comprehensive evaluation of the preliminary model. It was proposed that certain changes should be made, in particular with respect to the way in which the fate of organic nitrogen was modeled. These changes were subsequently adopted in the final version of the IAWPRC ASM1 (Henze *et al.*, 1987a and b).

The task group drew on a wide range of information in formulating the model. One research initiative that had a major influence on the model was the dynamic activated-sludge model developed by Marais and coworkers at the University of Cape Town (UCT) (Dold *et al.*, 1980, and van Haandel *et al.*, 1981). This dynamic model evolved out of the steady-state model of Marais and Ekama (1976). The steady-state model, in turn, constituted a development from a number of previous models for carbonaceous and nitrogenous material conversion and removal (Downing *et al.*, 1964; Lawrence and McCarty, 1970; McKinney, 1962; and McKinney and Ooten, 1969).

The ASM1 model, based on a mechanistic interpretation of the behavior of the organisms mediating the process reactions, has been shown to give a reliable description of the system response in wide ranges of system configurations (single and series reactor systems, aerated and nonaerated reactors, interreactor recycles), influent wastewater characteristics (chemical oxygen demand [COD], total Kjeldahl nitrogen [TKN], flow pattern), and operational parameters (sludge age, temperature, dissolved oxygen concentration). The model has had a significant effect on the development of design and operational procedures for activated-sludge systems. In full-scale plant operation it has also found application in assessing the effects of changes in waste flows and loads, operational modifications (such as changes in recycles), and proposed modifications to plant configuration. It has also proved valuable in operator training: through simulation exercises using the model, the operator acquires "instant" experience in the behavior to be expected with changes in inputs, system configuration, and operational strategies.

A restriction of ASM1 is that its scope is limited to systems incorporating carbonaceous energy removal and nitrification–denitrification (ND). The phenomenon of EBPR is not included. In EBPR activated-sludge systems, removal of phosphorus (P) is mediated by a group of organisms (poly-P organisms) that exhibit the propensity to store phosphorus as polyphosphate (poly-P) polymers called volutins (Buchan, 1981). Proposed explanations of the biochemical behavioral patterns associated with phosphorus release and uptake (and net phosphorus removal) have been presented in a number of models, such as that of Comeau *et al.* (1986), extended and modified by Wentzel *et al.* (1986), and that of Mino *et al.* (1987). The biochemical models are largely in agreement regarding the biochemical control mechanisms and have provided an explanation for the essential requirements for attaining EBPR—namely, an alternating anaerobic–aerobic sequence with the provision of short-chain fatty acids (SCFAs) during the anaerobic phase (also referred to as volatile fatty acids). In typical municipal wastewater, the SCFA content is minimal. In EBPR systems, the readily biodegradable COD component is transformed to SCFA by the non-poly-P organism mass, thereby making SCFA available to the poly-P organisms (Brodisch, 1985; Meganck *et al.*, 1985; and Wentzel *et al.*, 1985). Aside from this linkage, the poly-P and non-poly-P organisms in EBPR systems have been shown to act essentially independently. For this reason, Wentzel *et al.* (1989b) adopted the approach of developing "enhanced" cultures of poly-P organisms as the basis for studying the kinetics and stoichiometry of EBPR without the behavior being masked by that of the non-poly-P organisms.

Enhanced cultures of poly-P organisms were developed by Wentzel *et al.* (1989a and b) in continuous-flow activated-sludge systems (modified Bardenpho™ and UCT process configurations), with acetate as the only organic substrate. The sludge mass in these systems was shown to comprise the poly-P organism *Acinetobacter* spp. in excess of 90%. Based on observations of the continuous-flow systems and batch experiments using mixed liquor drawn from these systems, Wentzel *et al.* (1989a and b) developed a kinetic model for the enhanced culture EBPR system. The model provided a reasonable description of the response observed in a number of continuous-flow enhanced culture systems and the batch experiments with a single set of kinetic and stoichiometric parameters. The authors identified certain minor limitations in the proposed model and outlined problems that might be encountered when extending the model to mixed organism systems with municipal wastewater as influent. Despite any minor limitations, the enhanced culture model constituted a significant step toward the development of a general activated-sludge model capable of modeling the biological processes of carbonaceous energy removal, nitrification, denitrification, and EBPR.

Dold (1990 and 1991) merged the ASM1 model for non-poly-P heterotrophic organisms and autotrophic organisms (Henze *et al.*, 1987a and b) and the Wentzel *et al.* (1989b) model for poly-P organisms. In combining the models, certain extensions and modifications were incorporated in both the ASM1 and the Wentzel model components. Since the initial development, this NDEBPR model has been evaluated extensively against experimental data from laboratory-scale and full-scale treatment plants. This has led to further model refinement. The model is termed *general* in that it can be applied directly for the simulation of aerobic, anoxic–aerobic (ND), or anaerobic–anoxic–aerobic (NDEBPR) systems, predicting an appropriate balance between the three organism masses—the poly-P and non-poly-P heterotrophs and the autotrophs. This chapter presents the current model in some detail in the following sections.

The IAWQ task group has also extended ASM1 to include simulation of combined NDEBPR processes. The ASM2 model "is a compromise between complexity and simplicity, and between the many viewpoints on how the correct model should look like. It should be used as a conceptual platform for further model development" (Henze *et al.*, 1994).

**ASPECTS OF NITRIFICATION, DENITRIFICATION, ENHANCED BIOLOGICAL PHOSPHORUS REMOVAL MODEL DEVELOPMENT.** A mechanistic model for NDEBPR systems necessarily must account for a large number of biological processes to mimic the complex interactions that may affect the performance of a given treatment plant. This section briefly discusses certain key features of process behavior before presenting the general model in detail.

**Anoxic Growth of Poly-P Organisms.** A simplified representation of the behavioral patterns associated with phosphorus release and uptake in an EBPR system is shown in Figure 9.1. It is readily apparent that nitrate plays an important role in the performance of these systems. The current consensus interpretation is that nitrate entering the anaerobic zone will be used as an electron acceptor in the growth of non-poly-P heterotrophs. As a result, less substrate will be available for sequestration by the poly-P organisms, with the net effect that the amount of phosphorus removal will be reduced. The amount of substrate available to the poly-P organisms may be reduced to such an extent that these organisms are unable to sustain themselves in the system and the capacity for EBPR is lost.

This interpretation of the role of nitrate has proved valuable in the design and operation of EBPR systems. A basic principle for maximizing phosphorus removal is to "protect" the anaerobic reactor from ingress of nitrate as far as possible. Hence, as shown in Figure 9.1, the underflow recycle (nitrate rich) is directed to the anoxic reactor rather than directly to the anaerobic

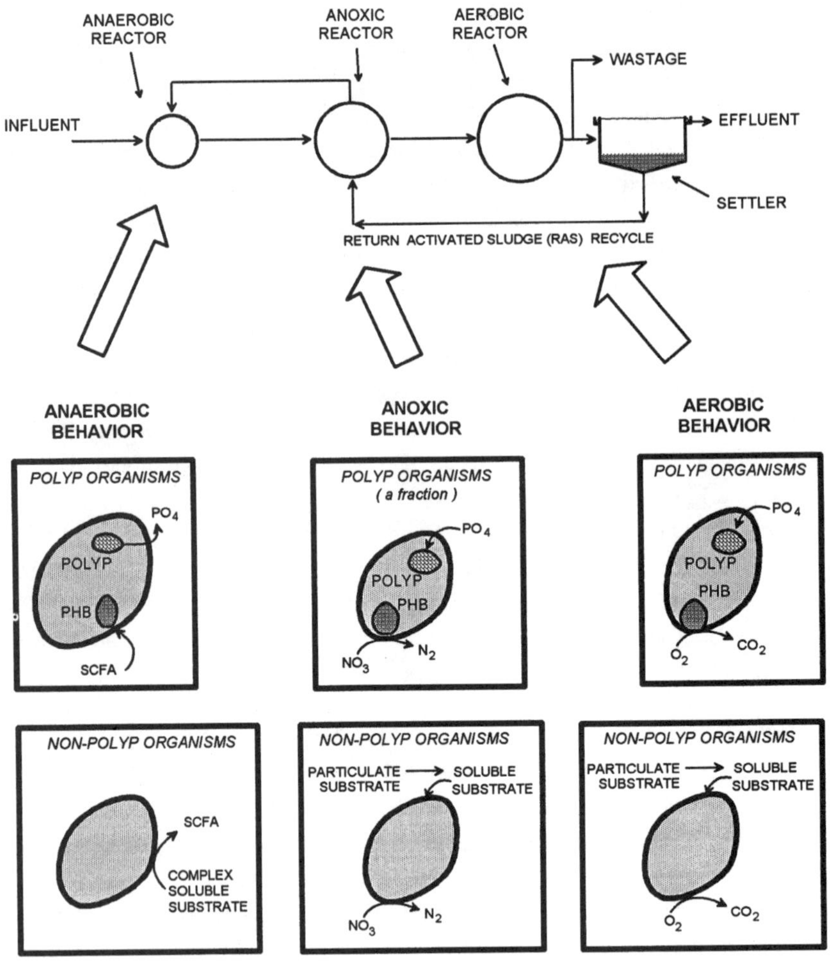


**Figure 9.1** **Schematic diagram outlining the roles of poly-P and non-poly-P heterotrophic organisms in a UCT/VIP™ process NDEBPR system.**

zone, and a recycle from the anoxic reactor (low in nitrate) transfers sludge to the anaerobic zone. However, implicit in this interpretation is the assumption that nitrate does not influence the behavior of the poly-P organisms directly. Denitrification by poly-P organisms was excluded from the Wentzel *et al.* (1989b) model because observations showed minimal denitrification occurred in laboratory systems comprising poly-P organisms. However, a recent review of experimental studies concerning denitrification behavior in EBPR activated-sludge systems (Barker and Dold, 1995b) indicates that phosphorus uptake by poly-P organisms does occur in anoxic zones of nutrient removal systems. Based on the results of microbiological studies, and many continuous and batch reactor experimental studies, a number of conclusions were drawn:

- Poly-P organisms are capable of concomitant denitrification and phosphorus uptake; however, indications are that not all poly-P organisms have this ability.
- Nitrate can serve as an electron acceptor for the oxidation of stored polyhydroxy butyrate (PHB); however, not all poly-P organisms capable of reducing nitrate appear able to use nitrite as an electron acceptor.
- Nitrate may not be as efficient as oxygen for phosphorus uptake. Batch tests indicate that more stored carbon (PHB, PHV) is used for a given amount of phosphorus taken up when nitrate is the electron acceptor in place of oxygen.
- Phosphorus uptake and PHB oxidation appear to occur simultaneously with phosphorus release and PHB storage when SCFAs are available under anoxic conditions.

In modeling denitrification by poly-P organisms, it is assumed that a fraction ($\eta_P$) of the poly-P organisms can use nitrate as an electron acceptor in the absence of oxygen for oxidation of stored PHB and uptake of phosphorus. In the model of Wentzel *et al.* (1989b), there are four aerobic growth processes; the four permutations reflect switches between ammonia and nitrate as the nitrogen source for synthesis and at high and low phosphorus concentration. The simplest means for incorporating anoxic growth of poly-P organisms is to allow the growth processes to occur under both aerobic and anoxic conditions. However, in an anoxic reactor of a continuous-flow system, growth is likely to occur in the presence of sufficient ammonia and soluble phosphorus. Therefore, in evaluating the possibility of anoxic growth, only the one aerobic growth process was duplicated for anoxic conditions, that is, growth with ammonia as the nitrogen source for synthesis and with no phosphorus limitation on the growth rate. The stoichiometric coefficient for phosphorus uptake, $f_{P,UPT}$, defined as the ratio of phosphorus taken up to PHB oxidized, was determined from batch experiments to be approximately 0.9 to 1.1 g P/g COD (Wentzel *et al.*, 1989b). With the inclusion of anoxic growth, there are now two stoichiometric parameters for phosphorus uptake, $f_{P,UPT1}$ and $f_{P,UPT2}$. Results of simulations indicate that the value for anoxic growth, $f_{P,UPT2}$, should be reduced to approximately 0.35 g P/g COD; the value for aerobic conditions, $f_{P,UPT1}$, remains unchanged at 0.95 g P/g COD. This was necessary to predict the observed phosphorus response.

**Conversion of Soluble Readily Biodegradable Chemical Oxygen Demand to Short-Chain Fatty Acids under Anaerobic Conditions.** The principal linkage between the poly-P and non-poly-P heterotrophic organism masses in EBPR systems treating municipal wastewater is the conversion of "complex" readily biodegradable soluble COD ($S_{BSC}$) to SCFAs ($S_{BSA}$). This process is the source of SCFA to sustain poly-P organism growth in the mixed-culture system, as the influent SCFA content typically is minimal. Wentzel *et al.* (1985) established the kinetics of this phenomenon and incorporated the process in the enhanced culture model for completeness even though all of the influent COD to the enhanced cultures was in the form of SCFA (acetate). The process was modeled as being first order with respect to the concentrations of $S_{BSC}$ and non-poly-P heterotrophs. In terms of stoichiometry, the process was assumed to yield one unit of $S_{BSA}$ for each unit of $S_{BSC}$ converted.

A study on COD and nitrogen mass balances in activated-sludge systems (Barker and Dold, 1995a) suggests that there is a "loss" of COD in activated-sludge systems incorporating anaerobic zones. Four different types of laboratory-scale system were studied: aerobic, anoxic, anoxic–aerobic, and anaerobic–anoxic–aerobic. The systems included a variety of configurations with differing wastewater characteristics and operating parameters. The results suggest that while good COD balances are to be expected in aerobic and anoxic–aerobic systems, systems incorporating anaerobic zones (that is, EBPR systems) tend to exhibit low COD balances (less than 80%). This "loss" of COD apparently is associated with the fermentation processes occurring in the anaerobic zone of EBPR systems treating municipal wastewater (anoxic-only systems also appear to exhibit a loss of COD, but to a lesser extent). Whether this COD loss is a direct result of fermentation (through the generation of gas that evolves during the actual fermentation process) or an indirect result (through the production of volatile compounds that are released from the system under aerated conditions) remains to be determined. Regardless, this disappearance of some 20% of the influent COD in EBPR systems translates into reduced sludge production and oxygen demand compared to non-EBPR systems. This feature alone strengthens the case for nutrient removal.

The following approach has been adopted for modeling of the readily biodegradable COD conversion process to incorporate the loss of COD because of fermentation:

- The process is assumed to be a fermentation process, that is, a growth process under anaerobic conditions. This is modeled using a Monod growth expression, with a yield of non-poly-P heterotrophic organisms ($Y_{H,ANA}$); correspondingly, the yield of fermentation products is $1 - Y_{H,ANA}$.
- Further, it is assumed that only a portion ($Y_{AC}$) of the fermentation products is $S_{BSA}$. The remainder is assumed to be COD lost to the system (perhaps as hydrogen gas).

The fermentation process is incorporated in the non-poly-P model component of the general model presented here because the process is mediated by the non-poly-P heterotrophic organism mass.

**Sequestration of Short-Chain Fatty Acids by Poly-P Organisms.** In the anaerobic sequestration of SCFA by poly-P organisms (for PHB storage, with associated phosphate release), it is assumed that the yield of PHB is

$Y_{PHB}$ units of PHB (as COD) per unit SCFA COD taken up. A value of 0.89 is suggested for $Y_{PHB}$ based on the assumption that 2 moles of PHB are formed per 2.25 moles of SCFA sequestered under anaerobic conditions (see biochemical model of Wentzel *et al.* [1986]). That is, the model incorporates COD loss in the sequestration reaction. This allows the general model to mimic the COD balances observed in the enhanced cultures.

**Nitrogen Source for Cell Synthesis.** In reviewing the initial IAWPRC model version, Dold and Marais (1986) postulated that under certain circumstances, nitrate, instead of ammonia-nitrogen, may serve as the nitrogen source for cell synthesis purposes. This postulate was confirmed from analysis of data collected during an extensive period, particularly in multiple-series reactor configurations operated at long sludge ages that exhibited high nitrification rates. The use of nitrate as a nitrogen source for poly-P organism synthesis when the ammonia concentration dropped to low levels was also observed by Wentzel *et al.* (1989a). On the basis of this information, two additional processes have been incorporated in the IAWPRC model version to give four growth processes: aerobic and anoxic growth of non-poly-P heterotrophs with either ammonia or nitrate as the nitrogen source for synthesis.

**Growth of Non-Poly-P Heterotrophs on Short-Chain Fatty Acids.** In the IAWPRC model, readily biodegradable soluble COD is used by the non-poly-P organisms (heterotrophs) in four possible growth modes: under aerobic or anoxic conditions with either ammonia or nitrate as the nitrogen source for synthesis purposes. For EBPR systems, it is necessary to distinguish between "complex" and SCFA readily biodegradable COD. Therefore, it is necessary to duplicate the four growth processes in the IAWPRC model to account for possible growth on the two components of the readily biodegradable COD for the mixed-culture system. With regard to growth on SCFA, it is likely that only one of the four processes would be of consequence: anoxic growth with ammonia as the nitrogen source. This is because SCFAs are removed in the unaerated zones at the "front end" of the continuous-flow systems and do not enter the aerobic zones in appreciable concentrations. However, for completeness, all four growth processes in the IAWPRC model (for "complex" COD) were duplicated in the general model for growth of non-poly-P organisms with SCFA as substrate. The same kinetic formulations and stoichiometry for growth on "complex" readily biodegradable COD and SCFA have been used.

**Hydrolysis/Solubilization of Slowly Biodegradable Chemical Oxygen Demand.** In the IAWPRC model, the biodegradable material is divided into a readily biodegradable fraction ($S_{BS}$) and a slowly biodegradable fraction ($S_{ENM}$). The readily biodegradable fraction is hypothesized to consist of material that can be absorbed readily by the organism and metabolized for energy and synthesis, whereas the slowly biodegradable fraction is assumed to be made up of particulate/colloidal material and complex organic molecules that require extracellular enzymatic breakdown before absorption and use. In the ASM1 model, the rate of solubilization under anoxic conditions is assumed to be reduced by a factor, $\eta_{SOL}$, compared to the rate under aerobic conditions. Under anaerobic conditions the rate is assumed to be zero. Recent research on enzymatic hydrolysis (Dold *et al.*, 1991, and San Pedro *et al.*, 1994) indicates that hydrolysis does occur under anaerobic conditions, and under anoxic conditions the rate of hydrolysis appears similar to that under aerobic conditions. This behavior requires further research. To provide flexibility in the model, two $\eta_{SOL}$ factors are incorporated: $\eta_{S,ANOX}$ and $\eta_{S,ANA}$.

**Decay of Poly-P Organisms under Anoxic Conditions.** The Wentzel (1989b) enhanced culture model did not consider anoxic behavior of poly-P organisms, and decay processes for these organisms were considered only for aerobic and anaerobic conditions. Processes for anoxic decay of poly-P organisms, stored polyphosphate lysis from anoxic decay, and PHB lysis from anoxic decay have been included here (three additional processes). Simulation results indicate that poly-P organisms under anoxic conditions have decay processes similar to those under aerobic conditions. This would seem reasonable because a significant number of the poly-P organisms can use nitrate as an electron acceptor.

**Conversion of Organic Nitrogen to Ammonia.** Because initial simulation results for mixed-culture EBPR systems indicated an overprediction of effluent TKN, the general model now assumes that both groups of heterotrophs (poly-P and non-poly-P) mediate the conversion of soluble organic nitrogen to ammonia.

**Yield of Heterotrophs in Anoxic Growth.** The IAWPRC model assumes a single yield coefficient ($Y_H$) for non-poly-P heterotrophs irrespective of whether oxygen or nitrate serves as the electron acceptor. It has long been surmised that the yield of organisms under anoxic conditions with nitrate as electron acceptor is lower than for aerobic growth. This does not have a very significant effect on IAWPRC model predictions of sludge production for ND systems treating municipal wastewater. For this situation the influent TKN:COD ratio typically is relatively low ($<0.12$ g N/g COD); therefore, the amount of nitrate generated is limited, and the organism mass generated by denitrification typically is small compared to the mass generated in aerobic growth. However, this is not the case in certain synthetic situations, for example anoxic-only systems with addition of sufficient quantities of nitrate to maintain the system organic substrate limited. McClintock *et al.* (1988) conducted side-by-side studies on five parallel sets of single-reactor systems operated over a range of sludge ages (1.5 to 15 days); in each set, one system was aerated while the other was not. The unaerated system was mixed and the soluble influent supplemented with an excess of nitrate (anoxic). In four of the systems, sludge production in the anoxic unit was significantly lower than in the corresponding aerobic one. However, sludge production in the 15-day sludge age systems was essentially the same. To account for the possibility of a lower anoxic yield, yield factors ($Y_{H,AER}$ and $Y_{H,ANOX}$) have been incorporated in the general model for both aerobic and anoxic growth, respectively, of non-poly-P heterotrophs. It should be noted that the expected decrease in sludge production has not been observed significantly in laboratory-scale anoxic and anoxic–aerobic systems treating municipal wastewater (Arkley and Marais, 1981, and Smyth, 1994).

**Switching Functions for Phosphorus Limitation.** The Wentzel *et al.* (1989a) model incorporates a switching function that allows the growth and phosphorus uptake processes to switch off when soluble phosphorus becomes limiting. Evidence from simulation of full-scale systems indicates that these two processes have different threshold values for switching on and off. To accommodate this, two additional switching functions for phosphorus limitation have been used, and the parameter ($K_{LP}$) has now been replaced by two parameters, $K_{LP,GRO}$ and $K_{LP,UPT}$. The parameter $K_{LP,GRO}$ is used in the switching functions for growth and has a value of approximately 0.005 g P/m$^3$. The parameter $K_{LP,UPT}$ represents the soluble phosphate limit for phosphorus uptake by poly-P organisms and has a value of approximately 0.25 g P/m$^3$.

**Releasable/Fixed Polyphosphate Components.** Experimental observations indicate that stored polyphosphate is portioned between a low- and a high-molecular-weight fraction (that is, $P_{PP\text{-}LO}$ and $P_{PP\text{-}HI}$) and that only the low-molecular-weight fraction can be released after being taken up. In modeling the phosphorus uptake/release processes, it is assumed that only a fraction ($f_{PP}$) of the phosphate stored as polyphosphate can be released in a subsequent anaerobic condition. The value for $f_{PP}$ of 0.94 suggested here was selected on the basis of simulating behavior in a range of EBPR systems and is in accordance with values reported in the literature (Mino *et al.*, 1984). The remainder of the stored polyphosphate is termed "fixed."

# GENERAL DYNAMIC MODEL FOR NUTRIENT REMOVAL SYSTEMS

A general mechanistic model for biological nutrient removal activated-sludge systems is presented in the following sections. The model is presented in two parts, following the structure of the previous discussion—a part for the non-poly-P and autotrophic organisms (relating principally to ND systems) and a part for the poly-P organisms (relating to EBPR behavior).

The model component (compound) symbols are defined in Table 9.2. The nomenclature adopted differs from that used by the IAWQ task group.

The model is presented in matrix format (two parts). To simplify presentation of the process rate equations, a number of switching functions are incorporated in the kinetic expressions; these are defined in Table 9.3. (The role of switching functions in activated-sludge modeling has been described by Billing and Dold [1988]). Briefly, switching functions serve a mathematical purpose to turn process rates on or off when concentrations of selected components are, respectively, above (SwitchYes) or below (SwitchNo) some low threshold magnitude. The switching functions also facilitate interpretation of the process model in the matrix format.

**Table 9.2 Definition of component symbols in the general model.**

| Component | Symbol | Definition | Units |
|---|---|---|---|
| 1 | $Z_H$ | Active non-poly-P heterotrophic biomass | g cell $COD/m^3$ |
| 2 | $Z_A$ | Active autotrophic biomass | g cell $COD/m^3$ |
| 3 | $Z_P$ | Active poly-P heterotrophic biomass | g cell $COD/m^3$ |
| 4 | $Z_E$ | Endogenous mass | g $COD/m^3$ |
| 5 | $S_{ENM}$ | Enmeshed slowly biodegradable substrate | g $COD/m^3$ |
| 6 | $S_{BSC}$ | Readily biodegradable "complex" substrate | g $COD/m^3$ |
| 7 | $S_{BSA}$ | Readily biodegradable SCFA substrate | g $COD/m^3$ |
| 8 | $S_{PHB}$ | Stored SCFA | g $COD/m^3$ |
| 9 | $S_{UP}$ | Particulate unbiodegradable matter | g $COD/m^3$ |
| 10 | $S_{US}$ | Unbiodegradable soluble substrate | g $COD/m^3$ |
| 11 | $P_{PP\text{-}LO}$ | Releasable stored poly-P | g $P/m^3$ |
| 12 | $P_{PP\text{-}HI}$ | Fixed stored poly-P | g $P/m^3$ |
| 13 | $P_{O4}$ | Soluble phosphate | g $P/m^3$ |
| 14 | $N_{BP}$ | Particulate biodegradable organic nitrogen | g $N/m^3$ |
| 15 | $N_{BS}$ | Soluble biodegradable organic nitrogen | g $N/m^3$ |
| 16 | $N_{O3}$ | Nitrate-nitrogen | g $N/m^3$ |
| 17 | $N_{H3}$ | Ammonia-nitrogen | g $N/m^3$ |
| 18 | $N_{US}$ | Unbiodegradable soluble nitrogen | g $N/m^3$ |
| 19 | $S_{O2}$ | Oxygen | g $(COD)/m^3$ |

**Table 9.3 Definition of switching functions in the general model.**

| Switching function | Equivalent | Definition |
|---|---|---|
| AirYesHet | 1 − AirNoHet | $\frac{S_O}{K_{O\ HET} + S_O}$ |
| AirNoHet | 1 − AirYesHet | $\frac{K_{O\ HET}}{K_{O\ HET} + S_O}$ |
| AirYesAut | 1 − AirNoAut | $\frac{S_O}{K_{O\ AUT} + S_O}$ |
| AirNoAut | 1 − AirYesAut | $\frac{K_{O\ AUT}}{K_{O\ AUT} + S_O}$ |
| $NH_3$Yes | 1 − $NH_3$No | $\frac{N_{H3}}{K_{NA} + N_{H3}}$ |
| $NH_3$No | 1 − $NH_3$Yes | $\frac{K_{NA}}{K_{NA} + N_{H3}}$ |
| $NO_3$Yes | 1 − $NO_3$No | $\frac{N_{O3}}{K_{NO} + N_{O3}}$ |
| $NO_3$No | 1 − $NO_3$Yes | $\frac{K_{NO}}{K_{NO} + N_{O3}}$ |
| $PO_4$GroYes | 1 − $PO_4$GroNo | $\frac{P_{O4}}{K_{LP\text{-}GRO} + P_{O4}}$ |
| $PO_4$GroNo | 1 − $PO_4$GroYes | $\frac{K_{LP\text{-}GRO}}{K_{LP\text{-}GRO} + P_{O4}}$ |
| $PO_4$UptYes | 1 − $PO_4$UptNo | $\frac{P_{O4}}{K_{LP\text{-}UPT} + P_{O4}}$ |
| $PO_4$UptNo | 1 − $PO_4$UptYes | $\frac{K_{LP\text{-}UPT}}{K_{LP\text{-}UPT} + P_{O4}}$ |
| $S_{BSA}$Yes | 1 − $S_{BSA}$No | $\frac{S_{BSA}}{K_{SSEQ} + S_{BSA}}$ |
| $S_{BSA}$No | 1 − $S_{BSA}$Yes | $\frac{K_{SSEQ}}{K_{SSEQ} + S_{BSA}}$ |
| PolyPYes | 1 − PolyPNo | $\frac{P_{PP\text{-}LO}}{K_{XP} + P_{PP\text{-}LO}}$ |
| PolyPNo | 1 − PolyPYes | $\frac{K_{XP}}{K_{XP} + P_{PP\text{-}LO}}$ |

**MODEL COMPONENT FOR NON-POLY-P HETEROTROPHS AND AUTOTROPHS.** The general model component proposed for describing the kinetic response of the non-poly-P heterotrophic and autotrophic organism masses is based on ASM1 (Henze *et al.*, 1987 a and b), with a number of modifications/extensions that have been previously described. This extended ASM1 model component of the general model is presented in matrix format in Table 9.4. Suggested values for the stoichiometric, kinetic, and switching

**Table 9.4 Model matrix for the non-poly-P organism and autotroph component of the general model.**

| | Component→ | 1 | 2 | 3 | 4 | 5 |
|---|---|---|---|---|---|---|
| | Process ↓ | $Z_H$ | $Z_A$ | $Z_P$ | $Z_E$ | $S_{ENM}$ |
| 1 | Aerobic growth of beterotrophs on $S_{BSC}$ with $N_{H3}$ | 1 | | | | |
| 2 | Anoxic growth of heterotrophs on $S_{BSC}$ with $N_{H3}$ | 1 | | | | |
| 3 | Aerobic growth of heterotrophs on $S_{BSC}$ with $N_{O3}$ | 1 | | | | |
| 4 | Anoxic growth of heterotrophs on $S_{BSC}$ with $N_{O3}$ | 1 | | | | |
| 5 | Aerobic growth of heterotrophs on $S_{BSA}$ with $N_{H3}$ | 1 | | | | |
| 6 | Anoxic growth of heterotrophs on $S_{BSA}$ with $N_{H3}$ | 1 | | | | |
| 7 | Aerobic growth of heterotrophs on $S_{BSA}$ with $N_{O3}$ | 1 | | | | |
| 8 | Anoxic growth of heterotrophs on $S_{BSA}$ with $N_{O3}$ | 1 | | | | |
| 9 | Decay of heterotrophs | $-1$ | | | $f_{EP,H}$ | $1-f_{EP,H}$ |
| 10 | Hydrolysis of stored/ enmeshed COD | | | | | $-1$ |
| 11 | Hydrolysis of organic nitrogen | | | | | |
| 12 | Ammonification | | | | | |
| 13 | Fermentation of $S_{BSC}$ to $S_{BSA}$ | $Y_{H,ANA}$ | | | | |
| 14 | Autotroph growth | | 1 | | | |
| 15 | Autotroph decay | | $-1$ | | $f_{EP,A}$ | $1-f_{EP,A}$ |

| 6 | 7 | 8 | 9 | 10 | 11 | 12 | 13 |
|---|---|---|---|---|---|---|---|
| $S_{BSC}$ | $S_{BSA}$ | $S_{PHB}$ | $S_{UP}$ | $S_{US}$ | $P_{PP\text{-}LO}$ | $P_{PP\text{-}HI}$ | $P_{O4}$ |
| $\frac{-1}{Y_{H,AER}}$ | | | | | | | $-f_{P,ZH}$ |
| $\frac{-1}{Y_{H,ANOX}}$ | | | | | | | $-f_{P,ZH}$ |
| $\frac{-1}{Y_{H,AER}}$ | | | | | | | $-f_{P,ZH}$ |
| $\frac{-1}{Y_{H,ANOX}}$ | | | | | | | $-f_{P,ZH}$ |
| | $\frac{-1}{Y_{H,AER}}$ | | | | | | $-f_{P,ZH}$ |
| | $\frac{-1}{Y_{H,ANOX}}$ | | | | | | $-f_{P,ZH}$ |
| | $\frac{-1}{Y_{H,AER}}$ | | | | | | $-f_{P,ZH}$ |
| | $\frac{-1}{Y_{H,ANOX}}$ | | | | | | $-f_{P,ZH}$ |
| | | | | | | | $f_{P,ZH} - f_{EP,H} \cdot f_{P,ZEH}$ |
| 1 | | | | | | | |
| $-1$ | $(1 - Y_{H,ANA})Y_{AC}$ | | | | | | $-f_{P,ZH} \cdot Y_{H,ANA}$ |
| | | | | | | | $-f_{P,ZA}$ |
| | | | | | | | $f_{P,ZA} - f_{EP,A} \cdot f_{P,ZEA}$ |

*(continued)*

**Table 9.4 Model matrix for the non-poly-P organism and autotroph component of the general model *(continued).***

| | 14 | 15 | 16 | 17 | 18 |
|---|---|---|---|---|---|
| | $N_{BP}$ | $N_{BS}$ | $N_{O3}$ | $N_{H3}$ | $N_{US}$ |
| 1 | | | | $-f_{N,ZH}$ | |
| 2 | | | $-\frac{(1-Y_{H,ANOX})}{2.86Y_{H,ANOX}}$ | $-f_{N,ZH}$ | |
| 3 | | | $-f_{N,ZH}$ | | |
| 4 | | | $-f_{N,ZH}$ $-\frac{(1-Y_{H,ANOX})}{2.86Y_{H,ANOX}}$ | | |
| 5 | | | | $-f_{N,ZH}$ | |
| 6 | | | $-\frac{(1-Y_{H,ANOX})}{2.86Y_{H,ANOX}}$ | $-f_{N,ZH}$ | |
| 7 | | | $-f_{N,ZH}$ | | |
| 8 | | | $-f_{N,ZH}$ $-\frac{(1-Y_{H,ANOX})}{2.86Y_{H,ANOX}}$ | | |
| 9 | $f_{N,ZH} - f_{EP,H} \cdot f_{N,ZEH}$ | | | | |
| 10 | | | | | |
| 11 | $-1$ | 1 | | | |
| 12 | | $-1$ | | 1 | |
| 13 | | | | $-f_{N,ZH} \cdot Y_{H,ANA}$ | |
| 14 | | | $\frac{1}{Y_A}$ | $-f_{N,ZA} - \frac{1}{Y_A}$ | |
| 15 | $f_{N,ZA} - f_{EP,A} \cdot f_{N,ZEA}$ | | | | |

| 19 | |
|---|---|
| $S_O$ | **Rate equation, $\rho_i$** |
| $-\frac{(1-Y_{H,AER})}{Y_{H,AER}}$ | $\mu_H \cdot \frac{S_{BSC}}{K_{SH}+S_{BSC}} \cdot Z_H \cdot AirYesHet \cdot NH_3Yes \cdot PO_4GroYes$ |
| | $\eta_{GRO} \cdot \mu_H \cdot \frac{S_{BSC}}{K_{SH}+S_{BSC}} \cdot Z_H \cdot AirNoHet \cdot NO_3Yes \cdot NH_3Yes \cdot PO_4GroYes$ |
| $-\frac{(1-Y_{H,AER})}{Y_{H,AER}}$ | $\mu_H \cdot \frac{S_{BSC}}{K_{SH}+S_{BSC}} \cdot Z_H \cdot AirYesHet \cdot NO_3Yes \cdot NH_3No \cdot PO_4GroYes$ |
| | $\eta_{GRO} \cdot \mu_H \cdot \frac{S_{BSC}}{K_{SH}+S_{BSC}} \cdot Z_H \cdot AirNoHet \cdot NO_3Yes \cdot NH_3No \cdot PO_4GroYes$ |
| $-\frac{(1-Y_{H,AER})}{Y_{H,AER}}$ | $\mu_H \cdot \frac{S_{BSA}}{K_{SH}+S_{BSA}} \cdot Z_H \cdot AirYesHet \cdot NH_3Yes \cdot PO_4GroYes$ |
| | $\eta_{GRO} \cdot \mu_H \cdot \frac{S_{BSA}}{K_{SH}+S_{BSA}} \cdot Z_H \cdot AirNoHet \cdot NO_3Yes \cdot NH_3Yes \cdot PO_4GroYes$ |
| $-\frac{(1-Y_{H,AER})}{Y_{H,AER}}$ | $\mu_H \cdot \frac{S_{BSA}}{K_{SH}+S_{BSA}} \cdot Z_H \cdot AirYesHet \cdot NO_3Yes \cdot NH_3No \cdot PO_4GroYes$ |
| | $\eta_{GRO} \cdot \mu_H \cdot \frac{S_{BSA}}{K_{SH}+S_{BSA}} \cdot Z_H \cdot AirNoHet \cdot NO_3Yes \cdot NH_3No \cdot PO_4GroYes$ |
| | $b_H \cdot Z_H$ |
| | $K_H \cdot \frac{S_{ENM}/Z_H}{K_X+S_{ENM}/Z_H} \cdot Z_H \cdot (AirYesHet + \eta_{S,ANOX} \cdot AirNoHet \cdot NO_3Yes + \eta_{S,ANA} \cdot AirNoHet \cdot NO_3No)$ |
| | $\rho_{10} \cdot \frac{N_{BP}}{S_{ENM}}$ |
| | $K_R \cdot N_{BS} \cdot (Z_H + Z_P)$ |
| | $K_C \cdot \frac{S_{BSC}}{K_{S,ANA}+S_{BSC}} \cdot Z_H \cdot AirNoHet \cdot NO_3No$ |
| $-\frac{(4.57-Y_A)}{Y_A}$ | $\mu_A \cdot \frac{N_{H3}}{K_{NH}+N_{H3}} \cdot Z_A \cdot AirYesAut$ |
| | $b_A \cdot Z_A$ |

function parameters are listed in Tables 9.5, 9.6, and 9.7. These are the values used in the case study presented later.

Space limitations do not allow for a detailed discussion of the model. The reader is referred to the literature cited, the publications of the IAWQ task group (Henze *et al.*, 1987a and b), and the excellent review by Grady (1989). The following paragraphs briefly summarize the essence of this part of the model.

- Growth of non-poly-P heterotrophs on "complex" readily biodegradable COD (processes 1 to 4)—growth of these organisms is modeled by the Monod relationship. There are four process equations to describe this, depending on whether growth occurs under aerobic or anoxic conditions and whether nitrate or ammonia is the nitrogen source for cell synthesis. In the absence of dissolved oxygen, only a portion of the non-poly-P heterotrophic organism population is capable of using nitrate, if available, as a terminal electron acceptor for the oxidation of organic matter. (Alternatively, all these organisms can use nitrate as electron acceptor, but at a reduced rate compared to aerobic conditions.) The process rate is adjusted by a factor ($\eta_{GRO}$) to account for the reduced rate of substrate removal under anoxic conditions. Different yield factors ($Y_{H,AER}$ and $Y_{H,ANOX}$) are incorporated for aerobic and anoxic growth, respectively. Switching functions enable the model to simulate the change from one electron acceptor to another or one nitrogen source to another. For the treatment of domestic wastewater, the growth of non-poly-P heterotrophs on complex readily biodegradable COD is responsible for a large portion of the removal of organic matter (and its oxygen requirement) and production of the bulk of the MLVSS.
- Growth of non-poly-P heterotrophs on SCFA COD (processes 5 to 8)—as above, there are four process equations to describe this behavior, depending on whether growth occurs under aerobic or anoxic conditions and whether ammonia or nitrate serves as the nitrogen source.
- Decay of heterotrophs (process 9)—this process is modeled according to the death-regeneration theory. The organisms are assumed to die at a certain rate, with a portion of the dead cells adding to the endogenous cell residue (nondegradable) and the rest becoming part of the pool of particulate biodegradable COD ($S_{ENM}$). The cellular nitrogen and phosphorus associated with the particulate biodegradable COD from decay become available as particulate organic nitrogen and soluble phosphate. Decay is assumed to occur under aerobic, anoxic, and anaerobic conditions.
- Hydrolysis of slowly biodegradable (enmeshed) COD (process 10)—the slowly biodegradable COD is assumed to be enmeshed in the sludge mass, and as it is broken down extracellularly, the products of breakdown are added to the pool of readily biodegradable (complex) COD. This process is modeled on the basis of Levenspiel's surface reaction kinetics and is assumed to occur under aerobic, anoxic, and

**Table 9.5 Stoichiometric parameters for the general model.**

| Symbol | Parameter | Value | Units |
|---|---|---|---|
| **Non-poly-P heterotrophs** | | | |
| $Y_{H,AER}$ | Yield (aerobic) | 0.666 | g cell COD yield/g COD used |
| $Y_{H,ANOX}$ | Yield (anoxic) | 0.666 | g cell COD yield/g COD used |
| $Y_{H,ANA}$ | Yield (anaerobic) | 0.100 | g cell COD yield/g COD used |
| $Y_{AC}$ | Fermentation $S_{BSA}$ yield | 0.40 | g $S_{BSA}$ COD/g $S_{BSC}$ COD |
| $f_{N,ZH}$ | Nitrogen content of active mass | 0.068 | g N/g COD active organisms |
| $f_{N,ZEH}$ | Nitrogen content of endogenous mass | 0.068 | g N/g COD endogenous residue |
| $f_{P,ZH}$ | Phosphorus content of active mass | 0.021 | g P/g COD active organisms |
| $f_{P,ZEH}$ | Phosphorus content of endogenous mass | 0.021 | g P/g COD endogenous residue |
| $f_{EP,H}$ | Fraction of active mass remaining as endogenous residue | 0.08 | g COD endogenous mass/g COD active mass |
| $f_{CV,H}$ | RatioCOD:VSS | 1.48 | g COD/g VSS |
| **Poly-P heterotrophs** | | | |
| $P_P$ | Yield | 0.639 | g cell COD yield/g COD used |
| $f_{P,UPT1}$ | Phosphorus uptake/COD utilized in aerobic growth | 0.95 | g P/g stored COD |
| $f_{P,UPT2}$ | Phosphorus uptake/COD utilized in anoxic growth | 0.35 | g P/g stored COD |
| $Y_{PHB}$ | PHB yield on sequestration of $S_{BSA}$ | 0.889 | g PHB COD/g $S_{BSA}$ COD |
| $f_{N,ZP}$ | Nitrogen content of active mass | 0.070 | g N/g COD active organisms |
| $f_{N,ZEP}$ | Nitrogen content of endogenous mass | 0.070 | g N/g COD endogenous residue |
| $f_{N,SEP}$ | Nitrogen content of soluble unbiodegradable COD | 0.070 | g N/g COD endogenous residue |
| $f_{P,ZP}$ | Phosphorus content of active mass (excluding poly-P content) | 0.021 | g P/g COD active organisms |
| $f_{P,ZEP}$ | Phosphorus content of endogenous mass | 0.021 | g P/g COD endogenous residue |
| $f_{EP,P}$ | Fraction of active mass remaining as particulate unbiodegradable endogenous residue | 0.25 | g COD endogenous mass/g COD active mass |
| $f_{ES,P}$ | Fraction of active mass remaining as soluble unbiodegradable residue | 0.20 | g COD/g COD active mass |
| $f_{P,REL}$ | Phosphorus release/SCFA COD uptake for sequestration | 0.49 | g P/g $S_{BSA}$ COD |
| $f_{CV,P}$ | Ratio COD/VSS | 1.42 | g COD/g VSS |
| $f_{PP}$ | Fraction of taken up P which can be released | 0.94 | g P/g P |
| **Autotrophs** | | | |
| $Y_A$ | Yield | 0.150 | g cell COD yield/g N used |
| $f_{N,ZA}$ | Nitrogen content of active mass | 0.068 | g N/g COD active organisms |
| $f_{N,ZEA}$ | Nitrogen content of endogenous mass | 0.068 | g N/g COD endogenous residue |
| $f_{P,ZA}$ | Phosphorus content of active mass | 0.021 | g P/g COD active organisms |
| $f_{P,ZEA}$ | Phosphorus content of endogenous mass | 0.021 | g P/g COD endogenous residue |
| $f_{EP,A}$ | Fraction of active mass remaining as endogenous residue | 0.08 | g COD endogenous mass/g COD active mass |
| $f_{CV,A}$ | Ratio COD:VSS | 1.42 | g COD/g VSS |

**Table 9.6 Kinetic parameters for the general model.**

| | | | | $\theta$ |
|---|---|---|---|---|
| **Non-poly-P heterotrophs** | | | | |
| $\mu_H$ | Maximum specific growth rate | 3.2 | $d^{-1}$ | 1.029 |
| $K_{S,H}$ | Half-saturation coefficient for growth | 5.0 | g COD/$m^3$ | 1.000 |
| $b_H$ | Organism decay rate | 0.62 | $d^{-1}$ | 1.029 |
| $\eta_{S,ANOX}$ | Anoxic solubilization factor | 1.00 | — | — |
| $\eta_{S,ANA}$ | Anaerobic solubilization factor | 0.50 | — | — |
| $\eta_{GRO}$ | Anoxic growth factor | 0.37 | — | — |
| $K_H$ | Maximum specific hydrolysis rate | 2.81 | $d^{-1}$ | 1.050 |
| $K_X$ | Half-saturation coefficient for hydrolysis | 0.15 | g COD/g COD | 1.000 |
| $K_C$ | Maximum specific fermentation growth rate | 4.0 | $d^{-1}$ | 1.029 |
| $K_{S,ANA}$ | Fermentation half-saturation coefficient | 5.0 | g COD/$m^3$ | 1.000 |
| $K_R$ | Ammonification rate | 0.08 | $m^3$/g COD·d | 1.029 |
| **Poly-P heterotrophs** | | | | |
| $\mu_{P1}$ | Maximum specific growth rate (no P limit) | 0.95 | $d^{-1}$ | 1.000 |
| $\mu_{P2}$ | Maximum specific growth rate (P limit) | 0.42 | $d^{-1}$ | 1.000 |
| $K_{SP1}$ | Half-saturation coefficient (no P limit) | 0.10 | g COD/g COD·d | 1.000 |
| $K_{SP2}$ | Half-saturation coefficient (P limit) | 0.05 | g COD/g COD·d | 1.000 |
| $\eta_P$ | Anoxic growth factor | 0.40 | — | — |
| $b_P$ | Organism endogenous mass loss rate | 0.04 | $d^{-1}$ | 1.000 |
| $b_{PP}$ | Poly-P cleavage rate for anaerobic "maintenance" energy generation | 0.03 | $d^{-1}$ | 1.000 |
| $K_P$ | Specific rate of SCFA sequestration | 6.00 | g COD/g COD·d | 1.000 |
| **Autotrophs** | | | | |
| $\mu_A$ | Maximum specific growth rate | 0.2–1.0 | $d^{-1}$ | 1.123 |
| $K_{NH}$ | Half-saturation coefficient for growth | 1.0 | g N/$m^3$ | 1.123 |
| $b_A$ | Organism decay rate | 0.04 | $d^{-1}$ | 1.029 |

**Temperature adjustment of kinetic parameters:**

$$K_T = K_{20} \cdot \theta^{(T-20)}$$

Where

$K_T$ = parameter value at $T$°C,

$K_{20}$ = parameter value at 20°C, and

$\theta$ = temperature adjustment coefficient.

**Table 9.7 Switching function parameters for the general model.**

| | | | |
|---|---|---|---|
| $K_{O,HET}$ | Aerobic/anoxic growth of heterotrophs | 0.002 | g $O_2/m^3$ |
| $K_{O,AUT}$ | Aerobic growth of autotrophs | 0.50 | g $O_2/m^3$ |
| $K_{NA}$ | Ammonia limit | 0.005 | g $N/m^3$ |
| $K_{NO}$ | Nitrate limit | 0.10 | g $N/m^3$ |
| $K_{LP,UPT}$ | Phosphate limit | 0.25 | g $P/m^3$ |
| $K_{LP,GRO}$ | Phosphate limit | 0.005 | g $P/m^3$ |
| $K_{XP}$ | Poly-P limit | 0.01 | g P/g COD |
| $K_{SSEQ}$ | SCFA limit | 5.00 | g $COD/m^3$ |

anaerobic conditions. The rate of hydrolysis under anoxic and anaerobic conditions is assumed to be a fraction of that under aerobic conditions ($\eta_{S,ANOX}$ and $\eta_{S,ANA}$, respectively).

- Hydrolysis of particulate organic nitrogen (process 11)—the breakdown of biodegradable particulate organic nitrogen to soluble organic nitrogen is assumed to occur at the same rate as that for carbonaceous slowly degradable matter described above.
- Ammonification of soluble organic nitrogen (process 12)—soluble organic nitrogen is converted to free and saline ammonia, a process mediated by the active non-poly-P and poly-P heterotrophs. The rate is assumed to be first order with respect to the total concentration of heterotrophs and the concentration of soluble biodegradable organic nitrogen.
- Fermentation of complex readily biodegradable COD (RBCOD) to SCFA (process 13)—non-poly-P facultative heterotrophs are responsible for the fermentation of complex RBCOD to SCFA under anaerobic conditions. This is modeled using a Monod growth expression, with a cell yield of $Y_{H,ANA}$ and a corresponding yield of fermentation products of $1 - Y_{H,ANA}$. It is assumed that only a portion ($Y_{AC}$) of the fermentation product is SCFA; the remainder is assumed to be COD lost to the system (possibly as hydrogen gas). This process represents the principal linkage between the poly-P and non-poly-P heterotrophic organism masses in EBPR systems treating municipal wastewater.
- Growth of autotrophs (process 14)—the autotrophic population is responsible for the oxidation of ammonia-nitrogen to nitrate (nitrification). It is assumed to be a single-step process, occurring under aerobic conditions only, with a cell yield of $Y_A$ (g cell COD/g $NH_3$-N). This process has a substantial effect on the prediction of oxygen demand; the effect on the total MLVSS is small because the yield of autotrophic nitrifiers is low. As with the growth of heterotrophs, ammonia-nitrogen is incorporated in the new cells.
- Decay of autotrophs (process 15)—this process parallels that for the heterotrophs, except that decay is assumed to occur at a much lower rate.

**MODEL COMPONENT FOR POLY-P HETEROTROPHS.** The general model component proposed for describing the kinetic response of the poly-P heterotrophic organism mass is similar to the enhanced culture model of Wentzel *et al.* (1989b), but with a number of modifications described earlier.

The poly-P organism component of the general model is presented in matrix format in Table 9.8. Suggested values of the stoichiometric, kinetic, and switching function parameters are listed in Tables 9.5, 9.6 and 9.7. These are the values used in the case study presented later.

Space limitations do not allow for a detailed discussion of this component of the model. The reader is referred to the literature previously cited. The following paragraphs briefly summarize this second part of the model.

- Aerobic growth of poly-P heterotrophs on PHB (processes 16 to 19)—there are four different process equations to describe poly-P organism growth on stored PHB, depending on whether soluble phosphorus is limiting or not and whether ammonia or nitrate serves as the nitrogen source for cell synthesis. The growth rate is dependent on the amount of stored substrate per unit poly-P organism mass. Switching functions enable the model to simulate a change to phosphorus limitation or a change in nitrogen source. Soluble phosphate uptake and storage as poly-P ($f_{P,UPT1}$ g P/g PHB-COD) occurs concomitant with growth. Only a fraction ($f_{PP}$) of the stored phosphate can be released subsequently.
- Anoxic growth of poly-P heterotrophs on PHB (process 20)—the model assumes that a fraction ($\eta_P$) of the poly-P organisms can use nitrate as an electron acceptor (in the absence of oxygen) for the oxidation of stored PHB and uptake of phosphorus. (Alternately, this factor $\eta_P$ can be regarded as a reduction in rate of growth and uptake compared to aerobic conditions.) The stoichiometric uptake of phosphate ($f_{P,UPT2}$ g P/g PHB-COD) differs from that for aerobic growth.
- Decay of poly-P heterotrophs (processes 21 to 32)—this process is modeled in a slightly different manner from that of the non-poly-P heterotrophs to account for the effect of electron acceptor on the poly-P organisms, as well as other differences such as the amount of ammonia, phosphorus, and stored PHB released because of cell decay. Twelve process rate equations in all are used to simulate aerobic, anoxic, and anaerobic decay of the poly-P organisms and the associated lysis of polyphosphate and PHB.
- Cleavage of poly-P for anaerobic maintenance (process 33)—the polyphosphate polymers stored by the poly-P organisms are used under anaerobic conditions to supply the maintenance energy requirements. Cleavage of polyphosphate is modeled as a first-order process with respect to the poly-P organism mass.

- Sequestration of SCFA by poly-P organisms (process 34)—cleavage of polyphosphate polymers supplies the energy requirements for the storage of the SCFA as PHB by the poly-P heterotrophic organisms. The rate of SCFA uptake is assumed to be zero order with respect to the SCFA concentration and first order with respect to the poly-P organism concentration. A stored PHB yield is $Y_{PHB}$ units (as COD) for each unit of SCFA COD taken up. The release of phosphorus because of polyphosphate cleavage is assumed to be directly proportional to the uptake of the SCFA. Switching functions allow the process to be switched off if either polyphosphate or SCFA becomes limiting.

**EVALUATION AND CALIBRATION OF THE GENERAL MODEL.** The general model has been applied to a wide range of systems operated during a range of conditions (sludge age and recycle rates), for example, aerobic systems (with and without nitrification), ND systems, EBPR systems, and NDEBPR systems. The model simulates the organism mass distribution and satisfactorily tracks the changes in a range of key parameters such as soluble phosphorus and nitrate concentrations. An important feature of the model evaluation is that a single set of kinetic and stoichiometric model parameters was used for all the simulations (one exception is the maximum specific growth rate of the nitrifiers, $\mu_A$ [see comments later in this chapter]). This provides a degree of support for the model structure and integrity.

An important requirement of the general model is the capacity to predict the appropriate active heterotrophic organism mass structure in a system with respect to the poly-P and non-poly-P organisms. Typically, three "normal" situations are possible:

- In completely aerobic systems the active organism mass should consist of only non-poly-P organisms and should not exhibit excess phosphorus removal. The general model predictions in this case should reduce to those of the modified ASM1 model.
- In EBPR systems incorporating unaerated zones and with only SCFA as substrate, the active heterotrophic organism mass should consist of only poly-P organisms and should exhibit excess phosphorus removal. This would correspond to the enhanced culture systems of Wentzel *et al.* (1989a). The general model predictions in this case should reduce to those of the enhanced culture poly-P model. In these systems, if the influent TKN:COD ratio is unusually high, it might be possible that a small mass of non-poly-P organisms is sustained through anoxic growth.
- In EBPR systems that have municipal wastewater as influent and exhibit phosphorus removal, the active organism mass should reflect the coexistence of both poly-P and non-poly-P organisms.

**Table 9.8 Model matrix for the poly-P organism component of the general model.**

| | Component→ | 1 | 2 | 3 | 4 | 5 |
|---|---|---|---|---|---|---|
| | Process ↓ | $Z_H$ | $Z_A$ | $Z_P$ | $Z_E$ | $S_{ENM}$ |
| 16 | Aerobic growth of poly-P organisms on $S_{PHB}$ with $N_{H3}$ | | | 1 | | |
| 17 | Aerobic growth of poly-P organisms on $S_{PHB}$ with $N_{O3}$ | | | 1 | | |
| 18 | Process 16 if $P_{O4}$ limited | | | 1 | | |
| 19 | Process 17 if $P_{O4}$ limited | | | 1 | | |
| 20 | Anoxic growth of poly-P organisms on $S_{PHB}$ with $N_{H3}$ | | | 1 | | |
| 21 | Aerobic decay of poly-P organisms | | | −1 | $f_{EP,P}$ | |
| 22 | $P_{PP\text{-}LO}$ lysis on aerobic decay | | | | | |
| 23 | $P_{PP\text{-}HE}$ lysis on aerobic decay | | | | | |
| 24 | $S_{PHB}$ lysis on aerobic decay | | | | | |
| 25 | Anoxic decay of poly-P organisms | | | −1 | $f_{EP,P}$ | |
| 26 | $P_{PP\text{-}LO}$ lysis on anoxic decay | | | | | |
| 27 | $P_{PP\text{-}HI}$ lysis on anoxic decay | | | | | |
| 28 | $S_{PHB}$ lysis on anoxic decay | | | | | |
| 29 | Anaerobic decay of poly-P organisms | | | −1 | $f_{EP,P}$ | |
| 30 | $P_{PP\text{-}LO}$ lysis on anaerobic decay | | | | | |
| 31 | $P_{PP\text{-}HI}$ lysis on anaerobic decay | | | | | |
| 32 | $S_{PHB}$ lysis on anaerobic decay | | | | | |
| 33 | Cleavage of poly-P for anaerobic maintenance | | | | | |
| 34 | Sequestration of SCFA by poly-P organisms | | | | | |

| 6 | 7 | 8 | 9 | 10 | 11 | 12 | 13 |
|---|---|---|---|---|---|---|---|
| $S_{BSC}$ | $S_{BSA}$ | $S_{PHB}$ | $S_{UP}$ | $S_{US}$ | $P_{PP\text{-}LO}$ | $P_{PP\text{-}HI}$ | $P_{O4}$ |
| | | $\frac{-1}{Y_P}$ | | | $f_{PP} \cdot \frac{f_{P,UPT1}}{Y_P}$ | $(1 - f_{PP}) \cdot \frac{f_{P,UPT1}}{Y_P}$ | $-f_{P,UPT1}/Y_P - f_{P,ZP}$ |
| | | $\frac{-1}{Y_P}$ | | | $f_{PP} \cdot \frac{f_{P,UPT1}}{Y_P}$ | $(1 - f_{PP}) \cdot \frac{f_{P,UPT1}}{Y_P}$ | $-f_{P,UPT1}/Y_P - f_{P,ZP}$ |
| | | $\frac{-1}{Y_P}$ | | | $-f_{P,ZP}$ | | |
| | | $\frac{-1}{Y_P}$ | | | $-f_{P,ZP}$ | | |
| | | $\frac{-1}{Y_P}$ | | | $f_{PP} \cdot \frac{f_{P,UPT1}}{Y_P}$ | $(1 - f_{PP}) \cdot \frac{f_{P,UPT2}}{Y_P}$ | $-f_{P,UPT2}/Y_P - f_{P,ZP}$ |
| | | | | $f_{ES,P}$ | | | $f_{P,ZP} - f_{EP,P} \cdot f_{P,ZEP}$ |
| | | | | | $-1$ | | 1 |
| | | | | | | $-1$ | 1 |
| | 1 | $-1$ | | | | | |
| | | | | $f_{ES,P}$ | | | $f_{P,ZP} - f_{EP,P} \cdot f_{P,ZEP}$ |
| | | | | | $-1$ | | 1 |
| | | | | | | $-1$ | 1 |
| | 1 | $-1$ | | | | | |
| | | | | $f_{EP,P}$ | | | $f_{P,ZP} - f_{EP,P} \cdot f_{P,ZEP}$ |
| | | | | | $-1$ | | 1 |
| | | | | | | $-1$ | 1 |
| | 1 | $-1$ | | | | | |
| | | | | | $-1$ | | 1 |
| | $-1$ | $Y_{PHB}$ | | | $-f_{P,REL}$ | | $f_{P,REL}$ |

*(continued)*

**Table 9.8 Model matrix for the poly-P organism component of the general model *(continued).***

| | 14 | 15 | 16 | 17 | 18 |
|---|---|---|---|---|---|
| | $N_{BP}$ | $N_{BS}$ | $N_{O3}$ | $N_{H3}$ | $N_{US}$ |
| 16 | | | | $-f_{N,ZP}$ | |
| 17 | | | $-f_{N,ZP}$ | | |
| 18 | | | | $-f_{N,ZP}$ | |
| 19 | | | $-f_{N,ZP}$ | | |
| 20 | | | $-\dfrac{(1 - Y_P)}{2.86 Y_P}$ | $-f_{N,ZP}$ | |
| 21 | | $f_{ES,P} \cdot f_{N,SEP}$ | | $f_{N,ZP}$ $- f_{EP,P} \cdot f_{N,ZEP}$ $- f_{ES,P} \cdot f_{N,SEP}$ | |
| 22 | | | | | |
| 23 | | | | | |
| 24 | | | | | |
| 25 | | $f_{ES,P} \cdot f_{N,SEP}$ | $-\dfrac{(1 - f_{EP,P} - f_{ES,P})}{2.86}$ | $f_{N,ZP}$ $- f_{EP,P} \cdot f_{N,ZEP}$ $- f_{ES,P} \cdot f_{N,SEP}$ | |
| 26 | | | | | |
| 27 | | | | | |
| 28 | | | | | |
| 29 | | $f_{ES,P} \cdot f_{N,SEP}$ | | $f_{N,ZP}$ $- f_{EP,P} \cdot f_{N,ZEP}$ $- f_{ES,P} \cdot f_{N,SEP}$ | |
| 30 | | | | | |
| 31 | | | | | |
| 32 | | | | | |
| 33 | | | | | |
| 34 | | | | | |

| 19 | |
|---|---|
| $S_O$ | **Rate equation, $\rho_i$** |
| $-\frac{(1-Y_P)}{Y_P}$ | $\mu_{P1} \cdot \frac{S_{PHB}/Z_P}{K_{SP1}+S_{PHB}/Z_P} \cdot Z_P \cdot \text{AirYesHet} \cdot \text{NH}_3\text{Yes} \cdot \text{PO}_4\text{UptYes}$ |
| $-\frac{(1-Y_P)}{Y_P}$ | $\mu_{P1} \cdot \frac{S_{PHB}/Z_P}{K_{SP1}+S_{PHB}/Z_P} \cdot Z_P \cdot \text{AirYesHet} \cdot \text{NO}_3\text{Yes} \cdot \text{NH}_3\text{No} \cdot \text{PO}_4\text{UptYes}$ |
| $-\frac{(1-Y_P)}{Y_P}$ | $\mu_{P2} \cdot \frac{S_{PHB}/Z_P}{K_{SP2}+S_{PHB}/Z_P} \cdot Z_P \cdot \text{AirYesHet} \cdot \text{NH}_3\text{Yes} \cdot \text{PO}_4\text{UptNo}$ |
| $-\frac{(1-Y_P)}{Y_P}$ | $\mu_{P2} \cdot \frac{S_{PHB}/Z_P}{K_{SP2}+S_{PHB}/Z_P} \cdot Z_P \cdot \text{AirYesHet} \cdot \text{NO}_3\text{Yes} \cdot \text{NH}_3\text{No} \cdot \text{PO}_4\text{UptNo}$ |
| | $\eta_P \cdot \mu_{P1} \cdot \frac{S_{PHB}/Z_P}{K_{SP1}+S_{PHB}/Z_P} \cdot Z_P \cdot \text{AirNoHet} \cdot \text{NO}_3\text{Yes} \cdot \text{NH}_3\text{Yes} \cdot \text{PO}_4\text{UptYes}$ |
| $-(1-f_{EP,P}-f_{ES,P})$ | $b_P \cdot Z_P \cdot \text{AirYesHet}$ |
| | $\rho_{21} \cdot \frac{P_{PP\text{-}LO}}{Z_P}$ |
| | $\rho_{21} \cdot \frac{P_{PP\text{-}HI}}{Z_P}$ |
| | $\rho_{21} \cdot \frac{S_{PHB}}{Z_P}$ |
| | $b_P \cdot Z_P \cdot \text{AirNoHet} \cdot \text{NO}_3\text{Yes}$ |
| | $\rho_{25} \cdot \frac{P_{PP\text{-}LO}}{Z_P}$ |
| | $\rho_{25} \cdot \frac{P_{PP\text{-}HI}}{Z_P}$ |
| | $\rho_{25} \cdot \frac{S_{PHB}}{Z_P}$ |
| | $b_P \cdot Z_P \cdot \text{AirNoHet} \cdot \text{NO}_3\text{No}$ |
| | $\rho_{29} \cdot \frac{P_{PP\text{-}LO}}{Z_P}$ |
| | $\rho_{29} \cdot \frac{P_{PP\text{-}HI}}{Z_P}$ |
| | $\rho_{29} \cdot \frac{S_{PHB}}{Z_P}$ |
| | $b_{PP} \cdot Z_P \cdot \text{AirNoHet} \cdot \text{PolyPYes}$ |
| | $K_P \cdot Z_P \cdot S_{BSA}\,\text{Yes} \cdot \text{PolyPYes}$ |

Dold (1990 and 1991) presented case studies that demonstrated the model application in each of these three situations. Of the three, the most rigorous evaluation of the general activated-sludge model is the third: a mixed-culture system in which poly-P and non-poly-P organisms coexist in balance (together with nitrifiers) (this is the situation for the system discussed and used as an example in the Application of the Activated-Sludge Models section).

# SIMPLIFIED STEADY-STATE MODEL

Wentzel *et al.* (1990) developed a steady-state model for NDEBPR systems. The model essentially is a combination of the steady-state model for non-poly-P organism behavior of Marais and Ekama (1976) and the enhanced culture model for poly-P organism behavior of Wentzel *et al.* (1989b). The main function of the model is to predict the amount of phosphorus removal and the total mass of volatile solids in an NDEBPR system operated at a given sludge age and with a specified anaerobic sludge mass fraction.

In the development of the steady-state model, a number of simplifying assumptions were made:

- The release of phosphorus by the poly-P organisms to supply anaerobic maintenance energy requirements is assumed to be negligible compared to the release of phosphorus for sequestration of SCFA.
- All of the SCFA from the influent or generated within the anaerobic reactor is taken up by the poly-P organisms.
- The PHB stored by the poly-P organisms in the anaerobic zone is assumed to be completely used in the subsequent aerobic zone.
- Anaerobic hydrolysis of particulate material to RBCOD is assumed to be negligible (that is, only RBCOD in the influent is assumed to be converted to SCFA for use by the poly-P organisms).
- The poly-P organisms are assumed to be incapable of using nitrate as an electron acceptor in the absence of oxygen.

The development of the steady-state model is outlined in detail by Wentzel *et al.* (1990). The equations for application with municipal wastewater as influent are generalized here in that both complex readily biodegradable COD and SCFA are included as possible substrates. In the following sections, the data required to use the model are summarized, and the step-wise, steady-state model algorithm is laid out.

In comparing the steady-state model to the general model, perhaps the most significant aspect is the prediction of sludge production. The steady-state model does not make provision for any loss of COD. Therefore, predictions of volatile solids concentrations from the steady-state model will be greater than for the general model presented earlier.

**DATA REQUIRED FOR THE STEADY-STATE MODEL.** Certain information regarding the influent wastewater characteristics (for COD content) and the NDEBPR system is required:

Wastewater characteristics:

- Total influent COD concentration ($S_{ti}$);
- Unbiodegradable soluble influent COD concentration ($S_{USi}$);
- Unbiodegradable particulate influent COD concentration ($S_{UPi}$);
- Total biodegradable influent COD concentration ($S_{Bi} = S_{Ti} - S_{USi} - S_{UPi}$);
- Total readily biodegradable influent COD concentration ($S_{BSi} = S_{BSCi} + S_{BSAi}$);
- Readily biodegradable complex influent COD concentration ($S_{BSCi}$); and
- Short-chain fatty acid influent COD concentration ($S_{BSAi}$).

Operating and system parameters:

- Influent flow rate ($Q$);
- Anaerobic mass fraction ($f_{XA}$);
- Aerobic mass fraction ($f_{AER}$);
- System sludge age ($\theta_X$);
- Number of equal-volume anaerobic reactors in series ($N$);
- Aerobic volume ($V_{AER}$);
- Recycle ratio to anaerobic reactor ($r$); and
- Nitrate concentration in the recycle to the anaerobic reactor ($N_{03,R}$).

The steady-state model also incorporates a number of model parameters; these are defined in Table 9.9. It should be noted that endogenous sludge production in the steady-state model is based on the endogenous respiration concept, rather than the death-regeneration approach used in the general model. Hence, the organism decay rate at 20°C is 0.24 $d^{-1}$ (see Dold *et al.*, 1980).

**STEADY-STATE MODEL DESIGN ALGORITHM.** The step-wise algorithm for the steady-state model is outlined in the following section.

Step 1: Determine the concentration of readily biodegradable COD available for conversion to SCFA using Equation 1; if the nitrate concentration in the recycle to the anaerobic zone is unknown, assume a value of 1 g/m$^3$:

$$\begin{aligned} S'_{BSCi} &= S_{BSCI} - r \cdot 8.6 \cdot N_{O3,R} \\ &= (S_{BSi} - S_{BSAi}) - r \cdot 8.6 \cdot N_{O3,R} \end{aligned} \tag{9.1}$$

**Table 9.9 Model parameters and symbol definitions for steady-state model.**

| Symbol | Definition | Value/units |
|---|---|---|
| $Y_P$ | Poly-P organism yield constant | 0.45 g VASS/g COD |
| $Y_H$ | Non-poly-P organism yield constant | 0.45 g VASS/g COD |
| $b_P$ | Specific endogenous mass loss rate constant for poly-P organisms | 0.04 $d^{-1}$ at 20°C |
| $b_H$ | Specific endogenous mass loss rate constant for non-poly-P organisms | 0.24 $d^{-1}$ at 20°C |
| $f_{EP,P}$ | Fraction of active poly-P mass remaining as particulate unbiodegradable endogenous residue | 0.25 g VSS/g VASS |
| $f_{EP,H}$ | Fraction of active non-poly-P mass remaining as particulate unbiodegradable endogenous residue | 0.20 g VSS/g VASS |
| $f_{P,XP}$ | Fractional phosphorus content of the biological active mass of poly-P organisms | 0.38 g P/g VASS |
| $f_{P,XI}$ | Fractional phosphorus content of inert solids in the system | 0.03 g P/g VISS |
| $f_{P,XEP}$ | Fractional phosphorus content of the poly-P endogenous mass | 0.03 g P/g VESS |
| $f_{P,XH}$ | Fractional phosphorus content of the biological active mass of non-poly-P organisms | 0.03 g P/g VASS |
| $f_{P,XEH}$ | Fractional phosphorus content of the non-poly-P endogenous mass | 0.03 g P/g VESS |
| $f_{P,SUPI}$ | Fractional phosphorus content of the influent unbiodegradable particulate COD | 0.03 g P/g VISS |
| $K$ | First-order conversion rate constant | 0.06 $d^{-1}$ |
| $f_{CV}$ | COD:VSS ratio | 1.48 g COD/g VSS |
| $S'_{BSCi}$ | Influent complex readily biodegradable COD concentration remaining after nitrate is exhausted | g COD/$m^3$ |
| $S_{BSC,N}$ | Readily biodegradable complex COD concentration in the $N$th anaerobic reactor | g COD/$m^3$ |
| $MS_{BSAi}$ | Influent mass of SCFA supplied per day ($QS_{BSAI}$) | g COD/d |
| $MS_{SEQ}$ | Mass of SCFA available per day for poly-P organisms | g COD/d |
| $MS_{B,H}$ | Mass of substrate available per day for growth of non-poly-P heterotrophic organisms | g COD/d |
| $MX_H$ | Total active mass of non-poly-P organisms in the system | g VASS |
| $MX_{E,H}$ | Endogenous mass from decay of non-poly-P organisms | g VESS |
| $MX_P$ | Total active mass of poly-P organisms in the system | g VASS |
| $MX_{E,P}$ | Endogenous mass from decay of poly-P organisms | g VESS |
| $MX_I$ | Total inert volatile mass in the system | g VISS |
| $X_V$ | Volatile suspended solids concentration in the aerobic zone | g VSS/$m^3$ |
| $\Delta P_H$ | Phosphorus removal due to non-poly-P organisms per unit volume influent | g P/$m^3$ |
| $\Delta P_P$ | Phosphorus removal due to poly-P organisms per unit volume influent | g P/$m^3$ |
| $\Delta P_I$ | Phosphorus removal due to wastage of inert material per unit volume influent | g P/$m^3$ |
| $\Delta P$ | Total phosphorus removal per unit volume influent | g P/$m^3$ |

Step 2: Use the following iterative process to calculate both the concentration of readily biodegradable COD remaining in the last anaerobic reactor and the mass of active non-poly-P heterotrophs present in the system per litre of influent flow:

- Assume an initial value for $S_{BSC,N}$ of 0 g COD/m$^3$.
- Calculate $MX_H/Q$ using Equation 9.2a:

$$\frac{MX_H}{Q} = \frac{\{S_{Bi} - [S_{BSCi} - (1 + r)S_{BSC,N}] - S_{BSAi}\}Y_H\theta_X}{(1 + b_H\theta_X)} \tag{9.2a}$$

- Using the value calculated above for $MX_H/Q$, calculate a value for $S_{BSC,N}$ using Equation 9.2b:

$$S_{BSC,N} = \frac{S_{BSCi}/(1 + r)}{\left[1 = \frac{Kf_{XA}MX_H}{N} Q/(1\ (1 = r)\right]^N} \tag{9.2b}$$

- Recalculate $MX_H/Q$ using the value calculated above for $S_{BSC,N}$.
- Repeat the iterative calculation of $MX_H/Q$ and $S_{BSC,N}$ until the values are near-constant.

Step 3: Determine the mass of SCFA taken up per day by the poly-P organisms:

$$MS_{SEQ} = QS_{BSAi} + [S_{BSCi} - (1 + r)S_{BSC,N}]Q \tag{9.3}$$

Step 4: Calculate the mass of substrate available per day for growth of the non-poly-P heterotrophs:

$$MS_{B,H} = QS_{Bi} - MS_{SEQ} \tag{9.4}$$

The sludge production and phosphorus removal can now be calculated separately for the two groups of organisms.

Step 5: Determine the total mass of active poly-P organisms in the system:

$$MX_P = \frac{Y_P MS_{SEQ}\theta_X}{(1 + b_P\theta_X)} \tag{9.5}$$

Step 6: Calculate the poly-P endogenous mass in the system:

$$MX_{E,P} = f_{EP,P}b_P MX_P\theta_X \tag{9.6}$$

Step 7: Using the sludge masses calculated above, estimate the phosphorus removal from the poly-P organisms:

$$\Delta P_{\mathrm{P}} = \left[ f_{\mathrm{P,XP}} \frac{MX_{\mathrm{P}}}{\theta_{\mathrm{X}}} + f_{\mathrm{P,XEP}} \frac{MX_{\mathrm{E,P}}}{\theta_{\mathrm{X}}} \right] \frac{1}{Q} \tag{9.7}$$

Step 8: Determine the total mass of active non-poly-P heterotrophs in the system, using the value for $MS_{\mathrm{B,H}}$ calculated in step 4:

$$MX_{\mathrm{H}} = \frac{Y_{\mathrm{H}} MS_{\mathrm{B,H}} \theta_{\mathrm{X}}}{(1 + b_{\mathrm{H}} \theta_{\mathrm{X}})} \tag{9.8}$$

(This calculation can be checked by multiplying the final value for $MX_{\mathrm{H}}/Q$ from Equation 2a by $Q$.)

Step 9: Calculate the non-poly-P endogenous mass in the system:

$$MX_{\mathrm{E,H}} = f_{\mathrm{EP,H}} b_{\mathrm{H}} MX_{\mathrm{H}} \theta_{\mathrm{X}} \tag{9.9}$$

Step 10: Estimate the phosphorus removal from the non-poly-P organisms, using the sludge masses calculated in steps 8 and 9:

$$\Delta P_{\mathrm{H}} = \left[ f_{\mathrm{P,XH}} \frac{MX_{\mathrm{H}}}{\theta_{\mathrm{H}}} + f_{\mathrm{P,XEH}} \frac{MX_{\mathrm{E,H}}}{\theta_{\mathrm{X}}} \right] \frac{1}{Q} \tag{9.10}$$

Step 11: Calculate the total inert sludge mass in the system:

$$MX_{\mathrm{i}} = QS_{\mathrm{UPi}} \theta_{\mathrm{X}} / f_{\mathrm{CV}} \tag{9.11}$$

Step 12: Determine the phosphorus removal occurring as a result of the wastage of inerts from the system:

$$\Delta P_{\mathrm{I}} = \left( \frac{f_{\mathrm{P,XI}} MX_{\mathrm{I}}}{\theta_{\mathrm{X}}} \right) \frac{1}{Q} \tag{9.12}$$

Step 13: Calculate the total system phosphorus removal by summing the phosphorus removals estimated in steps 7, 10, and 12:

$$\Delta P = \Delta P_{\mathrm{P}} + \Delta P_{\mathrm{H}} + \Delta P_{\mathrm{I}} \tag{9.13}$$

Step 14: The volatile suspended solids concentration in the aerobic zone can be estimated using the sludge masses previously calculated:

$$X_{\mathrm{V}} = \frac{f_{\mathrm{XAER}}(MX_{\mathrm{P}} + MX_{\mathrm{E,P}} + MX_{\mathrm{H}} + MX_{\mathrm{E,H}} + MX_{\mathrm{I}})}{V_{\mathrm{AER}}} \tag{9.14}$$

# APPLICATION OF THE ACTIVATED-SLUDGE MODELS

The general model and the steady-state model are now applied to an NDEBPR system. The example chosen is one of the laboratory-scale units (System 11a) operated by Wentzel *et al.* (1990). Figure 9.2 illustrates the configuration of this UCT/VIP™process. The 20°C system was operated at a 10-day sludge age with an influent of unsettled municipal wastewater input at a constant rate of 15 L/d. The characteristics of the influent wastewater in terms of total COD and TKN and unbiodegradable soluble and particulate COD are listed in Table 9.10. The wastewater was supplemented with phosphate to ensure that the influent concentration exceeded the removal capacity of the system. Average values of a number of system parameters monitored on a daily basis during steady-state operation are listed in Table 9.10. It should be noted that the effluent concentration of 10.1 g $P/m^3$ is slightly lower than the concentration in the last reactor; that is, there was a small uptake of phosphorus in the clarifier.

**APPLICATION OF THE GENERAL MODEL.** The predicted system response for certain key parameters is compared to the observed data in Table 9.10. The general model tracks the changes in soluble phosphorus and nitrate through the system and predicts the effluent TKN and COD concentrations with reasonable accuracy. What is most important is that the volatile suspended solids concentration and the oxygen use rates in the aerated reactors are also predicted reasonably. This is of particular significance given the model assumptions on fermentation under anaerobic conditions and the associated COD loss.

**APPLICATION OF THE STEADY-STATE MODEL.** The calculation procedure for application of the steady-state model algorithm is now outlined

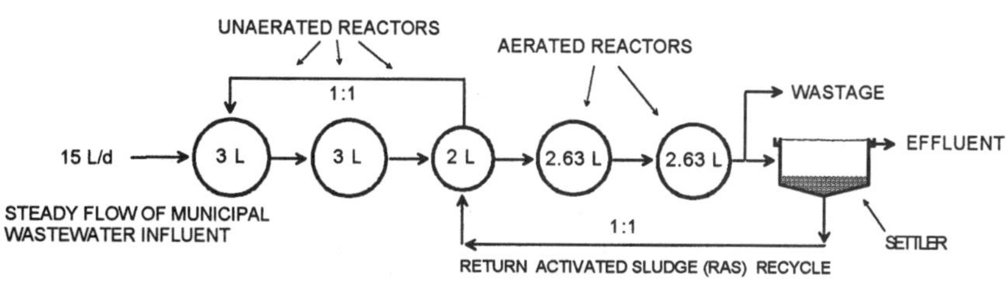


**Figure 9.2** UCT/VIP™ process NDEBPR system used by Wentzel *et al.* (1990) for the development of a mixed (poly-P + non-poly-P) organism culture (System 11a) (sludge age = 10 days; feed volume = 15 L/d; temperature = 20°C; influent = unsettled municipal wastewater).

**Table 9.10 Comparison of observed and predicted response of selected parameters in the NDEBPR mixed-culture system shown in Figure 9.2.**

| Parameter | Value | |
|---|---|---|
| | Observed | Predicted |
| Influent: | | |
| Total COD (g COD/m$^3$) | 506 | — |
| Readily biodegradable complex COD ($S_{BSC}$) (g COD/m$^3$) | 108 | — |
| Readily biodegradable SCFA COD ($S_{BSA}$) (g COD/m$^3$) | 0 | — |
| Unbiodegradable soluble COD ($S_{US}$) (g COD/m$^3$) | 53 | |
| Unbiodegradable particulate COD ($S_{UP}$) (g COD/m$^3$) | 46 | — |
| TKN (g N/m$^3$) | 51 | — |
| Ammonia (g N/m$^3$) | 38.3 | — |
| Phosphorus (g P/m$^3$) | 22.6 | — |
| Soluble phosphorus (g P/m$^3$): | | |
| Reactor 1 | 43.4 | 39.8 |
| Reactor 2 | 47.2 | 49.0 |
| Reactor 3 | 37.6 | 34.9 |
| Reactor 4 | 19.5 | 18.9 |
| Reactor 5 | 10.8 | 10.5 |
| Effluent | 10.1 | 10.5 |
| Overall phosphorus removal | 12.5 | 12.1 |
| Nitrate (g N/m$^3$): | | |
| Reactor 1 | 0.0 | 0.0 |
| Reactor 2 | 0.0 | 0.0 |
| Reactor 3 | 2.4 | 2.5 |
| Reactor 4 | 18.3 | 14.1 |
| Reactor 5 | 20.1 | 20.6 |
| Oxygen use rate (g $O_2$/m$^3$·h): | | |
| Reactor 4 | 40 | 40.5 |
| Reactor 5 | 23 | 25.1 |
| Volatile suspended solids (g VSS/m$^3$): | | |
| Reactor 5 | 1 418 | 1 478 |
| Effluent: | | |
| COD (g COD/m$^3$) | 61 | 61 |
| TKN (g N/m$^3$) | 4.7 | 4.6 |

for the same example. Model parameters are listed in Table 9.9 and wastewater characteristics in Table 9.10.

**Operating and System Parameters.** Operating and system parameters are as follows:

- Influent flow rate ($Q$ = 15 L/d);
- Anaerobic mass fraction ($f_{XA}$ = 0.29) ;
- Aerobic mass fraction ($f_{XAER}$ = 0.51) ;
- System sludge age ($\theta_X$ = 10 days);
- Number of equal-volume anaerobic reactors in series ($N$ = 2);
- Recycle ratio, that is, to anaerobic reactor ($r$ = 1);
- Aerobic volume ($V_{AER}$ = 5.25 L); and
- Nitrate concentration in the recycle to the anaerobic reactor ($N_{03,R}$ = 2.4 g N/m$^3$).

**Design Algorithm.** The design algorithm is as follows:

- Step 1: The concentration of readily biodegradable COD available for conversion to SCFA is determined using Equation 9.1, which yields a value of 87.4 g COD/m$^3$.
- Step 2: Assuming an initial value for $S_{BSC,N}$ of 0 g COD/m$^3$, calculate $MX_H/Q$ using Equation 9.2a. Then, using the value calculated for $MX_H/Q$, calculate a value for $S_{BSC,N}$ using Equation 9.2b. Recalculate $MX_H/Q$ using the value calculated for $S_{BSC,N}$ and repeat the iterative calculation until $MX_H/Q$ and $S_{BSC,N}$ converge. This method yields the following results:

  — $S_{BSC,N}$ = 5.2 g COD/m$^3$, and
  — $MX_H/Q$ = 437.2 g VASS/m$^3$·d.

- Step 3: The mass of SCFA taken up per day by the poly-P organisms is determined using Equation 9.3, which yields a value of 1 154.8 mg COD/d.
- Step 4: The mass of substrate available per day for the growth of the non-poly-P organisms is then calculated using Equation 9.4, which yields a value of 4 955.2 mg COD/d.
- Step 5: The total mass of active poly-P organisms in the system is calculated using Equation 9.5 and is equal to 3 711.8 mg VASS.
- Step 6: The poly-P endogenous mass in the system is determined using Equation 9.6, leaving a value of 371.2 mg VESS.
- Step 7: The phosphorus removal from the poly-P organisms is then estimated using the sludge masses previously calculated and Equation 9.7, and comes out to 9.48 g P/m$^3$ influent.

- Step 8: Using the value for $MS_{B,H}$ calculated in step 4 and Equation 9.8, the total mass of active non-poly-P heterotrophs in the system is determined to be 6 558.3 mg VASS.
- Step 9: The non-poly-P endogenous mass in the system is calculated, using Equation 9.9, to be 3 148.0 mg VESS.
- Step 10: The phosphorus removal because of the non-poly-P organisms is then estimated using the sludge masses calculated in steps 8 and 9 and Equation 9.10, and is calculated to be 1.94 g $P/m^3$ influent.
- Step 11: The total mass of inerts in the system is calculated from Equation 9.11, which yields 4 615.5 mg VISS.
- Step 12: Phosphorus removal resulting from the wastage of inerts is then determined using Equation 9.12, assuming $f_{P,XI} = 0.03$, and is equal to 0.92 g $P/m^3$ influent.
- Step 13: The total system phosphorus removal is estimated by summing the phosphorus removals calculated in steps 7, 10, and 12, using Equation 9.13, and is equal to 12.3 g $P/m^3$ influent. The total volatile sludge mass in the system is obtained by summing the sludge masses previously calculated. Hence, the total volatile suspended solids concentration in the aerobic zone ($X_V$), calculated using Equation 9.14, is equal to 1 788 g $VSS/m^3$.

# CONCLUSION

This chapter has outlined a general activated-sludge dynamic model with the capacity for modeling the biological processes of carbonaceous energy removal, nitrification, denitrification, and EBPR. A simplified steady-state analysis model suited for "pencil and calculator" application also has been presented.

The satisfactory results from use of the model provide a basis for further model evaluation and refinement. Improvements to the model no doubt will come about as the understanding of the complex interactions occurring within biological nutrient removal systems is expanded. Many aspects require further investigation, including the following:

- The mechanisms, kinetics, and temperature dependency of enzymatic hydrolysis of slowly biodegradable colloidal and particulate organic substrate—current understanding of this behavior is limited, particularly for anoxic and anaerobic conditions important to nutrient removal systems.
- The role of fermentation under anaerobic conditions—this has important consequences with regard to volatile sludge production and oxygen demand in EBPR systems.
- The temperature dependency of the poly-P organism kinetic behavior.
- The fate of "unbiodegradable" COD generated by the endogenous processes of poly-P organisms.

There is no doubt that modeling and simulation are assuming a prominent role in nutrient removal system design, operation, and control. It should be recognized that the model described here considers only the activated-sludge system. This is only one part of the wastewater treatment cycle. Other unit operations within the treatment plant (and operating procedures) are important to overall system performance. A comprehensive treatment plant simulator should incorporate all of these aspects, for example settling tanks, sludge fermenters, digesters, and handling of supernatant streams.

A final comment is worth making regarding the importance of a key aspect that is perhaps not emphasized explicitly in this chapter: namely, modeling parameters. In applying the model to predict system behavior, the model requires values for stoichiometric and kinetic parameters, such as yields and growth rates. In addition, a number of parameters that define the characteristics of the influent wastewater must be specified, for example the readily biodegradable COD content and the fraction of unbiodegradable COD that is soluble and particulate.

References are available that discuss measurement techniques for the stoichiometric and kinetic parameters (see, for exampe, Ekama *et al.*, 1986). However, several studies have shown that the stoichiometric and kinetic parameters, with one notable exception described below, do not change appreciably for different systems treating municipal wastewaters (Dold and Marais, 1986, and Gibson, 1991). For example, the decay rate estimated from the decline in oxygen use rate in a batch aerobic sludge digestion test is the same for sludge drawn from aerobic systems in Canada or South Africa (at least at 20°C). Therefore, it is proposed that the values for the stoichiometric and kinetic model parameters tabulated in this chapter (derived from previous studies on a range of treatment systems) should be acceptable initial estimates and should not require substantial adjustment in model calibration exercises.

The exception referred to above is the maximum specific growth rate of the nitrifiers, $\mu_A$. This parameter shows marked variations between systems treating different wastewater. Values ranging from 0.2 to 1.0 $d^{-1}$ (at 20°C) have been reported. This parameter has a significant effect on nutrient removal system design and performance, and obviously on model predictions as well. Therefore, special attention should be paid to measurement or estimation of $\mu_A$. In a sense, $\mu_A$ can be regarded as a wastewater characteristic.

Regarding the influent wastewater characteristics, studies have demonstrated that these do change, often appreciably, from one municipal waste to another. Wastewater characteristics have a significant effect on system performance, particularly for nutrient removal systems. A single characteristic such as the readily biodegradable COD fraction can determine whether or not a system designed for enhanced phosphorus removal will in fact remove phosphorus. Therefore, if the model is to provide reasonable predictions of system behavior, adequate knowledge of wastewater characteristics is extremely important. The Appendix outlines the basis for wastewater characterization as applicable to NDEBPR systems. It is likely that the importance of the measurement (or appropriate estimation) of wastewater characteristics will receive increasing attention through wider application of models such as the one presented here.

# REFERENCES


Arkley, M.J., and Marais, G.v.R (1981) The Effect of the Anoxic Zone on Sludge Production and Settleability in the Activated Sludge Process. Research Report W38, Dep. Civ. Eng., Univ. Cape Town, S. Afr.

Barker, P., and Dold, P.L. (1995a) COD and Nitrogen Mass Balances in Activated Sludge Systems. *Water Res.* (G.B.), **29**, 633.

Barker, P., and Dold, P.L. (1995b) Denitrification Behaviour in Biological Excess Phosphorus Removal Activated Sludge Systems. Accepted for publication in *Water Res.* (G.B.).

Billing, A.E., and Dold, P.L. (1988) Modelling Techniques for Biological Reaction Systems: Part 1. Mathematical Description and Model Representation. *Water SA*, **14**, 185.

Brodisch, K.E.U. (1985) Interaction of Different Groups of Micro-Organisms in Biological Phosphate Removal. *Water Sci. Technol.*, **17**, 139.

Buchan, L. (1981) The Location and Nature of Accumulated Phosphorus in Seven Sludges from Plants Which Exhibited Enhanced Phosphorus Removal. *Water SA*, **7**, 1.

Comeau, Y., *et al.* (1986) Biochemical Model for Enhanced Biological Phosphorus Removal. *Proc. UBC Conf. New Dir. Res. Waste Treat. Residuals Manage. Water Res.*, **20**, 1511.

Dold, P.L. (1990) A General Activated Sludge Model Incorporating Biological Excess Phosphorus Removal. *Proc. C.S.C.E. Annu. Conf.*, Hamilton, Ont., Can.

Dold, P.L. (1991) Modification du modele general des boues activees pour tenir compte de la dephosphatation. *Sciences et Techniques de L'eau*, **24**, 229.

Dold, P.L., and Marais, G.v.R. (1986) Evaluation of the General Activated Sludge Model Proposed by the IAWPRC Task Group. *Water Sci. Technol.*, **18**, 63.

Dold, P.L., *et al.* (1980) A General Model for the Activated Sludge Process. *Prog. Water Technol.*, **12**, 47.

Dold, P.L., *et al.* (1991) Hydrolysis of a(1-4) Glucan Bonds in Activated Sludge Mixed Bacterial Communities. *Environ. Technol.*, **12,** 871.

Downing, A.L., *et al.* (1964) Nitrification in the Activated Sludge Process. *J. Proc. Inst. Sew. Purif.*, **64**, 130.

Ekama, G.A., *et al.* (1986) Procedures for Determining Influent COD Fractions and the Maximum Specific Growth Rate of Heterotrophs in Activated Sludge Systems. *Water Sci. Technol.*, **18**, 91.

Gibson (1991) Activated Sludge System Modelling Using the IAWPRC Model. M. Eng. thes., Dep. Civ. Eng., McMaster Univ., Hamilton, Ontario, Can.

Grady, C.P.L. (1989) Dynamic Modeling of Suspended Growth Biological Wastewater Treatment Processes. In *Dynamic Modeling and Expert Systems in Wastewater Engineering.* G.G. Patry and D. Chapman (Eds.), Lewis Publishers, Chelsea, Mich.

Grady, C.P.L., *et al.* (1986) A Model for Single-Sludge Wastewater Treatment Systems. *Water Sci. Technol.*, **21**, 47.

Henze, M., *et al.* (1987a) Activated Sludge Model No. 1. IAWPRC Scientific and Technical Report No. 1, Int. Assoc. Water Pollut. Res. Control, London, G.B.

Henze, M., *et al.* (1987b) A General Model for Single Sludge Wastewater Treatment Systems. *Water Res.* (G.B.), **21**, 505.

Henze, M., *et al.* (1994) Activated Sludge Model No. 2. IAWQ Scientific and Technical Report No. 3, Int. Assoc. Water Qual., London, G.B.

Lawrence, A.W., and McCarty, P.L. (1970) Unified Basis for Biological Treatment Design and Operation. *J. Sanit. Eng. Div., Proc. Am. Soc. Civ. Eng.*, **96**, 757.

Marais, G.v.R., and Ekama, G.A. (1976) The Activated Sludge Process: Part 1—Steady State Behaviour. *Water SA*, **2**, 163.

McClintock, S.A., *et al.* (1988) Nitrate Versus Oxygen Respiration in the Activated Sludge Process. *J. Water Pollut. Control Fed.*, **60**, 342.

McKinney, R.E. (1962) Mathematics of Complete Mixing Activated Sludge. *J. Sanit. Eng. Div., Proc. Am. Soc. Civ. Eng.*, **88**, SA3, 87.

McKinney, R.E., and Ooten, R.J. (1969) Concepts of Complete Mixing Activated Sludge. *Trans 19th Sanit. Eng. Conf.*, Univ. Kansas, Lawrence, 32.

Meganck, M., *et al.* (1985) The Importance of Acidogenic Microflora in Biological Phosphorus Removal. *Water Sci. Technol.* , **17**, 199.

Mino, T., *et al.* (1984) Location of Phosphorus in Activated Sludge and Function of Intracellular Polyphosphates in Biological Phosphorus Removal Process. *Water Sci. Technol.*, **17**, 93.

Mino, T., *et al.* (1987) Effect of Phosphorus Accumulation on Acetate Metabolism in the Biological Phosphorus Removal Process. In *Advances in Water Pollution Control—Biological Phosphate Removal from Wastewaters.* R. Ramadori (Ed.), Pergamon Press, Oxford, Eng.

San Pedro, D., *et al.* (1994) Evaluation of the Rate of Hydrolysis of Slowly Biodegradable COD (SBCOD) Using Starch as Substrate Under Anaerobic, Anoxic and Aerobic Conditions. *Proc. Water Qual. Int. `94—IAWQ 17th Biennial Int. Conf.*, Budapest, Hungary.

Smyth, S.A. (1994) Sludge Production in Aerobic and Anoxic Activated Sludge Systems. M. Eng. thes., Dep. Civ. Eng., McMaster Univ., Hamilton, Ont., Can.

Van Haandel, A.C., *et al.* (1981) The Activated Sludge Process: Part 3—Single Sludge Denitrification. *Water Res.* (G.B.), **15,** 1135.

Wentzel, M.C., *et al.* (1985) Kinetics of Biological Phosphorus Release. *Water Sci. Technol.,* **17**, 57.

Wentzel, M.C., *et al.* (1986) Metabolic Behaviour of Acinetobacter spp. in Enhanced Biological Phosphorus Removal. *Water SA*, **12**, 209.

Wentzel, M.C., *et al.* (1989a) Enhanced Polyphosphate Organism Cultures in Activated Sludge Systems. Part II: Experimental Behaviour. *Water SA*, **15**, 71.

Wentzel, M.C., *et al.* (1989b) Enhanced Polyphosphate Organism Cultures in Activated Sludge Systems. Part III: Kinetic Model. *Water SA*, **15**, 89.

Wentzel, M.C., *et al.* (1990) Biological Excess Phosphorus Removal—Steady State Process Design. *Water SA*, **16**, 29.

# Chapter 10 Nutrient Removal by Aquatic Systems



Aquatic systems such as ponds, shallow reservoirs, and natural and constructed wetlands are considered low-cost alternatives for treating municipal, industrial, and agricultural effluents. Conventional treatment systems are energy intensive and operationally complicated compared to natural systems. Continual rises in energy costs make alternative low-energy treatment systems more economically viable. Wastewater treatment wetlands may also provide wildlife habitat and water storage. Wetland vegetation also may be harvested for biomass as an animal food, fiber, or fertilizer resource (Reddy and Smith, 1987). During the past two decades, a vast amount of literature has been published on wastewater treatment wetlands, including potential use of these systems to remove biochemical oxygen demand (BOD), nutrients, and to a limited extent metals and toxic organic compounds; wetland engineering assessment and design criteria; and treatment system optimization and cost effectiveness. Numerous studies on various aspects of constructed wetlands are available (Cooper and Findlater, 1990; Gersberg *et al.*, 1983; Hammer, 1989; Kadlec and Tilton, 1978; Kickuth, 1977; Lakshman, 1979; Mitsch and Cronk, 1992; Moshiri, 1993; Olson and Marshall, 1992; Reddy and Smith, 1987; Reddy *et al.*, 1982; Rogers *et al.*, 1985; Seidel, 1976; Spangler *et al.*, 1976; and Wolverton, 1987). Aquatic systems are now used for wastewater treatment throughout subtropical and temperate regions of the world. Several design manuals have been developed (Kadlec and Knight, 1995; Mitsch and Cronk, 1992; and Reed *et al.*, 1988). Because detailed design information for use of aquatic and wetland systems for waste is thoroughly covered in the cited references, this chapter will focus on processes involved in the removal of organic matter and nutrients (nitrogen and phosphorus).

Aquatic systems used for wastewater treatment can be grouped into two main categories: ponds and shallow reservoirs stocked with floating and submersed aquatic macrophytes, and constructed and natural wetlands dominated by emergent macrophytes.

# AQUATIC SYSTEMS

**PONDS AND SHALLOW RESERVOIRS.** Aquatic plants used in these systems float on the water surface (floating aquatic macrophytes) or grow in the water column, rooted in the sediment (submersed aquatic plants). These plants naturally grow in eutrophic lakes and streams. The high growth rates and nutrient assimilation capacities of these plants have encouraged water quality managers to consider them for reducing wastewater nutrient levels (Reddy and Sutton, 1984). The most typical floating aquatic plants used in treatment ponds are water hyacinth (*Eicchornia crassipes*), pennyworth (*Hydrocotyle umbellata*), and duckweed (*Lemna minor*), and the most typical submersed aquatic macrophyte used is elodea (*Egeria densa*) (see Figure 10.1).

**WETLAND SYSTEMS.** Wetlands are defined as the lands transitional between terrestrial and aquatic systems where the water table typically is at or near the surface of the land (Mitsch and Gosselink, 1986). Under current regulations, intentional use of natural wetlands for wastewater treatment is not permitted because of potential negative effects on flora and fauna. However, many natural wetlands may be used in polishing wastewater effluent. In this case, the wetland system is critically monitored to ensure no deterioration of water quality or of flora and fauna in the ecosystem. The effectiveness of natural processes in wetlands for wastewater treatment has led to the development of constructed wetlands that are managed and loaded at a higher rate than natural wetlands. Two types of constructed wetlands are used: free water surface (FWS) wetlands (see Figure 10.2, top) and subsurface flow (SSF) wetlands (Reed *et al.*, 1988) (see Figure 10.2, bottom).

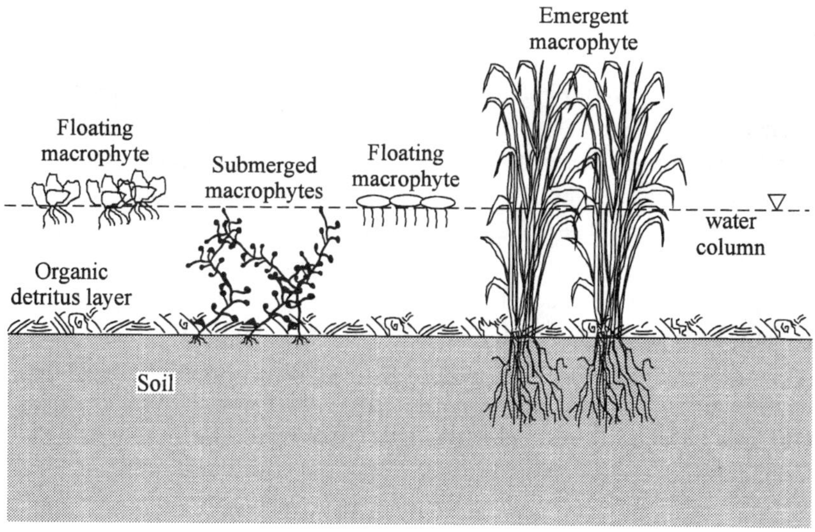


**Figure 10.1** **Schematic of a wastewater treatment pond stocked with aquatic macrophytes.**

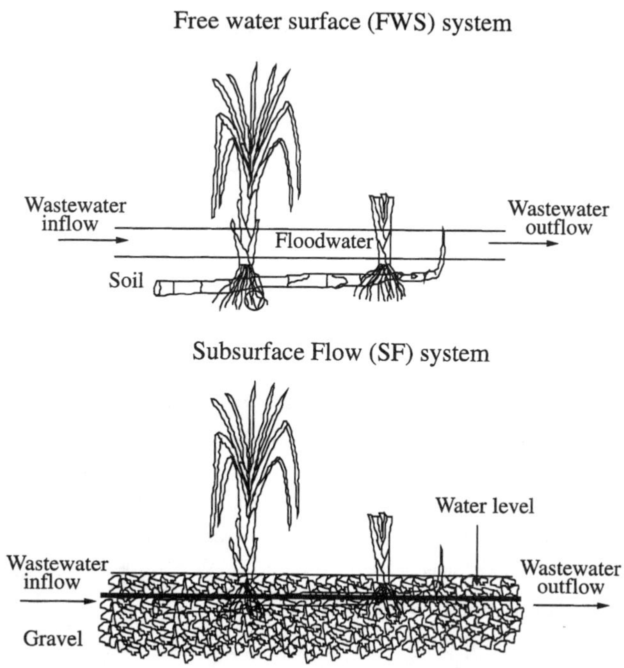


**Figure 10.2** **Schematic showing (top) free water surface and (bottom) subsurface flow wetland systems used for wastewater treatment.**

In FWS wetlands, the majority of the water flows over the surface of the soil (as in a natural wetland) at a managed floodwater depth. Water can be lost through evapotranspiration (ET) and infiltration. Percent water loss from ET may be significant depending on the hydraulic loading rates and climatic conditions. Wastewater applied to these systems only contacts the soil surface (see Figure 10.2, top). In SSF wetlands, water flows though a porous matrix of coarse sand or gravel in which the plants are rooted. Water flows through the bed instead of on top of the matrix; consequently, water may contact all surfaces of the matrix and plant roots. Water losses are from ET. In natural wetlands, the fauna, flora, and soils are determined by the hydroperiod. In constructed wetlands, the hydroperiod is controlled; flora may be maintained as monocultures or allowed to evolve with native vegetation. Different substrates such as native soil, sands, and gravel may be used as a rooting media for vegetation.

Constructed wetlands provide a number of advantages over natural wetlands for wastewater treatment. Water levels and flow rates can be easily managed. The soil/substrate and vegetation type can be selected for their ability to treat particular effluent constituents. Treatment cells are designed to site-specific conditions allowing easy and efficient management of the system. Constructed wetlands may be used in all climates, with colder climates (subfreezing temperatures) even facilitating vegetation maintenance (Brooks *et al.,* 1984; Small, 1976; and Wile *et al.*, 1985). However, under certain conditions, reduced performance during winter months has been reported (Wile *et al.*, 1985).

# *MAIN COMPONENTS OF WETLANDS AND AQUATIC SYSTEMS*

The three main components of wetlands and aquatic systems are hydrology, soils or sediments, and vegetation. The interactions among these components dictate the overall contaminant removal efficiency of the treatment systems.

**HYDROLOGY.** Hydrologic regime is the major regulating factor of all wetland and aquatic systems used for wastewater treatment. In ponds and reservoirs, hydrologic characteristics depend on configuration or geometry of ponds, wastewater loading rate, and water depth. Typical water depth in aquatic systems containing floating or submerged macrophytes ranges from 0.4 to 2 m, with an average water depth of approximately 1 m.

The FWS wetlands include both natural and constructed types and typically consist of channels or basins with natural soil. In certain areas, constructed FWS wetlands are lined with clay material to prevent downward movement of water. In FWS systems, water depth typically ranges from 0.1 to 1.0 m. The SSF wetlands include marshes (areas with semipervious bottoms), ponds, and trenches (with or without lining). Typically, soil or gravel is used as rooting media. In SSF systems, wastewater flows through a soil/gravel substrate, and the soil is not completely saturated with water; that is, there is no free-standing water on the soil or matrix surface. This inhibits mosquito production and reduces short-circuiting of water flow.

An accurate hydrologic budget is required to evaluate the efficiency of wetland systems for treating wastewater. Components of the budget include surface water inflow, precipitation, groundwater inflow, ET, surface water outflow, and infiltration. In constructed wetlands and ponds, the dominant surface inflows are the influent waste stream and precipitation. Infiltration losses typically are low because of the presence of liner or impervious soils. Accurate estimation of mass removal of nutrients requires accounting for evaporative losses. Evapotranspiration can cause significant water losses, which in turn can cause effluent constituent concentrations to increase. The influences of these factors are discussed in detail by Carter *et al.* (1978) and Kadlec (1989).

Hydraulic loading rates to constructed wetlands vary from 20 to 200 mm/d depending on treatment objectives, whereas the loading rates of natural marshes are in the range of 2 to 50 mm/d.

**SOILS/SUBSTRATE FOR ROOTING MEDIA.** Soils in natural wetlands can be broadly grouped into mineral soils and organic soils. Mineral soils typically have high bulk densities (1 to 1.8 g/cm$^3$), low carbon content (<10%), and pH ranging from acid to alkaline. Organic soils typically have lower bulk density (<0.3 g/cm$^3$), high organic carbon content (10 to 50%), and acidic to neutral pH. The type and texture of the soil affect physical, chemical, and biological mechanisms regulating the removal of waste constituents. Information on soil type for a given site can be obtained from the U.S. Department of Agriculture, Soil Conservation Service soil surveys. Free water surface wetlands used for wastewater treatment typically use native soils, whereas SSF wetlands use a gravel or rock matrix. In many systems, the soils are underlain by a plastic or clay liner to prevent infiltration and downward movement of contaminants.

**VEGETATION.** Herbaceous emergent aquatic plants are widely distributed on the shorelines of lakes and streams and in wetland ecosystems. Many of these plants are perennial and have a high capacity for biomass production and nutrient storage. The emergent aquatic plants most typically used in FWS and SSF wetlands include cattails (*Typha* spp.), reeds (*Phragmites* sp.), and bulrush (*Scirpus* spp.). Most of these plants occur in both southern and northern climates. During winter months, the aboveground portion of the plant dies back, and dormant roots and rhizomes provide energy for new spring growth. Cattails, reeds, and bulrushes currently are used by several small municipalities to treat domestic wastewater in the U.S., Canada, and many European countries.

Floating plants most typically used in aquatic treatment systems are water hyacinth, pennywort, and duckweeds. Elodea is a typical submersed macrophyte used in ponds for water treatment.

Forested wetlands may occur as hydrologically discrete stands, or they may be located along the interface of aquatic and terrestrial systems (on river floodplains and lake fringes). These wetlands can also be used for disposal of wastewater effluents (Boyt *et al.*, 1977; DeBusk and Reddy, 1987; and Dierberg and Brezonik, 1984).

# PHYSICOCHEMICAL CHARACTERISTICS

In natural and constructed wetlands, the dissolved oxygen of the water column may remain relatively high because of a low density of oxygen-consuming organisms; photosynthetic oxygen production by algae; or mixing of water by wind action. Oxygen demand in the water column is low compared to the soil; thus, oxygen diffusing through the water column is rapidly consumed at the soil surface. This results in two distinctly different soil layers: an oxidized or aerobic surface layer in which oxygen is present and an underlying reduced or anaerobic soil layer in which no free oxygen is present (see Figure 10.3). Aerobic and anaerobic biogeochemical processes regulate the removal or retention of nutrients by wetland soils.

Emergent macrophytes transport oxygen through stems and leaves into their root zone. This convective flow and diffusion of air into the root zone occurs in many wetlands plants as an adaptation to anaerobic soil conditions (Armstrong, 1964; Brix, 1988; Brix *et al.*, 1992; and Dacey, 1981). Oxygen not used during root respiration may leak into the adjacent soil, creating an aerobic zone around the roots. The excess oxygen in the rhizosphere oxidizes potentially toxic sulfides, reduced iron, and manganese (Engler and Patrick, 1975), and may be significant in the aerobic oxidation of soluble organic compounds and nitrification of ammonium-nitrogen (Reddy *et al.*, 1989).

Floating macrophytes also have the capacity to transport oxygen to the rhizosphere. The oxygen transport capacity of pennywort was found to be high (112 mg $O_2$/g [root]·d) compared to water hyacinth (55 mg $O_2$/g [root]·d) (Reddy *et al.*, 1990). The oxygen transport capacity of several emergent aquatic macrophytes was in the range of 4 to 37 mg $O_2$/g (root)·d. Various methods have been devised for measuring oxygen transport; results vary depending on units of qualification (see Table 10.1). Estimates of oxygen transport based on BOD removal values were calculated to be 29 to 45 g $O_2/m^2$·d; these values are higher than the values of 3 to 5 g/$m^2$·d calculated from $NH_4$-N oxidation (Burgoon, 1993). A range of 5 to 45 g/$m^2$·d has been

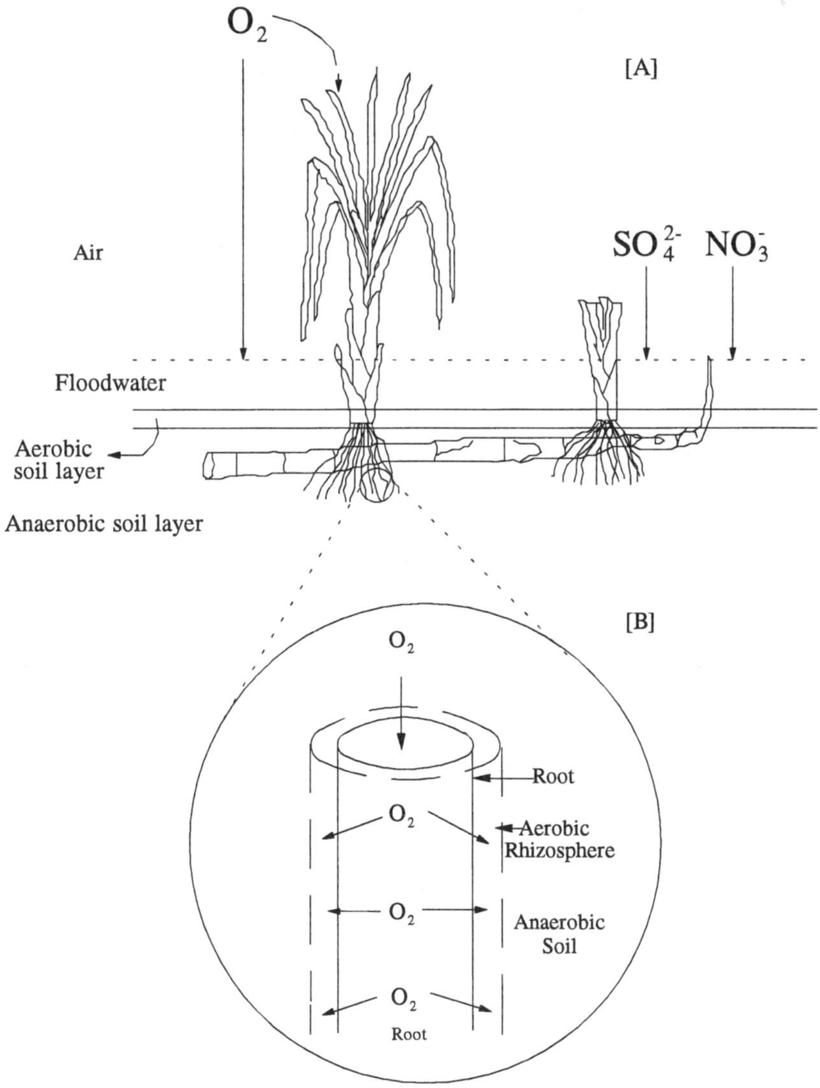


**Figure 10.3 Schematic showing aerobic soil–floodwater interface and root–soil–water interface in wetlands.**

recommended for design purposes of SSF treatment wetlands (Reed *et al.*, 1988). Field studies with and without plants give the most reliable estimate of oxygen available to bacteria in the rhizosphere. Although there is clear evidence of oxygen transport into the roots by diffusion and mass flow, at times oxygen transport may be insufficient to meet plant metabolic requirements (Brix, 1990), implying that there is not enough oxygen to significantly affect wastewater treatment.

Wetlands soils can be characterized by redox potential or oxidation-reduction potential (Eh), a measure of electron availability. The redox potential of

**Table 10.1** **Summary of estimated oxygen transport calculated from removal of BOD, organic matter, and nitrogen and accumulation of dissolved oxygen in the pore water in various field and laboratory research. Oxygen transport is presented using two types of units (Burgoon, 1993).**

| Reference | Plant | Age | System | Oxygen mg/L | Eh, mV | $O_2$ transport, $\mu g\ O_2/L{\cdot}h$ | $O_2$ transport, $g\ O_2/m^2{\cdot}d$ |
|---|---|---|---|---|---|---|---|
| Dunabin *et al.*, 1988 | *Typha domingensis* | 5 months | 10-L bucket | 3–4 | na[a] | 80–300 | na |
| Moorhead and Reddy, 1990 | *Hydrocotyle umbellata* | na | 635-mL flask | 0.5–7 | na | 700 (100) | na |
| Reed *et al.*, 1988 | *Phragmites communis* | Mature | na | na | na | na | 5–45 |
| Gersberg *et al.*, 1986 | *Scirpus validus* | Mature | 0.1 ha | na | na | na | 5–45 |
| Reddy *et al.*, 1990 | *Scirpus validus* | Small | 500-mL flask | 0.5–7 | na | 292 (35) | na |
| | *Scirpus pungens* | Small | " | " | na | 0 (27) | na |
| | *Typha* sp. | Small | " | " | na | 160 (48) | na |
| | *Sagittaria latifolia* | Small | " | " | na | 160 (90) | na |
| Burgoon, 1993 | *S. pungens* | Mature | 6 $m^2$ | 0 | <-200 | na | 29 |
| Burgoon, 1993 | *S. validus* | 6–12 mos | 2.2 L | 0 | <100 | 3 113 | 10 |

[a] Not available.

a soil is measured with an inert platinum electrode. Depending on soil type and intensity of soil reduction, a wide range of Eh values is observed in wetland soils (see Figure 10.4). Calcium, nitrogen, oxygen, selenium, manganese, and iron are the most important elements affected by soil redox conditions. Characteristic Eh values of well-drained soils or wetland soils during dry periods are typically greater than 300 mV. Nitrate is removed at Eh values between 200 and 300 mV. Lower Eh conditions result in the reduction of manganese, iron, sulfate, and carbon dioxide (see Figure 10.5). Depending on organic matter content and organic loadings of waterlogged soils, Eh values as low as 250 mV can occur (Mahapatra and Patrick, 1969). In wetlands used for wastewater treatment, these reduction processes can have significant effects on the metal solubility of copper, selenium, and arsenic (Dierberg *et al.*, 1987a, and Massacheleyn *et al.*, 1991) and on biodegradation of organic compounds (Dierberg *et al.*, 1987b; Haggblom *et al.*, 1993; and Pardue *et al.*, 1993). Figures 10.4 and 10.5 provide a general guide to the process of oxidation reduction in wetland soils. However, Eh measurements in natural systems represent mixed potential (because of concurrent redox processes) and may not be related quantitatively to any one redox couple (Ponnamperuma, 1972, and Stumm, 1966).

The pH of both acid and alkaline soils of constructed and natural wetlands tends to converge toward neutrality as a result of reduction reactions and consumption of protons. The pH affects aerobic and anaerobic bacterial growth, chemical precipitation–complexation reactions, adsorption reactions, and chemical solubility. Submersed macrophytes or algae in the water column may deplete carbon dioxide during photosynthesis and elevate pH val-

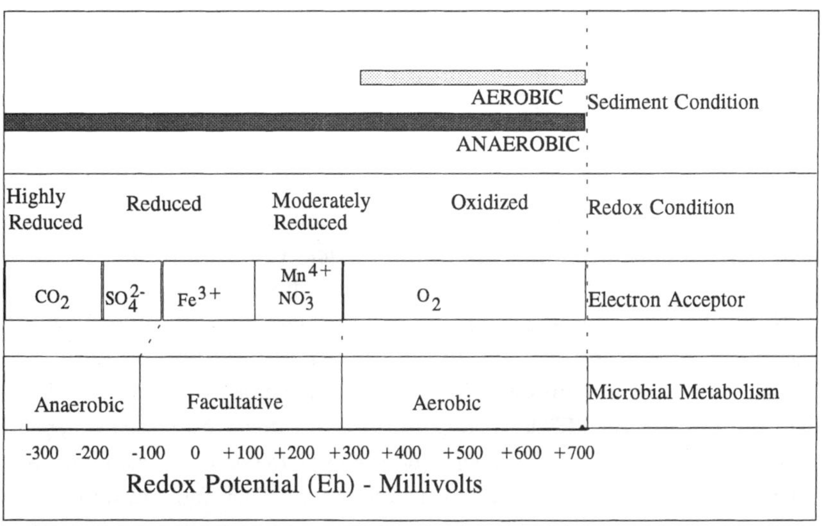


**Figure 10.4 Schematic showing the microbial processes functioning in a wetland at various redox potentials.**

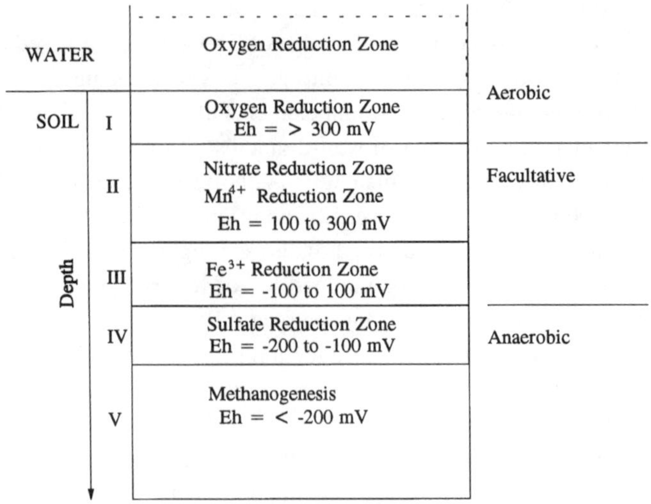


**Figure 10.5** **Schematic showing the reduction of various electron acceptors as a function of soil depth in a wetland.**

ues ($pH > 10$) during the daylight hours. The pH of the water column may decrease by 2 to 3 units after cessation of photosynthesis and because of the net increase of respiratory activity during the evening hours.

# *Nutrient storage/uptake capacity of aquatic plants*

The nutrient removal capacity of treatment wetlands is dependent on plant growth rates, wastewater composition, water depth, sediment characteristics, oxygen transfer capability of the plants into the root-zone (see Figure 10. 3), biochemical and physicochemical processes functioning at the root–water–sediment interface (Good and Patrick, 1987, and Reddy, 1983), plant density per unit area, plant harvesting, and climate.

Nutrient uptake rates and maximum growth rates define the nutrient storage capability of the plant. The potential rate of nutrient uptake by a plant is limited by its net productivity (growth rate) and the concentration of nutrients in the plant tissue (Reddy and DeBusk, 1987). Nutrient storage is dependent on plant tissue nutrient concentrations and on the potential for biomass accumulation (that is, the maximum standing crop). Therefore, desirable traits of a plant used for contaminant assimilation and storage would include rapid growth, high tissue nutrient content, and the capability to attain a high standing crop (biomass per unit area). Plant growth is directly proportional to nutrient supply at low to moderate nutrient loads; increased nutrient loads increase the growth of the plant but may result in luxury uptake of nutrients. Response to nutrient concentration or loads depends on the type of aquatic plant used in the treatment system. Maximum growth typically is attained at the lower range of nutrient supply, whereas maximum nutrient uptake rate is reached at much higher levels of nutrients.

Emergent macrophytes have the capability to grow in a wide range of substrates and in a variety of wastewaters. The influence of substrate and wastewater type can result in a wide range of nutrient composition in the plant tissue. In many soils, major plant nutrients (nitrogen and phosphorus) may limit plant growth. Addition of these nutrients through wastewater or fertilizer increases growth rate and tissue nutrient content (Cary and Weerts, 1984; Reddy and Portier, 1987; and Ulrich and Burton, 1985). In addition to the nutrients applied through wastewater, sources within the wetland such as the decomposition of soil organic matter, biological nitrogen fixation, and decomposition of detritus tissue can also contribute to overall nutrient supply to vegetation.

Maximum annual storage of nutrients by emergent macrophytes is in the range of 600 to 2 630 kg N/ha and 75 to 403 kg P/ha. Because emergent macrophytes have more supportive tissue than floating macrophytes, they may have greater potential for storing the nutrients over a longer period. More than 50% of nutrients were found to be stored in belowground portions of emergent plant tissue, which suggests that harvesting aboveground biomass alone may not be an effective nutrient removal mechanism. Frequent harvesting of emergent vegetation may not be necessary to achieve maximum nutrient removal, although harvesting aboveground biomass once a year may improve the overall nutrient removal efficiency of the wetland.

In natural stands, the aboveground portion of many aquatic plants dies back during winter months, while the belowground portions remain alive. The rhizome and roots are the central food storage organs of the plants during the winter months, and these parts keep the plants alive during winter dormancy. The shoots and rhizomes have aerenchyma cells, which are used to transport oxygen into the root zone. This mechanism keeps the plant alive under anaerobic soil conditions. At the end of the winter season, shoot regrowth occurs, and the growing season continues until the fall. Many plants approach maximum growth by August, while pollutant assimilation continues until the end of the growth season.

Many floating macrophytes are quite productive and their nutrient uptake rates are rapid. Plants with large biomass per unit area (such as water hyacinth) have the potential to store large amounts of nutrients as compared to duckweed, which occurs at a low standing crop biomass. Storage of nutrients in floating plants is short term; therefore, these plants need to be harvested periodically to maintain a high nutrient removal capacity for the treatment system.

Frequent harvesting of plant biomass may not be necessary to achieve maximum pollutant removal in FWS and SSF wetlands, although harvesting has the advantage of reducing the buildup of detritus on the soil surface, thus decreasing nutrient release during decomposition. Additionally, in some locations there exist beneficial uses for biomass, such as compost and biogas production (Smith and Frank, 1988).

Mechanical harvesting techniques have been developed for reed and cattails (Bagnall *et al.*, 1987, and Graneli, 1984). For recovering maximum amounts of biomass, plants should be harvested during the months of August to September. It should be noted that under many conditions, harvesting aboveground biomass may not be cost effective and may even be detrimental to wetlands. If plant management is necessary, techniques such as waste drawdown followed by fire can be used to control vegetation density. However, controlled burning may also have short-term negative effects on water quality and wildlife (Mitsch and Cronk, 1992).

# NUTRIENT REMOVAL PROCESSES

Removal mechanisms of carbon (expressed as BOD, chemical oxygen demand [COD], or total organic carbon), nitrogen (organic nitrogen, ammonium, and nitrate), and phosphorus (dissolved reactive phosphorus, dissolved total phosphorus, and total phosphorus) in aquatic systems are discussed in the following sections.

**CARBON REMOVAL PROCESSES.** Organic matter is decomposed by microorganisms through aerobic respiration, facultative anaerobic respiration, or anaerobic respiration. Organic carbon is used as an energy source by bacteria and converted to bacterial biomass, carbon dioxide, or methane. In FWS and SSF wetlands, bacteria may be attached to stalks of plants, detritus, or the rooting matrix. Respiratory pathways for degradation of the organic matter depend on the availability of electron acceptors

**Aerobic Respiration.** Oxygen is the terminal electron acceptor in aerobic systems and is reduced while organic carbon (source of electron donor) is oxidized. In FWS wetlands, oxygen is available from diffusion and in open areas through algal photosynthesis. In SSF wetlands, oxygen is available primarily through oxygen transport by plants. It has been estimated that 5 to 40 g $O_2/m^2 \cdot d$ is available through emergent macrophytes. Aerobic respiration is efficient, and potential energy available from oxidation of organic compounds is higher than for anaerobic respiration.

**Facultative Anaerobic Respiration.** In the absence of oxygen, the microbial community shifts to facultative anaerobic respiration and uses nitrate as an electron acceptor during the oxidation of organic matter. If nitrate is not present, manganic and ferric iron compounds are used. Rates of organic matter decomposed are directly correlated to the amounts of electron acceptors present. In most wastewater, concentrations of manganese and iron are too low to be significant in the oxidation of organic matter.

**Anaerobic Respiration.** In the absence of oxygen, nitrate, manganic, and ferric compounds, true anaerobes use sulfate and carbon dioxide as electron acceptors during the oxidation of substrate carbon. In an aquatic macrophyte system, these conditions will exist in the detrital layer and underlying sediment. Anaerobic decomposition of organic matter is probably the most significant pathway for BOD removal in FWS and SSF wetlands. The high surface area of detritus, sediment, and plant stalks facilitates high rates of decomposition by anaerobic bacteria. Gravel used in SSF wetlands may be low in surface area (80 to 150 $m^2/m^3$), but root surfaces and particles suspended in pore spaces may support the microbial populations that provide the high rates of BOD removal observed in gravel filters (Young and McCarty, 1969).

**NITROGEN REMOVAL PROCESSES.** Nitrogen reactions in wetlands effectively process inorganic nitrogen through nitrification and denitrification, ammonia volatilization, and plant uptake (see Figure 10.6). These processes aid in maintaining low levels of inorganic nitrogen in the water column. A significant portion of dissolved organic nitrogen is returned to water column during breakdown of detrital plant tissue or soil organic matter. The majority of this dissolved organic nitrogen is resistant to decomposition. Effluent leaving wetlands may therefore contain elevated levels of nitrogen in organic form. The potential export of dissolved organic nitrogen will, however, depend on the environmental conditions present in the water and soil. Organic nitrogen is present in the influent wastewater and in the detritus plant tissue. Depending on the metabolic activity of the bacteria, organic bound nitrogen can be mineralized to ammonium-nitrogen. The rate of conversion, however, depends on the nature of organic nitrogen and carbon:nitrogen ratio of the organic matter, electron acceptor availability, and environmental factors such as temperature. Mineralization also occurs in the underlying soil, and the ammonium-nitrogen formed can be subsequently released into the overlying water. Detailed reviews of the nitrogen processes functioning in the soil and the overlying water column of a wetland are presented by Bowden (1987), Johnston (1991), Reddy and D'Angelo (1994), and Reddy and Patrick (1984).

Nitrification involves the biological oxidation of ammonium to nitrate. The bacteria involved are chemoautotrophic and use oxygen as their electron acceptor, whereas ammonium is used as their substrate. In a floating aquatic macrophyte system, nitrification may occur in the water column and in the rhizosphere. Because the oxygen concentration of the water under floating plants typically is low, nitrification rates in these systems can be limited by oxygen supply.

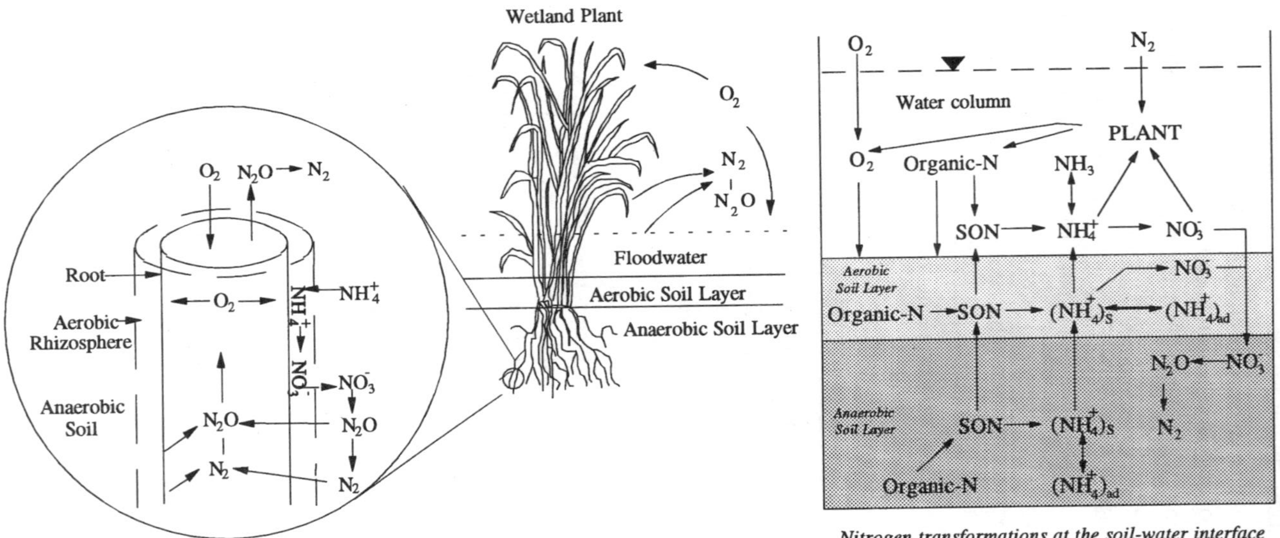


*Nitrogen transformations at the root-soil-water interface*

*Nitrogen transformations at the soil-water interface*

**Figure 10.6 Schematic showing the nitrogen transformations functioning in the soil and the water column of a wetland.**

In the absence of oxygen, facultative anaerobic microorganisms use nitrate as a terminal electron acceptor during their respiration (Firestone, 1982). During the denitrification process, nitrate is reduced to gaseous products such as nitrous oxide and nitrogen gas. Denitrification can potentially occur in the sediment, the water column devoid of oxygen, and the anoxic site of the rhizosphere. For example, in a water hyacinth system, detritus trapped in the root zone provides an ideal environment for denitrification.

Plant uptake of nitrogen is one of the main processes involved in the removal of nitrogen from wastewater. Removal of nitrogen through plant uptake will depend on the growth rate of the plant, culture density, and environmental parameters such as solar radiation and temperature. Aquatic plants are capable of assimilating both ammonium and nitrate. However, many aquatic plants appear to prefer ammonium over nitrate, even when both ions are present in the wastewater at the same time (Reddy and Tucker, 1983).

Vegetation can play a significant role by assimilating nitrogen into plant tissue and providing an environment in the root zone that supports nitrification–denitrification. Storage of nitrogen in plant biomass is highly variable, depending on the type of wetland (forested versus herbaceous). Nitrogen assimilation by herbaceous vegetation typically is short term; therefore, nitrogen cycles rapidly within the system. Unlike herbaceous macrophytes, forested wetlands provide long-term storage in the form of woody biomass. However, a significant portion of nutrients stored in the leaves is returned to the forest floor and eventually incorporated into the soil (Reddy and DeBusk, 1987). Forest litter decomposition is slow: the turnover periods are much longer. Plant uptake typically accounts for 16 to 75% of total nitrogen removal in wetlands (Reddy and DeBusk, 1987).

In addition to plant assimilation, pollutant removal in aquatic plant-based water treatment systems is affected by a number of biological, physical, and chemical processes functioning in the water, sediment, and rhizosphere (Good and Patrick, 1987, and Reddy, 1984). In studies on the fate of nitrogen added to floodwater, mass balances indicated that plant uptake accounted for 13 to 67% of the total nitrogen removal, while unaccounted-for nitrogen was assumed to be lost through nitrification and denitrification of $NH_3$ volatilization (Reddy, 1983, and Reddy and DeBusk, 1985).

Wetland plants have a unique capacity to transport oxygen to the root zone and create oxidized rhizosphere. The aerobic rhizosphere of wetland plants has been shown to support nitrification, whereas the adjacent anaerobic zone supports denitrification (Reddy *et al.*, 1989). These mechanisms explain overall nitrogen loss from wetlands not accounted for by plant uptake alone.

**PHOSPHORUS REMOVAL PROCESSES.** Phosphorus removal in wetlands and aquatic systems results from plant uptake, microbial immobilization into detritus plant tissue, retention by the underlying soil, and precipitation in the water column (see Figure 10.7). Because phosphorus is retained

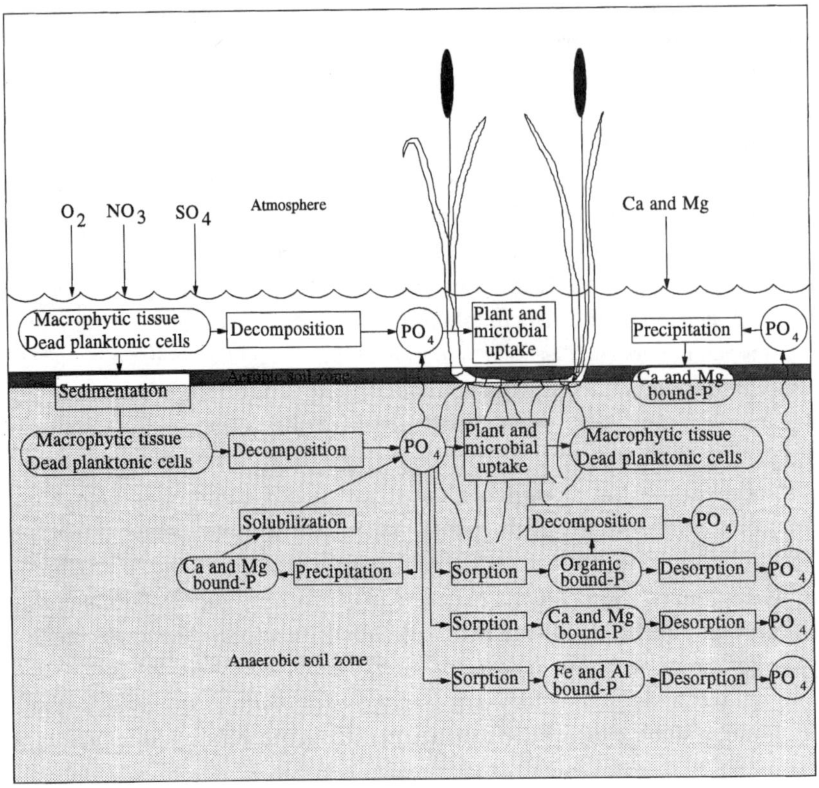


**Figure 10.7** **Schematic showing the phosphorus transformations in soil and water column of a wetland. (Reprinted from Reddy, K.R., and D'Angelo, E.M., Soil Processes Regulating Water Quality in Wetlands, In *Global Wetlands—Old World and New,* W. Mitsch [Ed.], Copyright 1994, with kind permission from Elsevier Science Ltd, The Boulevard, Langford Lane, Kidlington OX5 1GB, UK.)**

by the system components, the ultimate removal from the system can only be accelerated by harvesting the plants and dredging the sediments.

Phosphorus is present in organic and inorganic forms, of which inorganic forms are more mobile and more readily move from soil to the overlying water column. Unlike nitrogen and carbon, phosphorus cannot be lost from wetlands through metabolic processes (that is, there is no significant gaseous loss of phosphorus); thus, phosphorus tends to accumulate in the system. Wetland soils can function as a source or sink for the water column phosphorus. However, the amount of phosphorus retained or released by the soil depends on the phosphorus concentration of the overlying water column and associated biogeochemical processes functioning in the soil. These processes include adsorption/desorption reactions, precipitation, mineralization of organic phosphorus, and diffusion of phosphorus from the soil to the overlying water column and vice versa. It has been shown that inorganic phosphorus added at concentrations considerably greater than those present in the interstitial water of wetland soils may be retained by oxides and hydrous oxides of iron, aluminum, and calcium carbonate. At low phosphorus loadings, wetland soils have been found to release rather than retain phosphorus.

Several reviews have discussed the mechanisms of phosphorus retention in acid soils by aluminum and iron, and in alkaline soils by calcium and magnesium compounds (Berkheiser *et al.*, 1980; Bostrom *et al.*, 1982; Froelich, 1988; Khalid *et al.*, 1977; Logan, 1982; Reddy and D'Angelo, 1994; Sonzogni *et al.*, 1982; Syers *et al.*, 1973; and White, 1981). These reviews indicate that several phosphate minerals regulate the solubility of dissolved phosphorus in the interstitial waters, soils, and sediments. Some examples of these minerals include apatite, hydroxyapatite, fluorapatite, octocalcium phosphate, strengite, vivianite, variscite, and wavellite. Because of the complexity involved in mineral formation and solubility, it is difficult to attribute phosphorus retention to any single mineral.

Phosphorus retention by some wetland soils and sediments was shown to be directly correlated with aluminum (Harter, 1968; Richardson, 1985; and Shukla *et al.*, 1971). Results of batch incubation experiments (Reddy, 1990) conducted to determine the phosphorus buffering capacity of soils and sediments obtained from two wetland sites located in Lake Okeechobee Drainage Basin, Florida, indicated that phosphorus retention in these soils was proportional to oxalate extractable iron and aluminum (poorly crystalline and amorphous forms) as shown below:

$$S_{max} = 0.213\ 3 \text{ Oxalate } (Fe + Al) \qquad R^2 = 0.785 \tag{10.1}$$

Where

| | | |
|---|---|---|
| $S_{max}$ | = | phosphorus retention capacity, mmoles/kg; and |
| $(Fe + Al)$ | = | oxalate extractable iron and aluminum (amorphous and poorly crystalline forms), mmoles/kg. |

Oxidized forms of iron compounds such as ferric hydroxide, ferric oxyhydroxide, ferrous hydroxide, and amorphous oxides were shown to regulate soluble phosphorus concentrations in wetland soils and sediments. Hydrated iron oxides associated with aluminum and organic matter in the form of a gel complex were postulated to be the controlling sorption of inorganic phosphorus in lake sediments (Shukla *et al.*, 1971, and Williams *et al.*, 1971). Reduction of ferric oxyhydroxide to more soluble ferrous forms under low redox conditions was shown to increase overall phosphorus sorption relative to oxidized conditions (Patrick and Khalid, 1974). These results suggest that oxidation-reduction changes in wetland soils can significantly alter phosphorus solubility.

The conclusions made, based on previous research, are

- In acid soils, phosphorus is fixed as aluminum and ferric phosphates, if the activities of these cations are high;
- In alkaline soils, phosphorus fixation is governed by the activities of calcium and magnesium; and
- Phosphorus availability is greatest in soils with slightly acidic to neutral pH. The pH of wetland soils typically is near 7; thus, phosphorus retention by soils in wetland systems may be lower than that of upland soils.

Phosphorus retention by soils can only occur when inorganic phosphorus is in contact with the soil. In wetlands, phosphorus in the water column needs to diffuse into the soil where it can be retained by soil through adsorption and precipitation reactions. This happens only if the floodwater phosphorus concentration is higher than the soil pore-water phosphorus. Under these conditions, phosphorus retention by wetlands can be limited by diffusion. Similar to $NH_4$-N, phosphorus can be readily exchanged between soil and the overlying water column. The mobility of phosphorus between these two compartments depends on the concentration gradients, thickness of the aerobic soil–water interface, and physicochemical properties of the water column. In many wetlands, dissolved phosphorus concentration in soil pore water typically is higher than the phosphorus concentration in the overlying water column. Thus, net flux is from soil to overlying water column.

Uptake by algae and macrophytes may be a major phosphorus removal mechanism in these conditions. Water column phosphorus can be readily removed by algal uptake, followed by deposition of deal algal biomass on the soil surface. Similarly, vegetative uptake can provide either a short-term or long-term sink for phosphorus, depending on the type of vegetation present in wetland. Phosphorus stored in detrital plant tissue is rapidly released into the water column during decomposition. However, over long-term periods, significant portions of organic phosphorus may remain in the soil as a part of peat buildup in wetlands. These forms of phosphorus are relatively resistant to microbial breakdown and can be considered an important sink.

In calcareous systems, phosphorus released into the overlying water column can be precipitated as calcium-mineral-bound phosphorus, especially when the pH of the floodwater is altered diurnally by photosynthetic activities of algae. The steep gradients in pore-water phosphorus concentration result in the diffusive flux from soil to overlying water column, and the phosphorus released into the water column is either assimilated by planktonic organisms or precipitated with calcium minerals. In calcareous systems, coprecipitation of phosphorus with calcium carbonate was shown to be a dominant mechanism involved in the immobilization of soluble phosphorus. In the water column, where pH fluctuates diurnally, with high values during daytime (photosynthetic activity of algae) and low values during night (respiration by algae and bacteria), phosphorus can be precipitated and resolubilized (Diaz *et al.*, 1994).

# Design of Constructed Wetlands

Design of wetlands and aquatic plant systems for municipal wastewater treatment is relatively simple compared with that of conventional treatment systems. However, land costs and climatic limitations associated with these systems restrict the application of one design for all conditions. The design of constructed wetlands primarily consists of a shallow basin planted with aquatic plants typical to the site. A predetermined water depth optimal for plant growth is established and wastewater is provided at a specific hydraulic loading rate. However, wetlands are "open systems," which means the performance of the system is influenced by physical, chemical, biological, and climatic factors associated with site conditions. Many of these factors cannot be controlled or regulated. Thus, a "cookbook" approach to treatment wetland design cannot be developed without giving consideration to site-specific conditions and wastewater characteristics. Design information for wetlands and aquatic systems for wastewater treatment systems are described in detail by Lakshman (1979) and Reed *et al.* (1988). More recently, Kadlec and Knight (1995) have critically reviewed design criteria and proposed simple equations for designing FWS wetlands. In this chapter, we will summarize these design equations for carbon, nitrogen, and phosphorus removal by freshwater surface constructed wetlands.

More than 500 wetland systems currently are operating in North America for treating a variety of wastewaters, including municipal, industrial, mining operation, and stormwater. The largest FWS wetland treats approximately 49 200 $m^3$/d (13 mgd) and is located in Orlando, Florida. However, most wetland systems are small and serve towns with populations of less than 20 000. Many of these treatment wetlands are relatively young; therefore, a long-term database does not exist. In early research, natural wetlands were used to treat wastewater (Ewel and Odum, 1984). The Houghton Lake project in Michigan was designed and operated in 1978.

Several variables influence the design of wetlands for wastewater treatment. These include hydraulic loading (flow), depth of floodwater, area of wetland, soil type, vegetation, concentration of contaminant in the inflow water, and climatic conditions. Unlike conventional wastewater treatment systems, wetland systems are at their infancy, and the information available has not advanced enough to allow synthesis of short-term, small-scale, highly specialized studies into an accurate prediction of a fully developed and stabilized system (Kadlec, 1995).

Design equations developed by Kadlec (1993 and 1995) for major variables including inflow/outflow concentration, wetland area, detention time, hydraulic loading rate, and contaminant retention by wetlands are summarized in the following paragraphs. For additional information, the reader is referred to other sources (Kadlec and Knight, 1995, and Reed *et al.*, 1988).

**INFLOW/OUTFLOW CONCENTRATIONS.** Data available on constructed wetlands in North America and Europe primarily include information on inflows and outflows, with little or no information on other variables and internal processes regulating water quality. Using the North American Database (NADB), Kadlec and Knight (1995) developed simple empirical equations for predicting outflow concentration. Although these equations are easy to apply to estimate outflow concentrations, they should not be used in designing wetlands without considering other variables that control internal mass balances and processes. The equations presented are for FWS wetlands.

Biochemical oxygen demand:

$$C_o = 0.173C_i + 4.7 \qquad R^2 = 0.62; N = 440 \tag{10.2}$$

Where

$C_o$ = outflow concentration, mg/L (standard error in $C_o$ = 13. 6); and
$C_i$ = inflow concentration, mg/L;

Ranges over which this equation is applicable include 0. 27 $< q <$ 25.4 cm/d ($q$ = HLR, hydraulic loading rate); 10 $< C_i <$ 680 mg/L; 0.5 $< C_o <$ 227 mg/L.

Total nitrogen:

$$C_o = 0.409C_i + 0.122\,q \qquad R^2 = 0.48; N = 408 \tag{10.3}$$

Standard error in $C_o$ = 3.5; ranges over which this equation is applicable include 0.2 < HLR < 28. 6 cm/d; 0.4 $< C_o <$ 29.1 mg/L.

Ammonium-nitrogen:

$$C_o = 0.336q^{0.456}C_i^{0.728} \qquad R^2 = 0.44; N = 542 \tag{10.4}$$

Standard error in $C_o$ = 1.5; ranges over which this equation is applicable include 0.11 < HLR < 33.3; 0.04 $< C_i <$ 58.5 mg/L; 0.01 $< C_o <$ 58.4 mg/L.

Nitrate-nitrogen:

$$C_o = 0.093q^{0.745}C_i^{0.474} \qquad R^2 = 0.35; N = 0.553 \tag{10.5}$$

Standard error in ln $C_o$ = 1.6; ranges over which this equation is applicable include 0.017 $< q <$ 27.4 cm/d; 0.01 $< C_i <$ 24.5 mg/L; 0.01 $< C_o <$ 21.7 mg/L.

Total phosphorus:

$$C_o = 0.195q^{0.53}C_i^{0.91}$$
$$R^2 = 0.77;\ N = 373 \tag{10.6}$$

Standard error in ln $C_o$ = 1.0; ranges over which this equation is applicable include $0.02 < C_i < 0.20$ mg/L; $0.009 < C_o < 20$ mg/L; $0.1 < q_{\text{arg}} < 33$ cm/d.

The low correlation coefficient and high standard errors for the preceding equations demonstrate a high degree of the scatter in the NADB data. Considering this variability, application of these equations should be limited to the concentration ranges indicated.

**HYDRAULIC RETENTION TIME.** Increase in residence time of wastewater in the wetland often increases contaminant removal. The HRT can be estimated as follows:

$$\text{HRT} = \frac{AHn}{Q} \tag{10.7}$$

Where

$A$ = area of the wetland, m$^2$;
$H$ = water depth, m;
$Q$ = flow rate, m$^3$/d; and
$n$ = porosity or water column void fraction.

The value of $n$ for wetlands is typically > 0.95 (Kadlec, 1995). Vegetation density, detrital tissue accumulation, and concentration of suspended particles can affect the porosity in the water column.

**HYDRAULIC LOADING RATE.** The HLR is a function of both flow rate ($Q$) and wetland area ($A$) and is described as

$$q = Q/A$$

where $q$ = hydraulic loading rate, m/d.

The contaminant loading rate (CLR) is the product of hydraulic loading rate and concentration of the contaminant in the inflow water.

$$(\text{CLR})_I q_i C_i \tag{10.8}$$

Where

$\text{CLR}_I$ = inflow contaminant loading rate, g/m$^2$·d;
$q$ = inflow hydraulic loading rate, m/d; and
$C_i$ = inflow contaminant concentration, mg/L or g/m$^3$.

Similarly, outflow CLR can be described as

$$(\mathrm{CLR})_o = q_o C_o \tag{10.9}$$

Where

| | | |
|---|---|---|
| $(\mathrm{CLR})_o$ | = | outflow contaminant loading rate, g/m$^2$·d; |
| $q_o$ | = | outflow hydraulic loading rate (m/d) ($C_o$ = outflow contaminant concentration, mg/L or g/m$^3$). |

Net assimilation of contaminant by the wetland is given as

$$\mathrm{NCR} = (\mathrm{CLR})_I(\mathrm{CLR})_o \tag{10.10}$$

where NCR = contaminant removal rate, g/m$^2$·d.

A simple, time-averaged, steady-state, first-order model was used by Kadlec (1993) to describe contaminant assimilation rate:

$$J = kAC \tag{10.11}$$

Where

| | | |
|---|---|---|
| $J$ | = | contaminant removal rate, g/d; |
| $k$ | = | rate constant, m/d; |
| $C$ | = | contaminant concentration, g/m$^3$; and |
| $A$ | = | surface area of wetland, m$^2$. |

Equation 10.11 is rewritten as

$$\ln\left(\frac{C_o - C^1}{C_i - C^1}\right) = -\frac{k}{q} \tag{10.12}$$

where $C^1$ = background concentration (mg/L) ($C_o$, $C_i$, $k$, and $q$ are described in earlier equations).

By rearranging the above equation, and by converting all values on an annual basis, the area ($A$) needed for treatment can be calculated as follows:

$$A = \frac{Q}{k}\ln\left(\frac{C_i - C^1}{C_o - C^1}\right) \tag{10.13}$$

where $Q$ = water flow rate (m$^3$/d [365]); $k$ (m/a) and $A$ = area (m$^2$).

$$C_o = C^1 + (C_i - C^1)\exp\left(-\frac{kA}{Q}\right) \tag{10.14}$$

Average $k$ values estimated based on performance of several wetlands are as follows (Kadlec and Knight, 1995): BOD = 34 m/a; total N = 22 m/a; organic N = 17 m/a, and total P = 12 m/a. Coefficient of variation associated with those constants is approximately 50%. The average $C^1$ for wetlands has been reported at BOD = 7.7 mg/L; TN = 2 mg/L; organic N = 1.5 mg/L; and TP = 0.05 mg/L (Kadlec and Knight, 1995).

**DESIGN EXAMPLE.** Assume inflow wastewater concentrations ($C_i$) of BOD = 80 mg/L, total N = 30 mg/L, and total P = 10 mg/L; target concentrations are BOD = 15 mg/L, total N = 5 mg/L, total N = 5 mg/L, and TP = 1 mg/L. Design flow $Q$ = 1 000 $m^3$/d, water depth = 30 cm, and calculated HLR = 0.014 m/d. Using Equation 13, the areas required to achieve target concentrations are 24 600 $m^2$ (2.46 ha) for BOD; 37 000 $m^2$ (3.7 ha) for total N; and 71 400 $m^2$ (7.1 ha) for total P. For details on related calculations, see Kadlec and Knight (1995). It should be recognized that these calculations provide only estimates of preliminary sizing of the wetland to achieve targeted outflow concentration. The actual area should be determined from a pilot-scale project tailored to the wastewater treated. Design information collected on pilot projects can be used to optimize the treatment system for maximum efficiency.

The "$k$" values are a lumped constant that includes the cumulative effects of physical, chemical, and biological processes regulating contaminant retention in wetland. This model predicts annual average concentration. The uncertainty associated with these predictions could be in the range of 50 to 100%. This model also assumes negligible effects of rain and ET. The first-order model, however, accommodates several common observations made on the performance of wetlands for contaminant removal. These are summarized by Kadlec (1995) as follows:

- Increase in wetland area and reduction in hydraulic loading rate increases contaminant removal;
- Contaminants typically decrease in the direction of the flow, with values high at inlets;
- Contaminants are never reduced to zero levels; and
- Flow patterns in wetlands are intermediate between those of plug-flow and completely mixed reactors.

Final design of constructed wetlands should consider a number of factors including physical, environmental, and economic constraints (Mitsch and Cronk, 1992, and Richardson and Davis, 1987). Expectations of what wetlands can do should be recognized, and the effluent quality goals should be realistic. The simple approaches presented by Kadlec and Knight (1995) offer excellent first approximations for sizing the wetland and making preliminary estimates of effluent quality. Final wetland design should consider all site-specific conditions, including acceptable level of risk and regulatory requirements.

# REFERENCES

Armstrong, W. (1964) Oxygen Diffusion from the Roots of Some British Bog Plants. *Nature*, **204**, 801.

Bagnall, L.O., *et al.* (1987) Harvesting and Handling of Biomass. In *Aquatic Plants for Water Treatment and Resource Recovery*. K.R. Reddy and W.H. Smith (Eds.), Magnolia Publishing, Inc., Orlando, Fla.

Berkheiser, V.E., *et al.* (1980) Partitioning of Inorganic Orthophosphate in Soil-Water Systems. In *CRC Critical Reviews in Environmental Control*. CRC Press, Boca Raton, Fla.

Bostrom, B., *et al.* (1982) Phosphorus Release from Lake Sediments. *Arch. Hydrobiol. Beih.*, **18**, 5.

Bowden, W.B. (1987) the Biogeochemistry of Nitrogen in Freshwater Wetlands. *Biogeochemistry*, **4**, 313.

Boyt, F.L., *et al.* (1977) Removal of Nutrients from Treated Municipal Wastewater Vegetation. *J. Water Pollut. Control Fed.*, **49**, 789.

Brix, H. (1988) Light Dependent Variations in the Composition of the Internal Atmosphere of *Phragmites australis* (Cav.) Trin. ex Steudal. *Aquat. Bot.*, **30**, 319.

Brix, H. (1990) Gas Exchange Through the Soil-Atmosphere Interphase and Through Dead Culms of *Phragmites australis* in Constructed Reed Beds Receiving Domestic Sewage. *Water Res.* (G.B.), **24**, 259.

Brix, H., *et al.* (1992) Internal Pressurization and Corrective Gas Flow on Some Emergent Freshwater Macrophytes. *Limnol. Oceanogr.*, **37**, 1420.

Brooks, J.L., *et al.* (1984) Use of Peat for On-Site Wastewater Treatment: II. Field Studies. *J. Environ. Qual.*, **13**, 524.

Burgoon, P.S. (1993) Oxidation of Carbon and Nitrogen in the Root-Zone of Emergent Macrophytes Grown in Wetland Mesocosms. Ph.D. diss., Univ. Fla., Gainesville.

Carter, V., *et al.* (1978) Water Resources and Wetlands. In *Wetlands Functions and Values: the State of Our Understanding*. P.E. Greeson *et al.* (Eds.), Am. Water Resour. Assoc., Minneapolis, Minn.

Cary, P.R., and Weerts, P.G. (1984) Growth and Nutrient Composition of Typha Orientalis as Affected by Water Temperature and Nitrogen and Phosphorus supply. *Aquat. Bot.*, **19**, 105.

Cooper, P.F., and Findlater, B.C. (1990) *Constructed Wetlands in Water Pollution Control*. Pergamon Press, Oxford, Eng.

Dacey, J.W.H. (1981) Pressurized Ventilation in the Yellow Water-Lily. *Ecology*, **62**, 1137.

DeBusk, W.F, and Reddy, K.R. (1987) Removal of Floodwater Nitrogen in a Cypress-Mixed Hardwood Swamp Receiving Primary Sewage Effluent. *Hydrobiologia*, **153**, 79.

Diaz, O.A., *et al.* (1994) Solubility of Inorganic P in Stream Water as Influenced by pH and Ca Concentration. *Water Res.* (G.B.), **28,** 1755.

Dierberg, F.E., and Brezonik, P.L. (1984) The Effect of Wastewater on the Surface Water and Groundwater Quality of Cypress Domes. In *Cypress Swamps*. K.C. Ewel and H.T. Odum (Eds.), Univ. Fla. Press, Gainesville.

Dierberg, F.E., *et al.* (1987a) Removal of Copper and Lead Using a Thin-Film Technique. In *Aquatic Plants for Water Treatment and Resource Recovery*. K.R. Reddy and W.H. Smith (Eds.), Magnolia Publishing, Inc., Orlando, Fla.

Dierberg, F.E., *et al.* (1987b) Removal of Two Chlorinated Compounds from Secondary Domestic Effluent by a Thin Film Technique. *J. Environ. Qual.*, **16**, 321.

Engler, R.M., and Patrick, W.H., Jr. (1975) Stability of Sulfides of Manganese, Iron, Zinc, Copper, and Mercury in Flooded and Non-Flooded Soil. *Soil Sci.*, **119,** 217.

Ewel, K.C., and Odum, H.T. (1984) *Cypress Swamps*. Univ. Fla. Press, Gainesville.

Firestone, M.K. (1982) Biological Denitrification. *Agron. J.*, **22**, 289.

Froelich, P.N. (1988) Kinetic Control of Dissolved Phosphate in Natural Rivers and Estuaries: A Primer on the Phosphate Buffer Mechanism. *Limnol. Oceanogr.*, **33**, 649.

Gersberg, R.M., *et al.* (1983) Nitrogen Removal in Artificial Wetlands. *Water Res.* (G.B.), **17**, 1009.

Good, B.J., and Patrick, W.H., Jr. (1987) Root-Water-Sediment Interface Processes. In *Aquatic Plants for Water Treatment and Resource Recovery*. K.R. Reddy and W.H. Smith (Eds.), Magnolia Publishing, Inc., Orlando, Fla.

Graneli, W. (1984) Reed *Phragmites australis* (Cav.) Trin. ex Steudel as an Energy Source in Sweden. *Biomass*, **4**, 183.

Haggblom, M.H., *et al.* (1993) Influence of Alternate Electron Acceptors on the Anaerobic Biodegradability of Chlorinated Phenols and Benzoic Acids. *Appl. Environ. Microbiol.*, **59**, 1162.

Hammer, D.A. (1989) *Constructed Wetlands for Wastewater Treatment*. Lewis Publishers, Chelsea, Mich.

Harter, R.D. (1968) Phosphorus Adsorption Sites in Soils. *Soil Sci. Soc. Am. Proc.*, **33**, 630.

Johnston, C.A. (1991) Sediment and Nutrient Retention by Freshwater Wetlands: Effects on Surface Water Quality—CRC Critical Review. *Environ. Control*, **21**, 491.

Kadlec, R.H. (1989) Hydrologic Factors in Wetland Water Treatment. In *Constructed Wetlands for Wastewater Treatment*. D.A. Hammer (Ed.), Lewis Publishers, Chelsea, Mich.

Kadlec, R.H. (1993) Natural Wetland Treatment at Houghton Lake, Michigan: The First Fifteen Years. *Proc. 66th Annu. Conf. Water Environ. Fed.*, Anaheim, Calif.

Kadlec, R.H. (1995) Nutrient Removal in Constructed Wetlands: The Track Record. Paper presented at Interdisciplinary Conf. Animal Waste Land–Water Interface, Fayetteville, Ariz.

Kadlec, R.H., and Knight, R.L. (1995) *Treatment Wetlands*. Lewis Publishers, Boca Raton, Fla.

Kadlec, R.H., and Tilton, D.L. (1978) Wastewater Treatment via Wetlands Irrigation: Nutrient Dynamics. In *Environmental Quality Through Wetlands Utilization. Proc. Symp. Coordinating Council Restoration Kissimmee River Valley and Taylor Creek–Nubbin Slough Basin,* Tallahassee, Fla.

Khalid, R.A., *et al.* (1977) Phosphorus Sorption Characteristics of Flooded Soils. *Soil Sci. Soc. Am. J.*, **41**, 305.

Kickuth, R. (1977) Degradation and Incorporation of Nutrients from Rural Wastewaters by Plant Rhizosphere Under Limnic Conditions. In *Utilization of Manure by Land Spreading*, Comm. Eur. Commun., EUR 5627e, London, Eng.

Lakshman, G. (1979) An Ecosystem Approach to the Treatment of Wastewaters. *J. Environ. Qual.*, **8**, 353.

Logan, T.J. (1982) Mechanism for Release of Sediment-Bound Phosphate to Water and the Effects of Agricultural Land Management on Fluvial Transport of Particulate and Dissolved Phosphate. *Hydrobiologia*, **92**, 519.

Mahapatra, I.C., and Patrick, W.H., Jr. (1969) Inorganic Phosphate Transformations in Waterlogged Soils. *Soil Sci.*, **107**, 281.

Massacheleyn, P.H., *et al.* (1991) Arsenic and Selenium Chemistry as Affected by Sediment Redox Potential and pH. *J. Environ. Qual.*, **20**, 522.

Mitsch, W.J., and Cronk, J.K. (1992) Creation and Restoration of Wetlands: Some Design Consideration for Ecological Engineering. *Adv. Soil Sci.*, **17**, 217.

Mitsch, W.J., and Gosselink, J.G. (1986) *Wetlands*. Van Nostrand Reinhold, New York, N.Y.

Moshiri, G.A. (1993) *Constructed Wetlands for Water Quality Improvement*. Lewis Publishers, Ann Arbor, Mich.

Olson, R.K., and Marshall, K. (1992) The Role of Created and Natural Wetlands in Controlling Nonpoint Source Pollution. *Ecol. Eng.*, **1**, 1.

Pardue, J.H., *et al.* (1993) Assimilation of Hydrophobic Chlorinated Organics in Freshwater Wetlands: Sorption and Sediment-Water Exchange. *Environ. Sci. Technol.*, **27**, 875.

Patrick, W.H., Jr., and Khalid, R.A. (1974) Phosphate Release and Sorption by Soils and Sediments: Effect of Aerobic and Anaerobic Conditions. *Science*, **186**, 53.

Ponnamperuma, F.N. (1972) The Chemistry of Submerged Soils. *Adv. Agron.*, **24**, 29.

Reddy, K.R. (1983) Fate of Nitrogen and Phosphorus in Wastewater Retention Reservoir Containing Aquatic Macrophytes. *J. Environ. Qual.*, **12**, 137.

Reddy, K.R. (1984) Nutrient Transformations in Aquatic Macrophyte Filters Used for Water Purification. *Proc. Water Reuse Symp. III. Am. Water Works Assoc. Res. Found,*. San Diego, Calif.

Reddy, K.R. (1990) Phosphorus Retention Capacity of Stream Sediments and Associated Wetlands. Final Report submitted to South Fla. Water Manage. District, West Palm Beach.

Reddy, K.R., and D'Angelo, E.M. (1994) Soil Processes Regulating Water Quality In Wetlands. In *Global Wetlands—Old World and New*. W. Mitsch (Ed.), Elsevier, New York, N.Y.

Reddy, K.R., and DeBusk, W.F. (1985) Nutrient Removal Potential of Selected Aquaticmacrophytes. *J. Environ. Qual.*, **14**, 459.

Reddy, K.R., and DeBusk, W.F. (1987) Plant Nutrient Storage Capabilities. In *Aquatic Plants for Water Treatment and Resource Recovery*. M.A. Drew (Ed.), Magnolia Publishing, Inc., Orlando, Fla.

Reddy, K.R., and Patrick, W.H., Jr. (1984) Nitrogen Transformations and Loss in Flooded Soils and Sediments. *CRC Crit. Rev. Environ. Control*, **13**, 273.

Reddy, K.R., and Portier, K.M. (1987) Nitrogen Utilization by *Typha latifolia* L. As Affected by Temperature and Rate of Nitrogen Application. *Aquat. Bot.*, **27**, 127.

Reddy, K.R., and Smith, H.W. (1987) *Aquatic Plants for Water Treatment and Resource Recovery*. Magnolia Publishing, Inc., Orlando, Fla..

Reddy, K.R., and Sutton, D.L. (1984) Water Hyacinths for Water Quality Improvement and Biomass Production. *J. Environ. Qual.*, **13**, 1.

Reddy, K.R., and Tucker, J.C. (1983) Effect of Nitrogen Source on Productivity and Nutrient Uptake of Water Hyancinth (*Eichhornia crassipes*). *Econ. Bot.*, **37**, 236.

Reddy, K.R., *et al.* (1982) Water Treatment by an Aquatic Ecosystem: Nutrient Removal by Reservoirs and Flooded Fields. *Environ. Manage.*, **6**, 261.

Reddy, K.R., *et al.* (1989) Nitrification-Denitrification at the Plant-Root-Sediment Interface in Wetlands. *Limnol. Oceanog.*, **34**, 1004.

Reddy, K.R., *et al.* (1990) Oxygen Transport Through Aquatic Macrophytes: The Role in Wastewater Treatment. *J. Environ. Qual.*, **19**, 261.

Reed, S.C., *et al.* (1988) Natural Systems for Waste Management and Treatment. McGraw-Hill Book Co., New York, N.Y.

Richardson, C.J. (1985) Mechanisms Controlling Phosphorus Retention Capacity in Freshwater Wetlands. *Science*, **228**, 1424.

Richardson, C.J., and Davis, J.A. (1987) Natural and Artificial Wetland Ecosystems: Ecological Opportunities and Limitations. In *Aquatic Plants for Water Treatment and Resource Recovery*. K.R. Reddy and W.H. Smith (Eds.), Magnolia Publishing, Inc., Orlando, Fla.

Rogers, F.E.J., *et al.* (1985) *Wetlands for Wastewater Treatment with Special Reference to Municipal Wastewaters*. Witwatesrand Univ. Press, Johannesburg, S. Afr.

Seidel, K. (1976) Macrophytes and Water Purification. In *Biological Control of Water Pollution*. J. Tourbier and R.W. Pierson (Eds.), Univ. Pa. Press, Philadelphia.

Shukla, S.S., *et al.* (1971) Sorption of Inorganic Phoshate by Lake Sediments. *Soil Sci. Soc. Am. Proc.*, **35**, 244.

Small, M.M. (1976) Marsh/Pond Sewage Treatment Plants. In *Freshwater Wetlands and Sewage Effluent Disposal.* D.L. Tilson *et al.* (Eds.), Univ. Michigan, Ann Arbor.

Smith, W.H., and Frank, J.R. (1988) *Methane from Biomass: a Systems Approach.* Elsevier, New York, N.Y.

Sonzogni, W.C., *et al.* (1982) Bioavalability of Phosphorus Inputs to Lakes. *J. Environ. Qual.*, **11**, 555.

Spangler, K., *et al.* (1976) Experimental Use of Emergent Vegetation for the Biological Treatment of Municipal Wastewater in Wisconsin. In *Biological Control of Water Pollution.* J. Tourbier and W.P. Pearson (Eds.), Univ. Pa. Press, Philadelphia.

Stumm, W. (1966) Redox Potential as an Environmental Parameter: Conceptual Significance and Operational Limitation. Paper presented at 3rd Int. Conf. Water Pollut. Res., Water Pollut. Control Fed., Washington, D.C.

Syers, J.K., *et al.* (1973) Phosphate Chemistry in Lake Sediments. *J. Environ. Qual.*, **2**, 1.

Ulrich, K.E., and Burton, T.M. (1985) The Effects of Nitrate, Phosphate and Potassium Fertilization on Growth and Nutrient Uptake Patterns of *Phragmites australis* (Cav.) Trin. ex Steudel. *Aquat. Bot.*, **21**, 53.

White, R.E. (1981) Uptake and Release of Phosphate by Soil and Soil Constituents. In *Soils and Agriculture.* P.B. Tinker (Ed.), Halsted Press, John Wiley and Sons, New York, N.Y.

Wile, I., *et al.* (1985) Design and Use of Artificial Wetlands. In *Ecological Considerations in Wetlands Treatment of Municipal Wastewater.* P.J. Godfrey *et al.* (Eds.), Van Nostrand Reinhold Co., New York, N.Y.

Williams, J.D., *et al.* (1971) Fractionation of Inorganic Phosphate in Calcareous Lake Sediments. *Soil Sci. Soc. Am. Proc.*, **35**, 250.

Wolverton, B.C. (1987) Artificial Marshes for Wastewater Treatment. In *Aquatic Plants for Water Treatment and Resource Recovery.* K.R. Reddy and W.H. Smith (Eds.), Magnolia Publishing Inc., Orlando, Fla.

Young, J., and McCarty, P. (1969) The Anaerobic Filter for Wastewater Treatment. *J. Water Pollut. Control Fed.*, **41**, 160.

# Chapter 11
# Testing and Evaluation of Full-Scale Nutrient Removal Facilities



Nutrient removal from wastewater is becoming commonplace in many regions of the U.S. and abroad. To improve water quality, increasingly stringent nutrient limitations have been imposed on many dischargers. This challenge has led to considerable innovation in the field of nutrient removal in the last decade. The innovations and advances in this field are a direct result of research, piloting, testing, and evaluation of full-scale wastewater treatment facilities. This chapter provides an overview of the benefits, requirements, and techniques of testing and evaluation of full-scale nutrient removal facilities.

# BENEFITS OF FULL-SCALE TESTING

Bench- or pilot-scale testing can provide a wealth of data to aid in the modeling, design, and operation of nutrient removal systems. However, full-scale testing of nutrient removal technologies provides many benefits that bench- and pilot-scale testing cannot. Full-scale testing eliminates some uncertainties, such as "scale-up," encountered with bench and pilot studies. Under certain circumstances, full-scale testing may be simpler and less costly to implement than smaller-scale testing. Perhaps the most important benefit, however, is that full-scale testing typically provides a test of performance and problems that will be encountered under "real world" conditions. Full-scale tests are conducted outside of the controlled laboratory environment, and daily operation typically is performed by plant operators, not by scientists and engineers. The following are important benefits of full-scale testing of nutrient removal facilities:

- Validates or establishes process performance and capacity;
- Demonstrates new process configurations, enhancements, or operating modes;
- Compares alternatives for process selection or operation;
- Optimizes process performance ;
- Develops design and operating criteria;
- Trains operations and maintenance staff ;
- Verifies and calibrates process models; and
- Provides better definition of performance, thus may reduce design safety factors and tankage requirements.

# LIMITATIONS OF FULL-SCALE TESTING

Although there are significant benefits of full-scale testing of any waste treatment system, there are potential problems and limitations as well. Full-scale systems are considerably more difficult to control than bench- or laboratory-scale systems. Accuracy can be reduced by factors such as variation of flow and character of the wastewater; difficulties in measuring intraprocess flow splitting to tanks and clarifiers; and problems with accurate measurement of process variables such as air flow, solids concentrations, and recycle rates. These potential physical problems, combined with varying skill levels of operators and possible sampling and analytical difficulties, can produce potentially misleading data. The very reasons that make full-scale testing beneficial also make it a poor laboratory for precise measurement of process kinetics and growth rate determination.

Full-scale facility testing also can jeopardize process performance and effluent permit limitations. For example, making significant changes to the liquid treatment process can affect the character and settleability of the sludge, causing loss of solids in the effluent and possible operational problems such as biological foaming. Process changes and the incorporation of chemicals into the process train also can have a deleterious effect on solids handling (Murphy *et al.*, 1993).

Finally, full-scale testing can be taxing in terms of availability of plant staff and resources. Chemicals, energy, and facility modifications needed to accommodate the testing and the additional analyses typically required to analyze performance and operation can become costly.

# PROCEDURES FOR CONDUCTING FULL-SCALE TESTING

After the need for and potential benefits of full-scale nutrient removal testing are established and found to be acceptable, full-scale testing of a particular process or configuration may be warranted. This section provides an overview of the typical requirements and procedures for conducting full-scale testing. Planning, facilities and equipment, and process control and maintenance of testing conditions are addressed.

**TESTING PLAN.** A clear, concise testing plan is a prerequisite for a successful testing program. Well-defined and achievable goals and objectives must be established along with a plan of operation. After the goals and operating plans are established and the concepts are verified through process analysis, a sampling plan and testing schedule must be developed. "Reality checks" are required throughout the planning process to ensure that the testing program is adjusted to meet capabilities and financial constraints.

**SAMPLING PLAN.** The sampling program must provide the data needed to meet the objectives and goals of testing. Appropriate sample types must be taken at the appropriate point in the process train. Procedures for acquiring and preserving the samples must be carefully considered. Sufficient samples should be taken to ensure that primary experimental questions are answered and the data generated are statistically valid.

**PARAMETERS OF INTEREST.** Some of the physical, chemical, and biological parameters that are typically measured to characterize wastewater and process performance of nutrient removal systems are as follows:

- Biochemical oxygen demand (BOD), total and soluble;
- Chemical oxygen demand (COD), total and soluble;
- Volatile fatty acids (VFAs);
- Total Kjeldahl nitrogen—total and soluble, ammonia;
- Nitrite and nitrate;
- Total and soluble phosphorus;
- Solids—total and volatile;
- Alkalinity, pH;
- Temperature; and
- Dissolved oxygen.

Analyses of selected parameters at the appropriately sampled location throughout the plant or process can yield the data necessary to

- Characterize process influent and effluent;
- Determine process performance;
- Follow and establish the fate of selected parameters;
- Quantify mechanisms and rates of removal;
- Evaluate the stoichiometric relationship of various parameters (for example, alkalinity balance through a denitrification process can aid in the verification of process data and sampling and laboratory procedures); and
- Provide data needed to perform mass balances.

Because the character and quantity of organics dramatically affect rates in biological nutrient removal (BNR) systems, establishing the degradability of soluble and particulate forms of BOD and COD is beneficial. The relationship between many of these parameters (BOD/phosphorus, BOD/nitrogen) is useful for nutrient removal process selection, modeling, design, and operation. Procedures to determine biodegradability of the organic fractions and the relationship of these parameters to the nutrient removal processes are described in detail in Chapter 9.

**Sample Types and Locations.** Flow-weighted composite samples are used routinely for determining the average concentration of a particular constituent on a daily basis. Grab samples provide information on instantaneous conditions and can be useful in determining loading and performance variations. Grab samples also are of value in determining the fate of a constituent, or its concentration profile, through a tank or process. Grab samples also may be used for process control checks and determining averages of processes that do not vary significantly with time, such as certain solids-handling process recycles.

Sampling locations depend on the objectives of a program. Regulatory agencies typically require an assessment of the loading to and performance of a plant for routine parameters. A more extensive sampling program typically is required to acquire information for design, modeling, or process optimization. Examples of thorough sampling programs for analyses of nitrogen and phosphorus removal in a five-stage Bardenpho™ configuration are provided in the text by Randall *et al.* (1992).

**SCHEDULE.** Most systems require considerable time to equilibrate and produce consistent, reliable data. For example, a biological process typically requires three to four MCRTs to reach steady state for that operating condition (Lawrence and Brown, 1976). Implementing process changes to achieve a desired operating condition typically requires a series of gradual changes (typically no more than 10%) until the operating objective is achieved. In addition to providing adequate time for the testing considerations, the schedule should also provide flexibility and contingency to deal with unexpected problems and changes in the testing program and ensure safety at all times.

**TESTING FACILITIES AND EQUIPMENT.** Facilities and equipment used for nutrient removal testing can vary considerably depending on the objectives and goals of the testing program. In the case of biological treatment systems, physical options may include type of growth medium (fixed, suspended, or combination); environment (aerobic, anoxic, anaerobic, or combination); configuration of environments and hydraulic residence times; and control of substrate (for example, wastewater, recycle, methanol, or fermentation products). Major variables for chemical nutrient removal may include chemical type, water chemistry, point of application, and dose. Variables and process testing options are numerous. Equipment and facilities for testing may include chemical storage and feed facilities, pumping systems, mixing systems, tank configuration, and zone partitioning.

**PROCESS CONTROL AND MAINTENANCE OF TESTING CONDITIONS.** The extent of control of testing conditions should be compatible with testing objectives. For example, casual demonstration of a particular nutrient removal process may require considerably less control, data, and analysis than development of design criteria or establishing a capacity rating. Typically, test variables must be measured and maintained to the extent needed to meet the goals of the project with an appropriate level of confidence. A simple program may require only occasional testing of the effluent to determine whether a particular operating mode or process change is successful. Conversely, testing to determine the capabilities of facilities or to acquire a better understanding of a particular process or phenomenon may involve considerable effort.

Appropriate testing conditions help provide consistency, reproducibility, and reliability in the test data generated. To quantify or characterize performance, process variables need to be measured and controlled appropriately. The following are examples of process variables and conditions that may be required to be measured and accurately controlled during testing:

- Influent flow,
- Flow split to each unit process or reactor,
- Mixed liquor suspended solids and mixed liquor volatile suspended solids,
- Return sludge,
- Waste sludge,
- Internal recycles,
- Unit loading rates,
- Air flow,
- Oxidation-reduction potential, and
- Chemical dosage.

When possible, parallel treatment units should be used to conduct side-by-side testing. This affords the opportunity to investigate and compare alternative design/operating modes while eliminating variable conditions (such as loading, inhibitors, or temperature) that may affect or bias the outcome of testing done in a sequential manner. In addition, parallel testing units may provide an opportunity to control loading during the test period so that the experimental train can be operated at levels approaching or exceeding design conditions.

Where on-line instrumentation or flow measurement devices are not available, other methods can be used for estimating or determining average values. For example, estimates of flow or pumping capacity may be determined using volumetric methods such as either filling or drawing down a tank or using pump curves and discharge pressure readings. Open-channel flow may be measured using a weir or portable flume. Flow from chemical metering pumps can be measured at the pump using calibration columns or at the point of discharge using a bucket and stopwatch. Determination of flow splitting may be accomplished directly by flow measurement, through the use of a tracer, or by evaluation of operating data. The use of more sophisticated methods of plant evaluation and testing techniques, such as oxygen transfer, clarifier stress testing, and on-line monitoring, is presented in detail elsewhere (Daigger *et al.*, 1993).

# *EXAMPLES OF FULL-SCALE NUTRIENT REMOVAL SYSTEMS TESTING*

Examples of full-scale nutrient removal testing programs are provided in the following sections. Overviews of case studies for suspended-growth systems and fixed film systems are also provided. Suspended-growth systems are further divided into nitrogen removal systems, phosphorus removal systems, and combined systems. The full-scale testing case studies presented provide examples of various sizes of flow, technologies, and testing objectives.

**SUSPENDED GROWTH. Nitrogen Removal Process Testing.** Increased regulation of nitrogen discharges, particularly in coastal areas, has brought about a considerable number of full-scale testing programs of nitrogen removal techniques. Nitrogen removal can be costly to retrofit to existing secondary plants because of the large tankage requirements for nitrification. Full-scale testing can result in savings and improvements in technology because it provides valuable information to scientists, engineers, operators, and the regulatory community.

Examples of case studies of nitrogen removal testing at wastewater treatment facilities are shown in Table 11.1. The majority of the case studies presented in the table are demonstrations of nitrogen removal capability in existing tankage or relatively new process configurations. At two of the plants (Stamford, Connecticut, and Tallman Island, New York City), demonstration projects have been undertaken as part of the Long Island Sound Initiative. The Stamford plant was successful at reducing nitrogen in a number of operationally modified process configurations. Nitrogen reduction of close to 80% in warmer months was achieved without any plant modifications. At the 3.5-$m^3$/s (80-mgd) Tallman Island plant, nitrogen removal was accomplished in several step-feed configurations at relatively low hydraulic residence times. Some of the plant's plug-flow reactors were retrofitted with baffles and mixing equipment to create anoxic zones at various primary effluent feed points along the aeration basins (NEIETC, 1993).

Experiments at the Westport, Connecticut, wastewater treatment plant (WWTP) (also in the Long Island Sound watershed) using cyclic aeration in complete-mix mechanically aerated cells have been effective in removing an average of 60 to 70% of the influent nitrogen seasonally (Brink *et al.,* 1993). Many municipal WWTPs in the Long Island Sound area will be conducting low-cost retrofit projects for the removal of nitrogen (LISS, 1993).

In the Chesapeake Bay region of the U.S., biological nitrogen removal processes including the four-stage Bardenpho™ and Modified Ludzak-Ettinger configurations were tested at full scale and later implemented at Balti-

**Table 11.1 Nitrogen removal facilities testing.**

| | |
|---|---|
| **Stamford Water Pollution Control Facility—Stamford, Connecticut, U.S.** | |
| **Design flow (plant):** 75 000 $m^3/d$ (20 mgd)<br>**Test flow:** 68 000 $m^3/d$ (18 mgd)<br>**Process type:** Operational variations—anoxic/oxic, cyclic, step feed<br>**Testing objective:** Demonstration of nitrogen removal in various operating modes in existing facilities | **Features:** Operational variations in mechanically aerated 4-cell system with 6- to 7-hour HRT[a] resulted in total nitrogen removals of between 65 and 80%; cold weather nitrification/denitrification |
| **Tallman Island WWTP—New York, New York, U.S.** | |
| **Design flow:** 151 000 $m^3/d$ (40 mgd)<br>**Test flow:** 38 000 to 75 000 $m^3/d$ (10 to 20 mgd)<br>**Process type:** Step feed<br>**Testing objective:** Attempt nitrogen removal in limited tankage | **Features:** High-rate step feed process demonstration for nitrogen removal in low-strength wastewater ($BOD_5$:TKN[b] ratio $<2$); single and multiple anoxic zones achieved TN[c] removals of approximately 50 to 70% with HRT ranging from 4.2 to 7.1 hours |
| **Back River WWTP—Baltimore, Maryland, U.S.** | |
| **Design flow:** 680 000 $m^3/d$ (180 mgd)<br>**Test flow:** 102 000 $m^3/d$ (27 mgd)<br>**Process type:** Modified Ludzak–Ettinger (MLE) and four-stage Bardenpho™<br>**Testing objective:** Demonstrate several operating modes to determine process performance; establish design and operating criteria | **Features:** One-third of the existing plant was modified to operate in several nitrogen removal configurations; initial year of testing was promising with two technologies achieving TN less than 10 mg/L; plans call for the test facility to be operated for several years |
| **Parkway WWTP—Laurel, Maryland, U.S.** | |
| **Design flow:** 28 000 $m^3/d$ (7.5 mgd)<br>**Test flow:** 23 600 $m^3/d$ (6.0 mgd)<br>**Process type:** MLE and four-stage Bardenpho™<br>**Testing objective:** Demonstration, performance evaluation, and training | **Features:** Interim system operated during construction in an MLE and 4-stage Bardenpho™ modes; the removals exceeded 70% with 6- to 7-hour HRT; plant operates with a relatively low VSS[d] because of metal salt addition for P removal and treatment of water plant sludges |

*(continued)*

**Table 11.1 Nitrogen removal facilities testing (*continued*).**

**Chino Basin Municipal Water District Regional Plants No. 1 and No. 2—Rancho Cucamonga, California, U.S.**

| | |
|---|---|
| **Design flow:** NA<br>**Test flow:** NA<br>**Process type:** Anoxic/aerobic, mechanical surface aerators<br>**Testing objective:** Increase capacity and resolve operation problems | **Features:** Low-cost modifications to reliably nitrify and achieve 50% TN removal; modifications included reduction of BOD with $FeCl_3$ addition in primaries, equalizing recycles and septage, aerator motor modifications |

**Phoenix 91st Avenue Plant—Phoenix, Arizona, U.S.**

| | |
|---|---|
| **Design flow:** 567 000 $m^3/d$ (150 mgd)<br>**Test flow:** 115 000 $m^3/d$ (30 mgd)<br>**Process type:** Anoxic–aerobic<br>**Testing objective:** Upgrade to increase capacity and remove nitrogen; applied research | **Features:** High F:M[e] anoxic selectors for high-rate denitrification and improved SVI; successful denitrification in aerated anoxic zones; TN reductions of 75 to 80%; also conducted basic research at full scale—mass balances, kinetic and growth rate studies |

**Wolfsburg WWTP—Wolfsburg, Germany**

| | |
|---|---|
| **Design flow:** 14 500 $m^3/d$ (3.8 mgd)<br>**Test flow:** 14 500 $m^3/d$ (3.8 mgd)<br>**Process type:** Step-feed activated sludge<br>**Testing objective:** Demonstration of step feed for nitrogen removal | **Features:** Step-feed process achieved nitrogen removal ranging from 83 to 92% with a strong domestic wastewater (BOD—400 mg/L, TKN—55 mg/L); increased biological phosphorus also noted during operation; evaluated various increments of feed to three step-feed points |

**Kappala Sewage Treatment Plant—Liduingo, Sweden**

| | |
|---|---|
| **Design flow:** 150 000 $m^3/d$ (40 mgd)<br>**Test flow:** 25 000 $m^3/d$ (13 mgd)<br>**Process type:** Sludge reaeration with anoxic/aerobic contact zones<br>**Testing objective:** Year-round nitrification with 50% TN removal in winter | **Features:** Nine zones: 1 through 5 can be anaerobic, anoxic, or aerobic with internal recycle; use of sludge reaeration allows wintertime nitrification and approximately 50% TN reduction; 60 to 70% TN reductions are achieved in warmer months with other process combinations |

*(continued)*

**Table 11.1 Nitrogen removal facilities testing (*continued*).**

| | |
|---|---|
| **Landis WWTP—Vineland, New Jersey, U.S.** | |
| **Design flow:** 30 000 $m^3/d$ (8 mgd)<br>**Test flow:** 15 000 $m^3/d$ (4 mgd) (1 of 2 trains)<br>**Process type:** Modified Ludzak–Ettinger<br>**Testing objective:** Stress testing to verify design criteria and capacity | **Features:** On-line instrumentation for hydraulic measure, TSS, $O_2$ off-gas testing during 30-day peak month load test; high N load from food processing; plant met or exceeded design capacity |
| **Metro Wastewater Reclamation District Central Plant, North Complex—Denver, Colorado, U.S.** | |
| **Design flow:** 265 000 $m^3/d$ (70 mgd)<br>**Test flow:** 265 000 $m^3/d$ (70 mgd)<br>**Process type:** Anoxic/oxic<br>**Testing objective:** Capacity evaluation | **Features:** 60-day stress test to determine plant capacity; results used for process modeling and facilities planning |

[a] Hydraulic retention time.
[b] Total Kjeldahl nitrogen.
[c] Total nitrogen.
[d] Volatile suspended solids.
[e] Food-to-microorganism ratio.

more's Back River WWTP (Burns *et al.*, 1993) and the Washington Suburban Sanitary Commission's Parkway WWTP (Sadick *et al.*, 1992). Highlights from these testing programs, each of which can operate in various nitrogen removal configurations, are provided in Table 11.1. Plans are being made to conduct full-scale denitrification testing using methanol addition in suspended-growth reactors at the 13.5-$m^3/s$ (309-mgd) Blue Plains advanced WWTP in Washington, D.C. Nutrient-related testing has been conducted throughout the Chesapeake region for many years. The nutrient reductions mandated by the Chesapeake Bay Agreement should lead to further testing of both point and nonpoint nitrogen removal technologies.

On the western coast of the U.S., concern over ammonia toxicity and inorganic nitrogen are resulting in more stringent permit limitations. Full-scale testing at the Chino Basin Municipal Water District's Regional plants 1 and 2 confirmed that the plant could achieve a reasonable degree of nitrogen removal. This testing aided in the development of low-cost modifications to increase the capability and reliability of the plant to nitrify and achieve nitrogen removal (Drury *et al.*, 1992).

The Landis WWTP in Vineland, New Jersey, and the Metro Denver Wastewater Reclamation District Central plant in Colorado provide examples of testing to determine plant capacity for nitrogen removal under full-scale stress testing. On-line instrumentation was used to measure process flows, oxygen transfer, and clarifier performance in the Landis testing program. Results from the Metro Denver testing were used for model calibration and future facilities planning (Michael *et al.,* 1992).

The 91st Avenue WWTP in Phoenix, Arizona, has been the focal point of a number of full-scale tests over the past several years. Use of an oxic/anoxic configuration has been reported to effectively increase plant capacity by more than 40%. The plant is unique in that aeration is used in the anoxic zones for mixing and to partially satisfy oxygen uptake requirements. Work with aerated high food-to-microorganism (F:M) selectors for sludge settleability also has been conducted at this plant (Albertson and Hendricks, 1992). Full-scale testing also has been performed at the 91st Avenue plant to determine maximum nitrifier growth rate and kinetics (Stensel *et al.,* 1993).

In Europe, full-scale demonstration of nitrification and denitrification in relatively minimal tankage has been effectively demonstrated at the Wolfsburg WWTP in Germany (Kayser *et al.,* 1991). This testing focused on the demonstration and optimization of nitrification and denitrification in a step-feed mode of operation. Distribution of load among the step-feed points and the resulting effects on performance were primary objectives in this testing program.

Reactor basins at the Kappala WWTP in Liduingo, Sweden, were modified and configured as multiple operational zones capable of sustaining either aerobic or anoxic environments to test and evaluate various operational modifications. Operation in a contact stabilization mode permitted nitrification to occur in limited volume during the winter months at temperatures of 8 to 10°C (Palmgren, 1991).

**Phosphorus Removal Process Testing.** Although the role a particular nutrient plays in an aquatic environment is complex, conventional wisdom has focused on phosphorus reduction as a means of controlling eutrophication in freshwater systems (Randall *et al.*, 1992). Most phosphorus removal treatment methodologies use precipitation using iron or aluminum salts and, to a lesser extent, high pH treatment. During the last several years, however, enhanced biological phosphorus removal (EBPR) has become increasingly popular as an alternative to chemical precipitation. In addition to phosphorus removal to levels of 1 mg/L or less, the anaerobic selector effect of biological phosphorus systems can improve sludge settleability.

Most full-scale testing of phosphorus removal systems has focused on the themes of optimizing chemical addition, comparing EBPR to chemical addition, demonstrating and researching process performance, and producing and using fermentation products to enhance the performance of EBPR systems. Examples of such programs are presented in Table 11.2.

Metal salt addition for phosphorus control has been widely studied in the Chesapeake region of the U.S. Metal salt addition is used throughout this region. Plants along the Potomac River routinely achieve an effluent total phosphorus concentration of less than the required 0.18 mg/L. Many of the

**Table 11.2 Phosphorus removal process testing.**

| | |
|---|---|
| **Lower Potomac Pollution Control Plant—Fairfax County, Virginia, U.S.** | |
| **Design flow:** 137 000 $m^3/d$ (36 mgd)<br>**Test flow:** Various<br>**Process type:** Two-stage lime, metal salt addition<br>**Testing objective:** Evaluate lime versus ferric chloride; ferrous sulfite and impact on sludge production and handling | **Features:** In 1981, conducted full-scale side-by-side testing of ferric chloride versus two-stage lime treatment; following conversion to metal salts, full-scale testing was performed on process optimization and use of ferrous sulfite as an alternative to ferric chloride for phosphorus removal |
| **Rock Creek AWWTP—Hillsborough, Oregon, U.S.** | |
| **Design flow:** 75 700 $m^3/d$ (20 mgd)<br>**Test flow:** 53 000 $m^3/d$ (14 mgd)<br>**Process type:** Two-stage alum phosphorus removal<br>**Testing objective:** Verify pilot-scale tests and optimize full-scale operation | **Features:** 0.1 mg/L effluent TP[a] met using two-stage alum process; Al:P mole ratios of 1.4:1 in the primary and 15:1 in the tertiary were used; process optimization techniques were evaluated and refined costs were developed |
| **Cities of Ihtiman (Plant 1) and Omurtag (Plant 2) WWTPs—Ihtiman and Omurtag, Bulgaria** | |
| **Design flow:** Plant 1—12 000 $m^3/d$ (3.2 mgd); Plant 2—8 000 $m^3/d$ (2.1 mgd)<br>**Test flow:** Plant 1—10 000 $m^3/d$; Plant 2—68 000 $m^3/d$<br>**Process type:** Anaerobic/oxic<br>**Testing objective:** Demonstrate modified activated-sludge plants were capable of achieving effluent TP $< 1$ mg/L | **Features:** Parallel testing indicated that both plants were capable of achieving effluent phosphorus less than 1 mg/L 90% of the time when operated to maximize phosphorus removal; when operated in a nitrification mode, both systems achieved effluent TP of 2.6 mg/L or less 90% of the time |
| **Ypsilanti Community Utilities Authority—Ypsilanti, Michigan, U.S.** | |
| **Design flow:** NA<br>**Test flow:** 22 000 $m^3/d$ (5.8 mgd)<br>**Process type:** A/O™<br>**Testing objective:** Evaluate effectiveness of A/O™ retrofit with side-by-side testing and comparison with metal salt addition | **Features:** Parallel conventional activated-sludge and A/O™ trains operated to determine process performance with respect to TP, metals removal, and conventional parameters; also evaluated effects on $O_2$ usage and sludge dewatering; concluded that A/O™ could replace metal salt addition |

*(continued)*

**Table 11.2 Phosphorus removal process testing (*continued*).**

**Metropolitan WWTP—St. Paul, Minnesota, U.S.**

| | |
|---|---|
| **Design flow:** 950 000 $m^3d$ (250 mgd)<br>**Test flow:** 114 000 $m^3d$ (30 mgd)<br>**Process type:** Comparison of Enhanced Bio-P Removal (EBPR) using A/O™ and primary system metal salt addition<br>**Testing objective:** Evaluate performance and develop basis of design | **Features:** Used full-scale testing to investigate effects of chemical versus biological phosphorus removal; primary metal salt addition was shown to have a considerable negative effect on solids processing; design criteria and cost estimates were developed |

**Hyperion WWTP—Los Angeles, California, U.S.**

| | |
|---|---|
| **Design flow:** 110 000 $m^3/d$ (290 mgd)<br>**Test flow:** Various<br>**Process type:** Plug-flow activated sludge<br>**Testing objective:** Evaluate the impact of reduced MCRT on enhanced Bio-P removal | **Features:** As MCRT was reduced from 3 to 1.5 days, typical enhanced Bio-P removal was eliminated; effluent TP increased an average of 0.4 to 3.1 mg/L with reduced MCRT; struvite formation in digester piping was also eliminated |

**Orange Water & Sewer Authority Mason Farm Plant—Chapel Hill, North Carolina, U.S.**

| | |
|---|---|
| **Design flow:** 30 850 $m^3/d$ (8.15 mgd)<br>**Test flow:** 20 820 $m^3/d$ (8.15 mgd)<br>**Process type:** OWASA nutrified sludge<br>**Testing objective:** Demonstrate new biological phosphorus removal process | **Features:** VFA production from primary sludge fermenters mixed with RAS[b] in an anaerobic contractor; anaerobic contactor effluent blended with trickling filter process flow and introduced into aerobic cells for Bio-P uptake; subsequent anoxic zone removes nitrate |

**Northern, Bushkoppie, and Goudkoppies Works—Johannesburg, Republic of South Africa**

| | |
|---|---|
| **Design flow:** N/A<br>**Test flow:** N/A<br>**Process type:** Fermentation thickeners<br>**Testing objective:** Full-scale evaluation of elutriate from fermentation thickeners on Bio-P removal | **Features:** Acid fermentation and elutriation of fermentation products into the liquid process stream were shown to improve Bio-P removal at three plants; although increased nitrogen and phosphorus loadings were observed, significant savings in chemicals were realized; improved design of fermentation facilities also resulted |

[a] Total phosphorus.
[b] Return activated sludge.

facilities in the area have conducted full-scale tests to evaluate and optimize phosphorus removal. For example, at the Lower Potomac WWTP in Fairfax County, Virginia, several full-scale studies have been conducted related to phosphorus removal. In 1981, full-scale side-by-side testing was conducted at the Lower Potomac plant to compare multiple-point ferric chloride addition to the existing two-stage lime process (Canham *et al.,* 1981). As a result of the testing and substantial operations cost savings, the county switched to ferric chloride as the method of choice for phosphorus removal. In 1983, the Lower Potomac plant conducted full-scale tests using ferrous sulfate to determine phosphorus removal efficacy and its effects on solids handling (Hogge *et al.,* 1985). The Lower Potomac plant currently uses a multiple-point feed approach using both ferric chloride and ferrous sulfate. Full-scale metal salt optimization testing also has been conducted at the Blue Plains plant in Washington, D.C., and the Alexandria Sanitation Authority plant in Alexandria, Virginia.

In Oregon, concern over deteriorating water quality in the Tualatin River has led to total phosphorus limitations of 0.1 mg/L (seasonally). Several testing programs were conducted at the Unified Sewage Authority's Rock Creek plant. Full-scale tests demonstrated that a two-stage alum process was capable of reliably achieving a total phosphorus concentration of less than 0.1 mg/L. Testing also allowed process optmization and accurate cost estimates to be developed (Danzer, 1990, and Richwine *et al.,* 1992).

Testing of the Anaerobic/Oxic (A/O™) process has been conducted throughout the U.S. and abroad. Side-by-side testing of A/O™ systems and conventional chemical phosphorus removal has been conducted at Ypsilanti, Michigan (Kang *et al.,* 1991); the Metropolitan WWTP in St. Paul, Minnesota (Murphy *et al.,* 1993); and the wastewater treatment plants in the cities of Ihtiman and Omurtag, Bulgaria (Voutchkov, 1992). Typically, the use of the A/O™ process compared favorably to metal salt addition for meeting an effluent phosphorus concentration of less than 1 mg/L. Comparison of sludge production and handling characteristics was included in the Ypsilanti and Metropolitan plant studies. The features of each test program are presented in Table 11.2.

Full-scale testing of the secondary biological treatment process at the Hyperion WWTP in Los Angles, California, was conducted to examine the effect of mean cell residence time (MCRT) on EBPR (Mamais and Jenkins, 1992). The Hyperion plant has long plug-flow reactors that create an environment conducive to EBPR. When the MCRT of the system was reduced from 3 days to 1.5 days, enhanced removal of phosphorus ceased, thus indicating a washout time for the EBPR organisms of between 1.5 and 3 days under the operating conditions and temperature of the test.

Primary sludge fermentation products—VFAs—can be used to improve performance of biological phosphorus removal systems by providing readily degradable substrate for phosphorus-storing organisms. Full-scale testing of primary sludge fermentation and use of elutriated VFAs in the mainstream treatment train recently occurred at three plants in the Republic of South Africa (Lötter and Pitman, 1992). Significant phosphorus removal was noted at each of the plants tested. The results indicated that primary sludge fermentation is a viable alternative to chemical addition. Information obtained from these tests will be used for future plant retrofits.

In the U.S., full-scale testing was conducted by the Orange Water and Sewer Authority (OWASA) on an innovative process named the OWASA "Nutrified Sludge." In this process, primary fermentation products from fermentation reactors are mixed with return activated sludge in an anaerobic contactor. The contents are then blended with trickling filter effluent and introduced into a suspended-growth aerobic tank for bio-P uptake. The process was developed at OWASA's 30 000-$m^3$/d (8-mgd) Mason Neck plant in Chapel Hill, North Carolina (Kalb *et al.,* 1990). Full-scale testing of a primary sludge fermenter also was recently conducted in existing dissolved air flotation thickeners at the R.L. Sutton WWTP in Cobb County, Georgia (Daigger *et al.,* 1993). This testing provided data on the effects of MCRT on VFA production and on physical concerns and criteria for the design of a full-scale system.

**Combined Nitrogen and Phosphorus Process Testing.** Table 11.3 presents examples of plants where full-scale testing of nitrogen and phosphorus removal processes has occurred.

The Maryland Department of the Environment conducted extensive nutrient removal research and full-scale testing on six oxidation ditch systems in Maryland (Sen, 1992). The systems were reported to be capable of achieving effluent total phosphorus as low as 0.6 mg/L (with separate anaerobic tankage) and 4 mg/L total nitrogen. Multiparameter control approaches using dissolved oxygen, turbidity, alkalinity, and nitrogen forms were also investigated.

Full-scale performance testing of a sequencing batch reactor (SBR) was conducted at the Nottoway Correction Facility in Virginia. A similar test was conducted at the City of Ebensburg SBR facility in Pennsylvania. The testing was performed for regulatory acceptance and approval (Khararjian *et al.*, 1992).

The Eastern Service Area Facility in Orange County, Florida, and the Kelowna Works in British Columbia (Canada) are examples of five-stage Bardenpho™ plants where extensive full-scale testing has occurred. The Kelowna plant began operation in 1982. It was the first BNR facility located in Canada and the first to treat cold wastewater. Early experimentation with the use of fermentation products to improve nutrient removal also took place at Kelowna (Oldham, 1985). Parallel trains following primary clarification at Kelowna provide an ideal facility for process testing and optimization.

Compartmentalized tankage also provided an opportunity to investigate changes in hydraulic retention time in various zones. The flexibility in tankage and past optimization work recently led to the demonstration of a high-

**Table 11.3 Combined nitrogen and phosphorus removal facilities testing.**

**Bowie WWTP—Bowie, Maryland, U.S.**

**Design flow:** 7 600 $m^3$/d (2.0 mgd)
**Test flow:** 7 600 $m^3$/d (2.0 mgd)
**Process type:** Oxidation ditch
**Testing objective:** Applied research and demonstration of nitrogen and phosphorus removal

**Features:** Phased operating conditions and modifications to existing oxidation ditches were implemented to evaluate performance, develop control strategies, and conduct process research; similar work conducted at five other oxidation ditch plants in Maryland

**Nottoway Correction Facility Plant—Nottoway, Virginia, U.S.**

**Design flow:** 3 800 $m^3$/d (1 mgd)
**Test flow:** 570 $m^3$/d (0.15 mgd)
**Process type:** Sequencing batch reactors (ICEAS)
**Testing objective:** Demonstrate performance for state approval

**Features:** Three-month test of 4-hour cycle operation (1 hour fill, 2 hours aeration, 1 hour decant); 90% removal of TN and TP achieved

**Eastern Service Area WWT Facility—Orange County, Florida, U.S.**

**Design flow:** 45 400 $m^3$/d (12 mgd)
**Test flow:** 34 000 $m^3$/d (9 mgd)
**Process type:** 5-stage Bardenpho™
**Testing objective:** Rerating of plant capacity

**Features:** Stress testing of full-scale plant to evaluate nutrient removal performance, hydraulic capabilities, sludge settleability, and tertiary filter performance; concurrent kinetic bench studies were conducted; value of full-scale kinetic studies was limited

**Klagshamn Plant—Malmö, Sweden**

**Design flow:** 90 000 population equivalents
**Test flow:** NA
**Process type:** Combinations of chemical precipitation of phosphorus with predenitrification with wastewater and postdenitrification with methanol and ethanol
**Testing objective:** Evaluate process alternatives for retrofit

**Features:** On the basis of testing, process using precipitation of phosphorus with $FeCl_3$ and post-denitrification with methanol in summer and ethanol (faster denitrification rate) in winter was selected as most economical to meet annual average TN of 8 mg/L and TP of 0.3 mg/L

*(continued)*

**Table 11.3 Combined nitrogen and phosphorus removal facilities testing (*continued*).**

| Kelowna Treatment Plant—Kelowna, British Columbia, Canada | |
|---|---|
| **Design flow:** 22 500 $m^3/d$ (6 mgd)<br>**Test flow:** 22 500 $m^3/d$ (6 mgd)<br>**Process type:** 5-Stage Bardenpho™<br>**Testing objective:** Process optimization and upgrade of plant capacity | **Features:** First Canadian BNR works; parallel reactors after primary allowed side-by-side testing for process optimization and most recently comparison of a high-rate, three-stage system; past studies have included optimization of the production and application of fermentation products, tank configuration, RAS, and DO[a] control |
| **York River Plant—Yorktown, Virginia, U.S.** | |
| **Design flow:** 57 000 $m^3/d$ (15 mgd)<br>**Test flow:** 25 000 $m^3/d$ (6.5 mgd)<br>**Process type:** VIP/UCT™, A/O™, A2O™<br>**Testing objective:** Demonstrate and evaluate BNR process configurations | **Features:** Plant was configured to operate in four different BNR configurations sequentially; applied research was conducted and performance of each configuration was evaluated |

[a] Dissolved oxygen.

rate, three-stage configuration that will allow the plant capacity rating to be doubled.

Full-scale testing was recently conducted on the Eastern Service Area five-stage Bardenpho™ to measure the maximum treatment capacity of the biological treatment system (Reardon *et al.,* 1993). Hydraulic and process stress testing were conducted to establish firm capacity ratings and to develop an expansion plan for future upgrading. Concurrent kinetic testing was conducted in batch and continuous bench-scale reactors.

The Klangshamn plant in Malmö, Sweden, conducted testing to evaluate alternative process schemes for retrofitting the plant to meet new phosphorus and nitrogen standards. Ferric chloride precipitation was chosen for phosphorus removal. Postdenitrification with methanol or ethanol was found to be more cost effective than constructing and operating a predenitrification zone using wastewater as the carbon source. Bench-scale experiments indicated that denitrification rates using ethanol were considerably faster than with methanol. Ethanol, therefore, will be used during colder weather to optimize tank requirements (Andersson *et al.,* 1993).

Four alternative full-scale BNR configurations were tested at the Hampton Roads Sanitation District's York River plant, Virginia, in 1986. The process configurations included University of Cape Town (UCT™), VIP™, A/O™, and $A^2$/O™. The objectives of the program were to demonstrate the various BNR configurations and perform research in support of the Chesapeake Bay nutrient removal program. An emphasis of the York River plant testing program was evaluation of the mechanisms of EBPR and its performance under the effects of nitrification and relatively low-strength wastewaters (Waltrip, 1991).

**FIXED FILM PROCESSES.** The role of fixed film processes is continuously expanding in the field of nutrient removal from wastewater. Traditionally, fixed film processes were used for conventional secondary treatment. Recently, however, fixed film application has broadened to include separate-stage nitrification and denitrification; mixed or integrated media (combined fixed film in suspended growth); and compact, high-density treatment applications such as aerated biofilters and the fluidized beds. Because many of these applications are relatively new, much can be gained in terms of performance, design, and operating criteria from full-scale testing. Table 11.4 provides examples of full-scale tests or demonstration of fixed film integrated systems.

The first two projects listed in Table 11.4 are examples of integrated fixed film activated-sludge systems. Side-by-side full-scale testing of Ringlace (a multiple-stranded, polyvinyl chloride rope-type media) was conducted at the Annapolis WWTP in Maryland. Two single-sludge activated-sludge processes, both modified to obtain nitrogen removal, were operated in parallel. Ringlace media was installed in the aerobic zone of one of the process trains to compare performance. Enhanced volumetric rates of nitrification and denitrification were reported in the tank with Ringlace. The information obtained from the study will be used to establish design criteria and to aid in the development of an integrated fixed film activated-sludge system model (Sen *et al.,* 1993).

Another type of integrated system using polyethylene pellets in activated sludge was recently developed and tested in Japan. The proprietary system, named PEGASUS™, was demonstrated at a flow of 0.6 mgd. Nitrification and denitrification rates higher than those of single-sludge suspended-growth systems were reported (Takeshima *et al.,* 1993).

Full-scale tests of nitrifying trickling filters and an integrated fixed film activated-sludge system were conducted at the Sjölunda WWTP in Malmö, Sweden. Results of nitrification testing in crossflow trickling filter media showed high nitrification rates when temperatures exceeded 16°C. Nitrification efficiency decreased with temperatures below this threshold. Increasing the 3-m depth of media by 20% resulted in marginal improvement of nitrification. An identical trickling filter operated in parallel was used as a reference in these studies. In another test at the Sjölunda plant, parallel activated-sludge trains were modified to provide predenitrification with sludge recycle. One of the trains was outfitted with fixed beds with specific surface area of

**Table 11.4 Fixed film process testing.**

| Facility | Details | Features |
|---|---|---|
| **Annapolis WWTP—Annapolis, Maryland, U.S.** | **Design flow:** 37 850 $m^3$/d (10 mgd)<br>**Test flow:** 20 000 $m^3$/d (5.3 mgd)<br>**Process type:** Integrated fixed-film activated sludge (IFAS) system<br>**Testing objective:** Demonstration, applied research, development of design criteria | **Features:** Side-by-side tests with conventional activated sludge and IFAS using ringlace media; results indicated enhanced nitrification and denitrification volumetric rates; operational techniques investigated and data used for IFAS model development |
| **Japan Sewage Works Agency—Saitama, Japan** | **Design flow:** NA<br>**Test flow:** 2 250 $m^3$/d (0.6 mgd)<br>**Process type:** PEGASUS (Polyethylene Glycol Pellet Aided System)<br>**Testing objective:** Technology demonstration for nitrogen removal | **Features:** Pellet system in activated sludge achieved annual average TN removal of 70% to 8 mg/L with average HRT of 8 hours; nitrified recycle (at 2 to 5 times forward flow) by air-lift effect of aerobic zone, which raises water level high enough for recycle |
| **Sjölunda STP—Malmö, Sweden** | **Design flow:** 380 000 $m^3$/d (100 mgd)<br>**Test flow:** 121 000 $m^3$/d (32 mgd)<br>**Process type:** Nitrifying trickling filters and fixed bed submerged in activated-sludge basin<br>**Testing objective:** Evaluation of fixed-bed processes for improvement of nitrification and nitrogen removal | **Features:** Water temperatures down to 6°C; filters found to be temperature sensitive below 16°C; additional filter depth resulted in marginal improvement; submerged filters did not improve nitrification because of predation by protozoans |
| **Munich II WWTP—Munich, Germany** | **Design flow:** 216 000 $m^3$/d (57 mgd)<br>**Test flow:** 9 000 $m^3$/d (2.4 mgd)<br>**Process type:** Deep-bed denitrification<br>**Testing objective:** Evaluate cold weather performance, methanol breakthrough, and develop design criteria | **Features:** Filter cells modified to conduct testing; process can routinely achieve TN less than 10 mg/L with temperatures of 9°C; methanol breakthrough at feed rate of 3.5 mg/mg $NO_3$ and higher; continuous $NO_3$ analyzers used for process control |

*(continued)*

**Table 11.4 Fixed film process testing (*continued*).**

| Rambouillet WWTP—Paris, France | |
|---|---|
| **Design flow:** 12 000 $m^3/d$ (3.2 mgd)<br>**Test flow:** 1 000 $m^3/d$ (0.26 mgd)<br>**Process type:** Upflow aerated filter<br>**Testing objective:** Verify pilot; demonstrate prior to full-scale construction | **Features:** Evaluated nitrification, denitrification, and phosphorus removal in pilot and demonstration for development of performance, design, and operating criteria; 20 to 80% TN removal in 2- to 3-hour HRT |
| **Reno sparks WWTP—Reno, Nevada, U.S.** | |
| **Design flow:** 162 000 $m^3/d$ (43 mgd)<br>**Test flow:** 104 000 $m^3/d$ (28 mgd)<br>**Process type:** Trickling filter nitrification, fluidized-bed denitrification<br>**Testing objective:** Optimization of full-scale facilities | **Features:** Conducted full-scale testing to optimize hydraulic and process loading rates; trickling filter variables investigated included recirculation rates and natural and forced ventilation; for denitrification, methanol feed and bed control were objects of studies |

approximately 150 $m^2/m^3$. The purpose of the experiment was to determine whether the fixed bed could improve nitrification. The results of the test indicated that nitrification was not improved with the addition of the submerged media. Predation by protozoans in the downstream sections of the media where nitrification is thought to occur was presented as an explanation for this unexpected result. The SVIs of the integrated system were also found to be consistently lower than for the suspended-growth system (Andersson, 1993).

As part of a plant upgrade at the 57-mgd Munich II WWTP in Munich, Germany, filter cells were retrofitted with deep-bed denitrification filters for demonstration, development of design criteria, cost estimation, and comparison of treatment alternatives (Eichinger *et al.,* 1993). The filters were patterned after the units at the Howard F. Curren (Hookers Point) plant in Tampa, Florida. The results of the demonstration showed the process was effective at economically meeting the treatment objective of less than 10 mg/L total nitrogen at a water temperature of 9°C. Methanol dosage control using on-line nitrate analyzers was also investigated in the evaluation.

Full-scale demonstration of an upflow aerated filter with floating granular media (Biostr™) was used to verify basic kinetics and laboratory-scale performance at the treatment plant in Rambouillet, France (Toettrup *et al.,* 1993). One 22-$m^2$ filter unit was installed and operated in three modes: as tertiary nitrification only; as an anoxic zone where primary and nitrified effluents were mixed and treated; and finally as a complete nitrogen removal process using supplemental substrate addition. Following full-scale demonstration and verification of laboratory performance, several full-scale plants were constructed in Europe.

To meet stringent effluent requirements, nitrogen removal is accomplished at the Reno–Sparks Wastewater Treatment Facility in Sparks, Nevada, with nitrifying trickling filters and fluidized-bed denitrification reactors. Because operating information about these processes was somewhat limited, extensive full-scale optimization studies were undertaken at start-up of the plant. The nitrifying trickling-filter testing involved evaluation of liquid loading and recirculation rate along with an assessment of natural versus forced-draft ventilation. Forced-draft ventilation without recirculation was found to produce superior performance. Full-scale optimization of the fluidized-bed facility resulted in improved bed control and methanol feed techniques (Henderson and Porlein, 1989).

# REFERENCES


Albertson, O.E., and Hendricks, P. (1992) Bulking and Foaming Organism Control at Phoenix, AZ WWTP. *Water Sci. Technol.*, **26**, 461.

Andersson, B. (1993) Tentative Nitrogen Removal with Fixed Bed Processes in Malmö Sewage Treatment Plant. Int. Assoc. Water Pollut. Res. Control.

Andersson, B., *et al.* (1993) Nutrient Removal and Control at a Wastewater Treatment Plant in Malmö. *Proc. Eur. Water Pollut. Control Assoc. Int. Solid Waste Assoc. (EWPCA—ISWA) Symp.*, Munich, Ger.

Brink, W.P., *et al.* (1993) Full-Scale Complete Mix Nitrification/Denitrification Using Cyclical Aeration at Westport WPCF. *Proc. 66th Annu. Conf. Water Environ. Fed., Vol. 3, Liquid Treat. Processes,* Anaheim, Calif.

Burns, B.B., *et al.* (1993) Full-Scale Side-by-Side Testing of BNR Technologies. *Proc. 66th Annu. Conf. Water Environ. Fed., Vol. 3, Liquid Treat. Processes,* Anaheim, Calif.

Canham, R.A., *et al.* (1981) Full-Scale Evaluation of By-Product Ferric Chloride for Phosphorus Removal and Comparison with Two-Stage Lime Treatment. Paper presented at the 54th Annu. Conf. Water Pollut. Control Fed., Detroit, Mich.

Daigger, G.T., *et al.* (1993) Fermentation of Primary Sludge for Volatile Acid Production. 1993 Joint Can. Soc. Civ. Eng.–Am. Soc. Civ. Eng. Nat. Conf. Environ. Eng., Montreal, Que., Can.

Danzer, J. (1990) Rock Creek AWWTP Two-Stage Alum Process Phosphorus Removal Report. Prepared for Unified Sewage Authority, Oreg.

Drury, D.D., *et al.* (1992) How to Nitrify and Denitrify with Your Existing Facilities. *Proc. 65th Annu. Conf. Water Environ. Fed., Vol. 3, Liquid Treat. Processes Symp.,* New Orleans, La.

Eichinger, J., *et al.* (1993) Deep Bed Denitrification in Munich's (Ger.) Cold Wastewater. *Proc. 66th Annu. Conf. Water Environ. Fed., Vol. 3, Liquid Treat. Processes,* Anaheim, Calif.

Henderson, D.B., and Porlein, L.G. (1989) Pilot Testing and Full-Scale Operation of Fluidized Bed Biological Denitrification. Paper presented at the 62nd Annu. Conf. Water Pollut. Control Fed., San Francisco, Calif.

Hogge, A.H., *et al.* (1985) Full Scale Evaluation of Ferrous Sulfate for Phosphorus Removal and Its Impact on Solids Handling. Paper presented at the 58th Annu. Conf. Water Pollut. Control Fed., Kansas City, Mo.

Kalb, K.R., *et al.* (1990) Nutrified Sludge—An Innovative Process for Removing Nutrients from Wastewater. Paper presented at 63rd Annu. Conf. Water Pollut. Control Fed., Washington, D.C.

Kang, S.J., *et al.* (1991) Full Scale Demonstration of the A/O Process at the Ypsilanti Community Utilities Authority Wastewater Treatment Plant. Paper presented at 64th Annu. Conf. Water Environ. Fed., Toronto, Ont., Can.

Kayser, R., *et al.* (1991) Operational Results of the Wolfsburg Wastewater Treatment Plant. In *Design and Operation of Large Wastewater Treatment Plants, Proc. 6th IAWPRC Conf. Des. Oper. Large Wastewater Treat. Plants*, Prague, Czech.

Khararjian, H.A., *et al.* (1992) Removal in Intermittent Cycle Extended Aeration Systems (ICEAS®). Paper presented at 65th Annu. Conf. Water Environ. Fed., New Orleans, La.

Lawrence, A.W., and Brown, C.G. (1976) Design and Control of Nitrifying Activated Sludge Systems. *J. Water Pollut. Control Fed.*, **42**, 1779.

Long Island Sound Study Nutrient Removal Research Project (1993) Final Report. Prepared by New Engl. Interstate Environ. Training Cent./New Engl. Interstate Water Pollut. Control Comm.

Lötter, L.H., and Pitman, A.R. (1992) Improved Biological Phosphorus Removal Resulting from the Enrichment of Reactor Feed with Fermentation Products. *Water Sci. Technol.,* **26**, 943.

Mamais, D., and Jenkins, D. (1992) The Effects of MCRT and Temperature on Enhanced Biological Phosphorus Removal. *Water Sci. Technol.*, **26**, 955.

Michael, J.I., *et al.* (1992) Metro Wastewater Reclamation District North Complex Treatment Capacity Evaluation. *Proc. 65th Annu. Water Environ. Fed., Vol. 3, Liquid Treat. Processes Symp.,* New Orleans, La.

Murphy, S., *et al.* (1993) Full Scale Comparison of Primary Treatment, Solids Handling and Solids Incineration With and Without Chemical Precipitation for Phosphorus Removal. *Proc. 66th Annu. Conf. Water Environ. Fed., Vol. 3, Liquid Treat. Processes,* Anaheim, Calif.

New England Interstate Environmental Training Center (1993) Long Island Sound Nutrient Removal Research Project—Field Report.

Oldham, W.K. (1985) Full-Scale Optimization of Biological Phosphorus Removal at Kelowna, Canada. *Water Sci. Technol.*, **17**, 243.

Palmgren, T. (1991) Nitrogen Reduction—Volume Removal. *Water Sci. Technol,* **25,** 4/5, 233.

Randall, C.W., *et al.* (1992) *Designated Retrofit of Wastewater Treatment Plants for Biological Nutrient Removal.* Technomic Publishing Co., Inc., Lancaster, Pa.

Reardon, R.D., *et al.* (1993) Determining Nitrification and Denitrification Kinetics at a Full Scale Nutrient Removal POTW. *Proc. 66th Annu. Conf. Water Environ. Fed., Vol. 3, Liquid Treat. Processes,* Anaheim, Calif.

Richwine, R.D., *et al.* (1992) Phosphorus Removal at the Rock Creek Facility to Meet a 0.07 mg/L Total Phosphorus Standard. *Proc. 65th Annu. Conf. Water Environ. Fed., Vol. 3, Liquid Treat. Processes Symp.,* New Orleans, La.

Sadick, T.E., *et al.* (1992) Operational Experiences with BNR Facilities in the Chesapeake Bay and Long Island Sound Regions of the U.S.A. *Proc. Int. Conf. Nutrient Removal*, Leeds, G.B.

Sen, D. (1992) Case History: Design and Operational Features Determining Biological Nitrogen and Phosphorus Removal in Oxidation Ditches. Paper presented at Nutrient Control: Issues, Theory, Practical Solutions, Va. Water Pollut. Control Assoc., Richmond.

Sen, D., *et al.* (1993) Full Scale Evaluation of Nitrification and Denitrification on Fixed-Film Media (Ringlace) for Design of Single-Sludge Nitrogen Removal System. *Proc. 66th Annu. Conf. Water Environ. Fed., Vol. 3, Liquid Treat. Processes,* Anaheim, Calif.

Stensel, H.D., *et al.* (1993) Evaluation of Anoxic-Aerobic Treatment at the Phoenix 91st Avenue Plant. *Proc. 66th Annu. Conf. Water Environ. Fed., Vol. 3, Liquid Treat. Processes,* Anaheim, Calif.

Takeshima, T., *et al.* (1993) PEGASUS: An Innovative High-Rate BOD and Nitrogen Removal Process for Municipal Wastewater. *Proc. 66th Annu. Conf. Water Environ. Fed., Vol. 3, Liquid Treat. Processes,* Anaheim, Calif.

Toettrup, H., *et al.* (1993) The Treatment Trilogy of Floating Filters: From Pilot to Prototype to Plant. Paper presented at 2nd Int. Conf. Biofilm Reactors, Paris, Fr.

Voutchkov, N.S. (1992) Full Scale Phosphorus Removal Study at Two Modified Activated Sludge Plants. *Proc. 65th Annu. Conf. Water Environ. Fed., Vol. 3, Liquid Treat. Processes Symp.,* New Orleans, La.

Waltrip, G.D. (1991) Full-Scale Nutrient Removal Demonstration Project at York River Treatment Plant with the VIP Process. Paper presented at 45th Annu. Conf. Va. Water Pollut. Control Assoc., Williamsburg.

# *Chapter 12*
# *Costs for Nutrient Removal*



Considerable skill, knowledge, and experience are required to prepare accurate estimates of costs. Cost estimates are important considerations, and one should not be misled by the offhanded manner in which they can sometimes be treated. This chapter is intended to provide the reader information on the factors that influence the cost of nutrient control at typical wastewater treatment facilities. It cannot be a substitute for a carefully prepared estimate of cost based on a sound and thorough engineering evaluation. Often, what appear to be minor issues can have considerable cost effects for reasons that are not always well understood. Site-specific conditions that will be presented later can completely invalidate estimates derived from cost curves or empirical factors. This is especially true for nutrient control processes in which the influent characteristics, the climate, and the permit limits can have a dramatic effect on the selection and cost of the best or most desirable technology.

# ESTIMATING PROCEDURES

The technical literature appears to recognize several major categories of cost estimates depending on the state of development, definition of scope, and effort expended on the estimate. These are summarized in Table 12.1. The definitions by the American Association of Cost Engineers provide good guidelines. They consider three main categories, as follows:

- Order-of-magnitude estimates—this is an approximate estimate made without detailed engineering data. Examples include an estimate from cost-capacity curves, an estimate using scale-up or scale-down factors,

**Table 12.1 Types of capital cost estimates.**

| Estimate type | Cost basis | Purpose | Expected reliability |
|---|---|---|---|
| Order of magnitude | Cost capacity curves<br>Scale factors<br>Approximate ratios | Conceptual planning | +50% to −30% |
| Study | Flow diagrams<br>Layout drawings<br>Material and energy balances<br>Equipment type and size | Process selection<br>General evaluations<br>Guidance for further investigation | ±30% |
| Budget | Same as above<br>Soil and property surveys<br>Preliminary design of foundations, structures, architectural, electrical and instrumentation | Facility owner budget<br>Project authorization | +30% to −15% |
| Definitive | Plot plans and elevations<br>Piping diagrams<br>Instrumentation diagrams<br>One-line electrical diagrams<br>Structural sketches<br>Equipment data sheets and quotes | Improved estimate of project costs | +15% to −5% |
| Estimate of probable construction costs | Detailed surveys, engineering drawings and specifications | Control construction costs | ±5% |

and an approximate ratio estimate. It is typically expected that an estimate of this type would be accurate within +50% or −30%.

- Budget estimate—budget in this case applies to the facility owner's budget estimate and not to the budget as a project control document. A budget estimate is prepared with the use of flowsheets, layouts, and equipment details. It is typically expected that an estimate of this type would be accurate within +30% or –15%.
- Definitive estimate—as the name implies, this is an estimate prepared from defined engineering data. As a minimum, the data must include fairly complete plot plans and elevations, piping and instrument diagrams, one-line electrical diagrams, equipment data sheets and quotations, structural sketches, soil data, sketches of major foundations, building sketches, and a complete set of specifications. The "maximum" definitive estimate would be made from "Approved for Construction" drawings and specifications. It is expected that a definitive estimate would be accurate within +15% to -5%.

Capital costs for nutrient control projects include both the construction cost and other project-related costs. Project costs typically include many of the engineering, legal, and financial services listed in Table 12.2. The magnitude of these costs will vary with the size and complexity of the individual project and will typically range from approximately 14 to 20% of the construction cost.

**Table 12.2 Typical project (nonconstruction) costs.**

| Category | Typical percentage of construction cost | |
|---|---|---|
| | Low complexity | High complexity |
| Engineering services | | |
| Basic | 5–20 | 10–25 |
| Special | 0–2 | 0–2 |
| Soil surveys | 0.5 | 1.0 |
| Resident inspection | 2–6 | 6–12 |
| Total | 8–14 | 14–20 |
| Administration/legal services | | |
| Basic | 0–1.5 | 0–2 |
| Environmental | 0–1 | 0–5 |
| Total | 0–2 | 4 |
| Financial | | |
| Basic[a] | 0–12 | 0–15 |
| Interest during construction | 0–20 | 0–20 |
| Contingency | 5 | 10 |
| Total | 20–30 | 30–40 |

[a] Includes 10% debt service reserve.

# FACTORS

Attempts to accurately predict the costs of wastewater treatment facilities of any type are hampered by the fact that these costs are strongly influenced by a number of factors that are not directly related to the plant capacity, types of unit processes, or desired effluent quality. Some of these factors are beyond the control of the facility owners, including those that result from regulatory requirements and the market forces of a free economy. Many of these uncontrollable, subjective cost factors include the following:

- Regulatory design standards and mandated schedules,
- Conventions of engineering practice,
- Competitiveness of construction and supplier markets,
- Local labor and material costs,
- Construction time constraints,
- Condition of existing facilities,
- General economic climate,
- Extent to which cost is a controlling factor,
- Climatic conditions,
- Building code requirements,
- Fire code requirements,
- Seasonal and wet-weather load variations, and
- Land costs.

There are, however, a number of decisions made by the owner and the design professionals during the project design that have a significant effect on overall facility costs. Often these decisions are not made deliberately with full knowledge of their cost consequences, but rather are a consequence of habit, custom, and corporate standards. Some of these types of design cost decisions include

- Architectural design.
- Choice of bidding procedures.
- Materials selection.
- Corrosion control:
    - Coatings, and
    - Cathodic protection.
- Piping:
    - Materials,
    - Restraint, and
    - Details.

- Method of effluent disposal or reuse.
- Site work:

    - Plant layout,
    - Lighting,
    - Landscaping,
    - Roadways, and
    - Stormwater/site drainage.

- Structural design:

    - Buoyancy control, and
    - Water stops.

- Sludge treatment and disposal.
- Control systems.
- Monitoring systems.
- Electrical design:

    - Power distribution,
    - Emergency power, and
    - Length of wire runs.

- Mechanical design:

    - Equipment quality,
    - Installation details, and
    - Lubrication systems.

- Redundancy.
- Auxiliary systems:

    - Compressed air,
    - Fuel supply,
    - Chemical feed,
    - Service water, and
    - Tank/equipment drains.

- Support facilities:

    - Administrative space,
    - Maintenance facilities, and
    - Laboratory facilities.

Even a cursory review of the caveats and cost factors should sufficiently prove that obtaining a reliable estimate of costs is a difficult task, even for experienced professionals.

# IMPLICATIONS OF DESIGN FOR NUTRIENT CONTROL

The costs to implement nutrient control have been found to be highly variable and dependent on the influent wastewater characteristics (Kunihiro *et al.*, 1992, and Morales *et al.*, 1991). This is especially true at existing facilities. Some existing plants have been able to obtain relatively high nutrient removal efficiencies at minimal costs because of favorable influent characteristics and the design and configuration of the existing plant. Others have required significant capital expenditures.

The greatest cost effects will likely occur when nitrification is required, particularly for treatment plants in cold climates. At present, nitrification increases capital costs by requiring additional reactor volume and a larger aeration system. Increases in the operating costs for nitrification are mainly a result of increased power for aeration and, in some areas, supplemental alkalinity; however, this may be partially offset by increased oxygen transfer efficiency and, for long solids retention time (SRT) systems, possibly reduced sludge production.

The incremental cost for nitrogen removal, after nitrification is required, is not as large. Additional capital costs are required for anoxic reactors and, depending on the denitrification technology used, pumps and mixers or chemical feed systems. For suspended-growth processes, in some areas of the country, the capital costs for nitrogen removal may be partially offset by a reduction in aeration power costs and supplemental alkalinity. When the total Kjeldahl nitrogen (TKN):chemical oxygen demand (COD) ratio of the influent to the biological reactors is not favorable, additional capital and operating costs will be incurred to provide an external carbon source.

The capital costs for both chemical and biological phosphorus removal are relatively small. For enhanced biological phosphorus removal (EBPR), this is particularly true when the influent wastewater has a favorable biochemical oxygen demand ($BOD_5$):total phosphorus (TP) ratio. The operating costs, however, will vary widely depending on the influent characteristics and the method of removal being used. This is discussed further in the following sections.

The site-specific process design for nutrient control does affect the relative cost of a wastewater treatment plant (WWTP). The owner and design engineer should be cognizant of the ways in which effluent standards and the process design criteria for nutrient control can affect the capital and operating costs of a WWTP. This is especially important during the selection of a nutrient control technology and during final design. During process selection, design criteria can significantly alter the relative costs of different technologies in comparison with each other. Thus, it is important to use reasonable, accurate, and comparable design criteria. Failure to do this can result in selection of a process that will either not perform as intended or cost more to operate.

Overly conservative choices of process design criteria can also have the indirect effect of increasing the effective safety factor and, hence, the cost of a facility. Some of the basic design criteria, for which conservative assumptions are often made, can result in an improper design with respect to nutrient control. These often hidden safety factors can be included in

- The choice of design flow rate and wastewater strength;
- The selection of stoichiometric and kinetic coefficients such as the sludge yield, nitrifier growth rates, and temperature correction factors; and
- The choice of design effluent criteria.

**EFFECT OF REACTION RATE SAFETY FACTORS.** To size biological reactors, it is typically accepted design practice to calculate the reactor volume required based on the mass of substance to be removed and an appropriate reaction rate (or growth rate) for removal of the substance. This requires some engineering judgment on the part of the designer because the actual site conditions will likely vary in a significant range depending on a host of conditions that typically cannot be predicted precisely beforehand. To ensure proper operation in the range of possible conditions, it is prudent practice to use design safety factors. The use of safety factors, however, will increase the tank volumes beyond the absolute minimum requirement and will therefore increase the construction cost. There is sometimes a considerable degree of cost and competitive pressure to reduce the design safety factors to some perceived absolute minimum. The owner and designer should know what safety factors are being used to assess the associated costs or cost savings and the inherent risk when choosing a given safety factor.

**Nitrification.** Nitrification reactors typically are sized using an SRT or unit nitrification rate approach. In the SRT approach, the minimum theoretical SRT for nitrification is multiplied by a safety factor to obtain the design SRT. The estimated effect on the construction cost of the nitrification reactors caused by varying the safety factor between 1.0 and 2.0 increases the capital cost of the nitrification reactor approximately 50% as compared to no safety factor; however, this is typically 4 to 10% of the overall plant cost. The use of a safety factor to reduce the nitrification rate in the unit rate approach will have a similar cost effect because both design approaches are basically the same. The effects of a safety factor on the unit nitrification rate in the sizing of attached-growth systems for nitrification are somewhat different because there is still disagreement within the engineering community on the best design approach, and a higher degree of empiricism is required at this time.

**Denitrification.** Anoxic reactors for denitrification typically are sized using empirical unit rate design approaches, as presented in Chapters 4 and 7. Discrepancies do exist in some of the typically used empirical equations for predicting specific denitrification rates (SDNR) that will cause the design assumptions and use of safety factors to have different effects, depending on the approach used. Researchers in South Africa typically predict SDNRs using an equation that is zero order with respect to nitrate and carbon substrate concentrations and first order with respect to the estimated concentration of active biological mass. Their choice of rate constants is, however, dependent on the type of substrate present (that is, one constant is recommended when readily biodegradable substrate [RBCOD] is present, and another is recommended in the absence of RBCOD). The published South African denitrification rate constants represent the mean of a large number of individual rate determinations. In contrast, many plants in the U.S. have been sized using empirical equations that predict the SDNR as a function of the carbon substrate (applied $BOD_5$). Two different equations are used depending on the presence or absence of exogenous carbon substrate. Typically, engineering practice in both the U.S. and South Africa has been to not use safety factors on the denitrification rate, although the RSA guidelines suggest a safety factor between 1.1 and 1.2 be used to reduce the design SDNRs when doubt exists about the applicability of the suggested rates.

**EFFECT OF PREAERATION ANOXIC ZONES.** The need to remove nitrogen will impose additional capital costs and operating requirements on a WWTP. Under some circumstances it can be more economical to operate a suspended-growth nutrient control facility for both nitrification and nitrogen removal, even though only ammonia removal is required. Preaeration anoxic zones reduce operating costs for aeration power and chemicals for alkalinity supplements, and longer SRTs can result in reduced sludge production. These effects can offset the increased capital costs for anoxic reactors and equipment and additional power costs for mixing and recycle pumping. The capital cost items will include the preaeration anoxic reactors, the anoxic tank mixers, the mixed liquor recycle pumping and piping, and any chemical feed systems. Any reduction in operating costs will result from the recovery of oxygen and alkalinity that occurs with denitrification. The magnitude of the potential operating cost savings will vary according to the site-specific conditions. When there is a potential economic benefit from nitrogen removal in an activated-sludge process, it is suggested that the utility carefully investigate the expected costs at its facilities rather than assume that there will or will not be an advantage. The potential for savings will be greatest for wastewater with lower TKN:COD ratios and wastewater with insufficient natural alkalinity. The magnitude of the operating costs associated with supplemental alkalinity addition is affected by the type of chemical used to pro-

**Table 12.3 Comparative chemical costs for supplementing alkalinity.**

| | Hydrated lime | Quicklime | Soda ash | Caustic soda | Sodium bicarbonate | Magnesium hydroxide |
|---|---|---|---|---|---|---|
| Alkalinity added, mg $CaCO_3$/L per mg/L chem | 1.35 | 1.79 | 0.94 | 1.25 | 0.60 | 1.71 |
| Typical purity, % | 92 | 92 | 60 | 50 | 99 | 55 |
| Specific gravity | NA | NA | NA | 1.525 3 | NA | 1.49 |
| Density, lb/gal at 20°C | NA | NA | NA | 6.364 | NA | 12.5 |
| Typical bulk density, lb/cu ft[a] | 25 | 60 | 65 | NA | 60 | 93.7 |
| Truckload weight, lb[b] | 66 000 | 66 000 | 66 000 | 48 000 | 66 000 | 47 000 |
| Daily use[c,d] | | | | | | |
| lb 100%[e]/d[f] | 618 | 466 | 877 | 667 | 1 390 | 488 |
| tons[g] 100%[c]/d | 0.31 | 0.23 | 0.44 | 0.33 | 0.70 | 0.24 |
| lb bulk/d | 671 | 506 | 1 479 | NA | 1 404 | NA |
| tons bulk/d | 0.336 | 0.253 | 0.739 | NA | 0.702 | NA |
| cu ft bulk/d[h] | 27 | 8 | 23 | NA | 23 | NA |
| gal 100%/d[i] | NA | NA | NA | NA | NA | NA |
| gal solution·n/d | NA | NA | NA | 20 970 | NA | 70 |

[a] lb/cu ft × 16.02 = $kg/m^3 \cdot d$.
[b] lb × 0.453 6 = kg.
[c] Alkalinity required as calcium carbonate = 100 mg $CaCO_3$/L.
[d] Plant flow = 1 mgd.
[e] 100% = pure chemical.
[f] lb/d × 0.453 6 = kg/d.
[g] tons × 1.016 = MG.
[h] cu ft/d × $(2.832 \times 10^{-2})$ = $m^3$/d.
[i] gal/d × $(3.785 \times 10^{-3})$ = $m^3$/d.

vide the alkalinity. This is shown in Table 12.3. It is apparent that the use of certain chemicals can be much more expensive than others in many cases. For example, the use of lime can result in a large cost advantage over caustic soda when a significant amount of alkalinity is required. The individual designer and facility owner must evaluate the perceived advantage of ease of use of a specific compound in relation to the expected cost. The person doing such cost comparisons should include the expected changes in the influent TKN and alkalinity that will result from solids-handling sidestreams especially when anaerobic digestion is used. Any possible changes in solids-handling cost is harder to generalize and should be estimated on a case-by-case basis. The use of supplemental chemicals can increase sludge production slightly because of inerts and impurities in the chemical. Typically, sludge quantities will not increase because of the formation of calcium or magnesium carbonates. At the relatively low pH values (≤7–8) in the aeration tanks and low alkalinity doses, bicarbonate will be the principal anion, and the solubility of the typical calcium salts will not be exceeded.

**BIOLOGICAL VERSUS CHEMICAL PHOSPHORUS REMOVAL.** Design practices for EBPR are still evolving. The design procedures presented in Chapters 5 and 8 are still largely empirical and based on anaerobic hydraulic detention time or mass fraction. As yet, there is no well-defined concept of safety factor for this function because the relative size of the required anaerobic zone is small (0.5 to 2.5 hours nominal hydraulic retention time), the cost of being conservative is not great, and apparently it is not harmful to process efficiency up to a certain point. Randall's work (Randall *et al.*, 1992) has shown that anaerobic hydraulic retention times greater than 1.5 to 2.0 hours can reduce the readily biodegradable COD enough that the denitrification rates in a downstream anoxic reactor will be reduced.

The costs for biological or chemical methods for phosphorus removal can be heavily affected by the $BOD_5$:TP ratio of the influent wastewater. The economics will shift from chemical to biological phosphorus removal as the concentration of phosphorus to be removed increases and as the $BOD_5$:TP ratio increases. All biological systems remove some phosphorus as required for bacterial growth. As the concentration of phosphorus to be removed increases, a point will be reached at which typical metabolic needs are insufficient to remove the required amount. Additional measures are then required. When the incremental concentration of enhanced phosphorus removal needed is small, the costs for chemical removal can be less than those associated with EBPR. This is a result of the low capital cost for chemical feed systems, the low dosage of chemical required, and the small incremental increase in sludge production. With increasing requirements for phosphorus removal, the required chemical doses and sludge production become larger until a point is reached when it is cheaper to use EBPR. For large amounts of phosphorus removal, this will remain true even if supplemental chemical addition is required to meet the absolute discharge standard. Supplemental chemical addition may take the form of commercially available metal salts or the on-site production of volatile fatty acids. Higher influent $BOD_5$/phosphorus ratios and higher effluent phosphorus concentrations favor the use of EBPR. The use of primary sedimentation will tend to increase the cost of phosphorus removal because it will decrease the influent $BOD_5$ concentration and, hence, the quantity of phosphorus removed through typical sludge production. The use of biological phosphorus removal will also be favored at locations where the incremental cost of sludge handling and disposal is high.

**EFFECT OF NUTRIENT REMOVAL (LONG SOLIDS RETENTION TIME OPERATION) ON AERATION POWER COSTS.** It has been suggested that nitrification, or more directly the long SRTs required for nitrification, will result in a significant increase in the oxygen mass-transfer coefficient in process water as compared to clean water (U.S. EPA, 1989). This results in an increase in the oxygen-transfer efficiency and helps to offset the increased power cost associated with nitrification. Many studies of the effect of organic loading on oxygen transfer have shown some inverse correlation between the two parameters. It must be noted, however, that a wide variety of things affect the value of the oxygen-transfer efficiency (Stenstrom and Gilbert, 1981), and the relative effects of different parameters have been so far impossible to completely identify and quantify. It is suggested that it would be unwise to rely too much on this phenomenon in any economic analysis of the cost of nitrification.

**SELECTION OF OPTIMUM MIXED LIQUOR SUSPENDED SOLIDS CONCENTRATION.** For suspended-growth nutrient removal processes, the design mixed liquor suspended solids (MLSS) concentration selected can have a significant effect on the capital cost of the aeration tanks and final clarifiers. It is generally accepted that there is an optimum MLSS that will reduce the combined capital cost of aeration tanks and clarifiers when the clarifier capacity is thickening-limited (van Haandel, 1992). A final clarifier is thickening-limited whenever the MLSS concentration is greater than some minimum concentration; when it is less than the concentration at the minimum in the total solids flux curve; and when the return sludge concentration is greater than the critical concentration (van Haandel, 1992). It can be shown that the optimum MLSS will increase with the higher SRTs typically required for nutrient removal (Pincince *et al.,* 1997). Nutrient removal processes can improve sludge settling properties when the anaerobic or anoxic zones function as biological selectors. This improvement in settling velocity will also change the optimum MLSS for minimum capital cost. As sludge settling properties improve, the cost of the clarifiers will decrease and the optimum MLSS will also increase (Pincince *et al.*, 1997).

**EFFECTS OF SEASONAL NUTRIENT REMOVAL ON CAPITAL AND OPERATING COSTS.** Seasonal nutrient control limits can affect both the capital and operating costs. Capital costs will be affected to the extent that the method of nutrient control is affected by seasonal changes in wastewater temperature and changes in the wastewater strength resulting from seasonal changes in infiltration and inflow. Suspended-growth nitrification and all denitrification processes are directly and severely affected by decreases in wastewater temperature as presented in Chapters 4, 6, and 7. Attached-growth nitrification typically is not temperature limited, and EBPR has been reported to sometimes improve with decreasing temperature. The capital costs for activated-sludge nitrification will increase with decreasing design temperature to the extent that the tank costs increase with the required increases in aeration reactor volume. The choice of design temperature will not have an effect on the incremental operating costs for nitrification; however, actual operating costs will change with temperature in response to changes in oxygen demand and sludge production. With cooler wastewater temperature the sludge production typically will increase while oxygen consumption will decrease.

The capital costs for denitrification will also increase with decreasing design wastewater temperature as a result of decreases in the denitrification rate. Operating costs should not be expected to change with changes in temperature.

When chemical addition is required for nutrient control, the annual operating costs will change in direct proportion to the number of days that chemical supplements are required. This can affect the selection of a nutrient control technology particularly for phosphorus removal.

# *COSTS FOR EXISTING BIOLOGICAL NUTRIENT REMOVAL FACILITIES*

Historical construction and operating costs for nutrient control WWTPs can provide useful information to environmental planners, engineers, and utility managers. At the very least they provide an approximate measure of the costs of these types of technologies in comparison with more conventional methods. There are problems in using historical data for conceptual or preliminary level planning. As discussed previously, costs are influenced by many factors, and when examining historical costs it may be impractical to know all the things that have influenced a particular cost. Historical costs also tend to represent the way the profession used to do things and do not take into account recent improvements in processes and equipment. To compare historical costs from different times, it is necessary to use cost indices to try to adjust the numbers to a common base. Improved accuracy can be obtained by using other factors to adjust prices from different locations for regional differences in the costs of labor and supplies. These indices have many inherent assumptions that may or may not apply to a particular facility. Thus, the adjustment process will introduce additional unknowns and inaccuracies into original cost figures.

**SURVEY.** A survey of full-scale WWTPs using BNR processes was conducted in 1991 and 1992 (Reardon, 1994, and Reardon and Maurer, 1991) and updated in 1995. This work collected cost design information on BNR facilities that could be identified from the literature, patent and license holders, and personal experience. Technical information requested included the type of BNR process, permit limits, and basic design criteria. Cost data requested included bid prices, schedules of values, actual construction costs, and annual operating expenses or budgeted operations and maintenance expenses. A total of 272 BNR plants were identified in North America. Of the total, 163 have a design capacity under 18 900 $m^3/d$, 80 are between 19 000 and 76 000 $m^3/d$ capacity, and 29 have a design capacity greater than 76 000 $m^3/d$. The frequency of use of the typical BNR processes found in this survey is summarized in Table 12.4.

**Capital Costs.** All the data collected on capital costs and design criteria are summarized in Tables 12.5 and 12.6. The capital cost data were analyzed

**Table 12.4 Summary of BNR WWTPs by process type.**

| Process type | Number in operation[a] | Process type | Number in operation[a] |
|---|---|---|---|
| Modified Bardenpho™ | 34 | Oral simpre | 10 |
| Modified Ludzack–Ettinger | 32 | Multiple sludge | 9 |
| | | PhoStrip™ | 8 |
| Sequencing batch reactor In-channel oxidation | 28 | Other anoxic/aerobic processes | 8 |
| Ditch denitrification | 26 | Bardenpho™ | 7 |
| Schreiber | 23 | Phase isolation ditch | 7 |
| A/O™ | 20 | Intermittent aeration | 4 |
| Denitrification filters | 20 | UCT/VIP™ | 3 |
| $A^2/O$™ | 18 | Immobilized cells | 3 |
| Oxidation ditch with integral anoxic zone | 15 | Submerged packed bed | 2 |
| Total 277 | | | |

[a] As of 1995.

**Table 12.5 Summary of design SRT, MLSS, and low temperature criteria.**

| | Design SRT, days | | | Design MLSS, mg/L | | | Design low temperature, °C | | |
|---|---|---|---|---|---|---|---|---|---|
| **Process type** | **Low** | **Average** | **High** | **Low** | **Average** | **High** | **Low** | **Average** | **High** |
| A/O™ | 5 | 8.5 | 15 | 2 500 | 3 150 | 4 000 | 6 | 10 | 12 |
| $A^2$/O™ | 2.7 | 9.3 | 17 | 1 400 | 2 150 | 4 000 | 9 | 14.6 | 20 |
| Bardenpho™ | 10 | 19.5 | 35 | 2 200 | 3 915 | 5 200 | 10 | 16.4 | 20 |
| Modified Ludzak–Ettinger | 8 | 18.3 | 25 | 3 200 | 4 290 | 5 200 | 18 | 19.4 | 20 |
| Other | 5 | 15.8 | 25 | 1 800 | 2 633 | 3 500 | 10 | 13.4 | 20 |

against a number of parameters in attempts to develop some reasonable correlations. The best fits with linear regression were found for total capital cost versus design flow. Regression lines were fit to the total capital cost data using equations of the following forms:

$$C = K_1 Q_D^M + K_2 \tag{12.1}$$

Or

$$C = \frac{1}{K_1/Q + K_2} \tag{12.2}$$

Where

$C$ = capital cost, dollars;
$K_1$ = empirical base cost factor;
$M$ = empirical cost exponent;
$K_2$ = empirical cost factor; and
$Q$ = design flow.

**COMPARISONS WITH PREVIOUS WORK.** The correlation between capital costs and the BNR plant design capacity was compared with similar curves from other studies by updating the other historical curves with the ENR construction cost index. It can be seen that the construction costs for BNR WWTPs are not significantly different from those of WWTPs designed and constructed during the U.S. Environmental Protection Agency construction grants program of the 1970s.

The conclusions that can be drawn from this survey with respect to capital costs are

- Actual capital costs for WWTPs using BNR technology are comparable to historical capital costs for secondary treatment. It appears that the advances in BNR technology indeed allow municipal utilities to do

**Table 12.6 Summary of reported BNR capital costs.**

| Process type | Number of plants | $R^2$ | Best-fit equation | Upper 90 percentile equation |
|---|---|---|---|---|
| Two-stage (A/O™ and modified Ludzak–Ettinger) | 20 | 0.800 | $\$ = 2.0 \times 10^4 Q^{0.647}$ | $\$ = 902\ 3 \times Q^{0.779}$ |
| Four- & five-stage (Bardenpho™) | 17 | 0.765 | $\$ = 1/[(1.02 \times 10^{-3}/Q) + 4.1 \times 10^{-8}]$ | $\$ = 1/[(7.62 \times 10^{-4}/Q) + 6.14 \times 10^{-8}]$ |
| Upgraded plants | 7 | 0.779 | $\$ = 240Q + 3.34 \times 10^6$ | $\$ = 355Q + 7.51 \times 10^6$ |
| Expanded and upgraded plants | 23 | 0.615 | $\$ = 395Q + 5.70 \times 10^6$ | $\$ = 513Q + 8.97 \times 10^6$ |
| New plants | 44 | 0.950 | $\$ = 636Q + 2.88 \times 10^6$ | $\$ = 9\ 690Q + 5.70 \times 10^6$ |

[a] $Q$ = Design capacity in $m^3/d$.
[b] ENR Index = 5 000.

Table 12.7 **Summary of reported BNR operations and maintenance costs.**

| **Reported unit operating costs, \$/100 m³ (\$/1 000 gal)** | | | |
|---|---|---|---|
| **Cost component** | **Average** | **Maximum** | **Minimum** |
| Chemical cost | 1.43 (0.054) | 5.92 (0.224) | 0.002 2 (0.000 08) |
| Labor cost | 8.40 (0.318) | 32.50 (1.23) | 2.09 (0.079) |
| Maintenance cost | 1.84 (0.070) | 5.07 (0.192) | 0.21 (0.008) |
| Miscellaneous cost | 8.61 (0.326) | 34.61 (1.31) | 0.72 (0.027) |
| Power cost | 5.13 (0.194) | 13.58 (0.514) | 0.32 (0.012) |
| Total cost[b] | 22.56 (0.854) | 57.07 (2.16) | 6.66 (0.252) |

[a] ENR Index = 5 000.
[b] Note: Totals do not equal sum of components.

more for a similar amount of capital invested in wastewater reclamation facilities.

- Construction costs for WWTPs, from an overall perspective, appear to be more significantly influenced by other site-specific criteria than the need for nutrient control.

**OPERATION AND MAINTENANCE COSTS.** Operation and maintenance costs were collected from existing full-scale treatment facilities using biological nutrient control processes, in a similar manner to what was done for nutrient control construction costs. To the extent that it was convenient, the respondents segregated the costs by power, labor, and chemicals in addition to providing overall operating costs. These data are presented in Table 12.7.

# REFERENCES


Kunihiro, C.I., *et al.* (1992) Retrofitting for Biological Nutrient Removal—How Much Will Your Plant Cost. *Water Environ. Technol.*, **4**, 7, 67.

Morales, L.M., *et al.* (1991) Cost Impacts of Upgrading Wastewater Treatment Plants to Provide Nutrient Removal. Paper presented at 64th Annu. Water Pollut. Control Fed., Toronto, Ont., Can.

Pincince, A.B., *et al.* (1997) Minimizing Costs of Activated-Sludge Systems. *Water Environ. Res.*, **69,** 326.

Randall, C.W., *et al.* (1992) *Design and Retrofit of Wastewater Treatment Plants for Biological Nutrient Removal.* Technomic Publishing Co., Lancaster, Pa.

Reardon, R.D. (1994) Cost Implication of Design for Nutrient Removal. Paper presented at 67th Annu. Water Environ. Fed. Tech. Exposition Conf., Chicago, Ill.

Reardon, R.D., and Maurer, D.E. (1991) Water Quality, What Price? BNR Costs—Revisited. Paper presented at 1991 Fla. Water Res. Conf., Pensacola.

Stenstrom, M.K., and Gilbert, R.G. (1981) Effects of Alpha, Beta and Theta Factor Upon the Design, Specification and Operation of Aeration Systems. *Water Res.* (G.B.), **15**, 643.

U.S. Environmental Protection Agency (1989) *Design Manual Fine Pore Aeration Systems*. EPA-625/1-89-023, Off. Res. Dev., Cincinnati, Ohio.

van Haandel, A.C. (1992) Activated Sludge Settling Part II: Settling Theory and Application to Design and Optimization. *Water SA*, **18,** 3.

# Appendix: Wastewater Characterization



Studies have demonstrated that influent wastewater characteristics change, often appreciably, from one municipal wastewater to another. Differences in characteristics typically are a function of a combination of socioeconomic factors; water usage; degree of inflow and infiltration; use of garbage grinders; presence, magnitude, and nature of any industrial content; size, retention time, and storage capacity within the wastewater collection system; and presence of a phosphate detergent ban.

# ORGANIC MATERIAL

Characterization of the carbonaceous material in municipal wastewater is in terms of the chemical oxygen demand (COD). The advantage of selecting COD as the parameter for quantifying the "strength" of organic material in the influent, as opposed to biochemical oxygen demand (BOD) or total organic carbon (TOC), is that it provides a consistent basis for description of the activated-sludge process.

**BIODEGRADABLE AND NONBIODEGRADABLE FRACTIONS.** The first division of the influent COD is into biodegradable COD and non-biodegradable COD. Each of these fractions is divided further into two sub-fractions.

**Biodegradable Subfractions.** The biodegradable material is divided into a readily biodegradable fraction ($S_{BS}$) and a slowly biodegradable fraction ($S_{ENM}$). The readily biodegradable fraction is hypothesized to consist of material that can be absorbed readily by the organism and metabolized for energy and synthesis. The slowly biodegradable fraction is assumed to be made up of particulate/colloidal material and complex organic molecules that require extracellular enzymatic breakdown before absorption and use. The apportionment between readily and slowly biodegradable COD has a significant influence on process dynamic behavior (for example, oxygen demand) and is of crucial importance in the design of biological nitrogen and phosphorus removal systems. For purposes of modeling biological excess phosphorus removal systems, it is necessary to distinguish a further division of the readily biodegradable material ($S_{BS}$). The readily biodegradable fraction is hypothesized to consist of complex readily biodegradable COD ($S_{BSC}$) and readily biodegradable COD in the form of short-chain volatile fatty acids ($S_{BSA}$).

**Nonbiodegradable Subfractions.** The nonbiodegradable material is divided into an nonbiodegradable soluble fraction ($S_{US}$) and an nonbiodegradable particulate fraction ($S_{UP}$). Both components are hypothesized to be unaffected by biological action in the system. In an activated-sludge system, the $S_{US}$ leaves through the secondary clarifier overflow at a concentration equal to that in the influent. The $S_{UP}$ is enmeshed in the sludge mass and accumulates in the system. At steady state, the mass of $S_{UP}$ entering the system in the influent will be balanced by the mass leaving via the sludge wastage stream. From a mass-balance standpoint, the mass of $S_{UP}$ in the system will equal the influent mass per day multiplied by the system sludge age. In systems operated at a long sludge age, the $S_{UP}$ can constitute a significant fraction of the mixed liquor settleable solids concentration.

**QUANTIFYING THE DIVISION.** From certain practical considerations, the division of the influent wastewater COD into fractions typically is defined by fractional parameters: $f_{US}$, $f_{UP}$ and $f_{BS}$, where

$f_{US}$ = fraction of the total influent COD that is nonbiodegradable soluble;
$f_{UP}$ = fraction of the total influent COD that is nonbiodegradable particulate; and
$f_{BS}$ = fraction of the total influent COD that is readily biodegradable.

Using these fractions, the division of the total influent COD ($S_{ti}$) into the various subfractions may be expressed as follows:

$$S_{USi} = f_{US} S_{ti} \tag{A.1}$$

$$S_{UPi} = f_{UP} S_{ti} \tag{A.2}$$

$$S_{BSi} = f_{BS} S_{ti} \tag{A.3}$$

$$S_{ENMi} = (1 - f_{BS} - f_{US} - f_{UP}) S_{ti} \tag{A.4}$$

The division of influent organic material as outlined here does not incorporate heterotrophic biomass as a component of influent wastewater. Studies in Europe indicate that this component may be significant.

# NITROGENOUS MATERIAL

Characterization of the nitrogenous material in the influent ($N_{ti}$) is in terms of the total Kjeldahl nitrogen (TKN). (In a limited number of cases, mainly in Europe, appreciable influent concentrations of nitrate/nitrite have been encountered in municipal wastewater. This has a significant effect on nutrient removal system design and performance. For these cases the influent nitrate/nitrite concentration must be included as an additional wastewater characteristic, separate from the TKN.) For design and modeling purposes, it is necessary to specify the magnitudes of the various fractions of the influent TKN.

**FREE AND SALINE AMMONIA AND ORGANICALLY BOUND FRACTIONS.** The first division of the influent TKN is into free and saline ammonia ($N_{H3}$) and organically bound TKN. The organically bound TKN is divided further into two fractions, biodegradable and nonbiodegradable, each with soluble and particulate portions.

**Biodegradable Organically Bound Total Kjeldahl Nitrogen.** The biodegradable organic nitrogen is divided into a soluble fraction ($N_{BS}$) and a particulate fraction ($N_{BP}$).

**Nonbiodegradable Organically Bound Total Kjeldahl Nitrogen.** This component of the organic nitrogen is divided into a nonbiodegradable soluble fraction ($N_{US}$) and a nonbiodegradable particulate fraction ($N_{UP}$). The nonbiodegradable fractions are hypothesized to be unaffected by biological action in the system. In an activated-sludge system, the soluble nonbiodegradable material leaves through the secondary clarifier overflow at a concentration equal to that in the influent. The particulate nonbiodegradable fraction is material associated with the nonbiodegradable particulate COD in the influent and leaves the system through the sludge wastage stream.

**QUANTIFYING THE DIVISION**. Again, from practical considerations the nitrogen fractions of the influent wastewater TKN are defined by fractional constants: $f_{NA}$, $f_{NUS}$, $f_{NUP}$, and $f_{NOP}$, where

$f_{NA}$ = fraction of the total influent TKN that is free and saline ammonia;
$f_{NUS}$ = fraction of the total influent TKN that is nonbiodegradable soluble;
$f_{NUP}$ = fraction of the total influent TKN that is nonbiodegradable particulate; and
$f_{NOP}$ = fraction of the biodegradable organic TKN that is particulate.

The division of the influent TKN ($N_{ti}$) into the various subfractions may be expressed as follows:

$$N_{H3i} = f_{NA} N_{ti} \tag{A.5}$$

$$N_{USi} = f_{NUS} N_{ti} \tag{A.6}$$

$$N_{UPi} = f_{NUP} N_{ti} \tag{A.7}$$

The biodegradable organic nitrogen, $N_{OBi}$, is given by the total influent TKN less the sum of $N_{H3}$, $N_{US}$, and the nonbiodegradable particulate organic nitrogen associated with the nonbiodegradable particulate influent COD:

$$N_{OBi} = (1 - f_{NA} - f_{NUS} - f_{NUP}) N_{ti} \tag{A.8}$$

The soluble ($N_{BS}$) and particulate ($N_{BP}$) influent biodegradable organic nitrogen concentrations are, respectively,

$$N_{BSi} = (1 - f_{NOP}) N_{OBi} \tag{A.9}$$

$$N_{BPi} = f_{NOP} N_{OBi} \tag{A.10}$$

# PHOSPHORUS CONTENT

Characterization of the phosphorus content of the influent ($P_{ti}$) is in terms of the total phosphorus (TP) concentration. In the context of nutrient removal, division of the TP into subfractions has not received as much attention as the COD and TKN. The approach in modeling and design has been to assume that a portion of the influent TP is associated with the nonbiodegradable influent COD (approximately 10 to 15% of TP) and that the remainder can be considered orthophosphate. Limited data on influent soluble reactive phosphorus (SRP—typically considered to be orthophosphate) would indicate this approach to be reasonable. That is, measurements of influent SRP are similar to TP less phosphorus associated with the nonbiodegradable influent COD (assuming a phosphorus content of approximately 0.02 mg P/mg non-biodegradable particulate COD).

Several factors support the need for a more rigorous accounting of phosphorus fractions, considering the increasing emphasis on low effluent phosphorus concentrations (<0.3 mg P/L). For example, a large portion of the total soluble phosphorus in aquatic systems can be made up of soluble unreactive phosphorus (SUP). If SUP is an appreciable fraction of the influent TP, this likely would affect the attainable minimum effluent phosphorus concentration for biological nutrient removal systems. A suitable methodology of accounting for the various influent phosphorus fractions would be analogous to the division of influent TKN.

# NITRIFICATION RATE

The sequential biological oxidation of ammonia-nitrogen to nitrite and then to nitrate (nitrification) plays an important role in many nutrient removal systems. The microorganisms responsible for nitrification are autotrophic bacteria. The oxidation of saline ammonia first to nitrite ions and then to nitrate results from two specific genera of autotrophic bacteria, *Nitrosomonas* and *Nitrobacter*. *Nitrosomonas* oxidizes free and saline ammonia to nitrite, and *Nitrobacter* oxidizes nitrite to nitrate. The first step (conversion of ammonia to nitrite) typically is the rate-limiting factor in the two-step nitrification process. As a result, nitrification typically is considered a single-step process mediated by one group of nitrification bacteria.

The growth rate parameter for the nitrifiers ($\mu_A$) defines the maximum specific growth rate of this surrogate organism mass. Variations in values for $\mu_A$ appear to correlate with the extent of the industrial component in the wastewater; it is possible that with increased industrial input there is an increased possibility of inhibition of nitrifying organisms. Another factor influencing the rate between wastewaters may be differences in pH. Consequently, the maximum specific growth rate of nitrifiers should perhaps be regarded as a wastewater characteristic.

# MEASUREMENT OF WASTEWATER CHARACTERISTICS

In recent years a number of experimental protocols for wastewater characterization have been published. However, little of this research has been based on North American wastewater.

Methods for determination of the fractional parameters characterizing a domestic wastewater COD and TKN content divide into three groups:

- Measurement using standard analytical techniques applied directly to the wastewater (for $f_{NA}$);
- Estimation through analysis of data observed in continuous-flow activated-sludge systems (for $f_{US}$, $f_{UP}$, and $f_{BS}$); and
- Monitoring of fill-and-draw systems or batch tests mixing influent wastewater with activated sludge (for $f_{US}$, $f_{UP}$, and $f_{BS}$).

A number of procedures have also been proposed for determining the maximum specific nitrifier growth rate ($\mu_A$).

# Index

## A

# B

# C

# D

# E



# F

# G

# H



# I



# J



# K



# L

# M

# N

# O



# P

# R



# S

# *T*

# U



# V



# W

# Y